Modern Digital Design and Switching Theory

Eugene D. Fabricius

California Polytechnic State University
San Luis Obispo, California

CRC Press
Boca Raton London New York Washington, D.C.

Library of Congress Cataloging-in-Publication Data

Fabricius, Eugene D.
  Modern digital design and switching theory / by Eugene D. Fabricius.
      p. cm.
  Includes bibliographical references and index.
  ISBN 0-8493-4212-0
  1. Digital electronics. 2. Switching theory 3. System design.
  I. Title.
  TK7868 D5F33 1992
  621.381'5—dc20   91-48255
                                                                                        CIP

This book contains information obtained from authentic and highly regarded sources. Reprinted material is quoted with permission, and sources are indicated. A wide variety of references are listed. Reasonable efforts have been made to publish reliable data and information, but the author and the publisher cannot assume responsibility for the validity of all materials or for the consequences of their use.

Neither this book nor any part may be reproduced or transmitted in any form or by any means, electronic or mechanical, including photocopying, microfilming, and recording, or by any information storage or retrieval system, without prior permission in writing from the publisher.

The consent of CRC Press LLC does not extend to copying for general distribution, for promotion, for creating new works, or for resale Specific permission must be obtained in writing from CRC Press LLC for such copying.

Direct all inquiries to CRC Press LLC, 2000 Corporate Blvd., N.W., Boca Raton, Florida 33431.

Trademark Notice: Product or corporate names may be trademarks or registered trademarks, and are only used for identification and explanation, without intent to infringe.

To Sherrie, Carol, and Brian,
inheritors of a digital world

CONTENTS

PROLOGUE

An Historical Overview

Humans have long been interested in increasing their native computing power. Circa 3000 BCE the abacus was invented in China. This is the first known mechanical calculating machine. In 1617 CE, John Napier of Scotland invented a tool referred to as Napier's Bones. In 1642 CE, Blaise Pascal of France invented a desk calculator, or arithmetic machine. Gottfried Leibniz of Germany invented what he called a Stepped Reconer in 1671. Pascal's and Leibniz's machines were mechanical devices which used gears to do arithmetic.

The first modern approach to computer design was the automated loom designed by the Frenchman Joseph Jacquard in 1801. His loom used punched cards to control the pattern and revolutionized the weaving industry. In 1833, the Englishman, Charles Babbage invented the Analytical Engine, which calculated and printed mathematical tables. He was the first to conceive the idea of a stored-program computer, in which all the numbers and instructions were entered before any calculations were made. The Electric Tabulating Machine, invented in 1887 in the U.S.A. by Herman Hollerith, was the first true data-processing machine. It was used to tabulate the 1890 census.

The age of electronics began in 1904 with the invention of the first vacuum tube, a diode called a Fleming valve by its inventor, J. A. Fleming. Electronic computing became possible with the invention of the triode vacuum tube, called an audion, by Lee DeForest in 1906; but it was 1938 before the first true electronic digital computer, composed of vacuum tubes, was built by John Atanasoff. The electronic numerical integrator and computer, called ENIAC, was the first large-scale electronic digital calculating machine, composed of 18,000 vacuum tubes and weighing 30 tons. It occupied 1500 square feet, or about 130 square meters, and consumed 130,000 watts of power, but it could multiply two numbers in a couple of milliseconds, three orders of magnitude faster than any previous calculating machine.

Although ENIAC could perform high-speed mathematical operations, it had to pause for each instruction to be entered in real time by its human operators. John von Neumann revived the stored-program concept first used by Babbage, which allowed the machine to read instructions from its memory without waiting for human intervention. This stored-program concept allows todays computers to operate at speeds approaching nanoseconds (10^{-9} s).

In 1948, the International Business Machines computer, the IBM 604, became the first general-purpose digital computer. In 1949, the first stored-program computer, called the electronic delay storage automatic calculator, or EDSAC, went on line, and in 1954, IBM introduced the IBM 650. Thus, the period from roughly 1938 to 1954 defines the first generation of digital electronic computing machines.

The invention of the transistor by John Bardeen and Walter H. Brattain, under the direction of William P. Shockley, at Bell Labs in 1947 (patented in 1948) earned the three of them the 1956 Nobel Prize in Physics and opened the era of second-generation digital computers. The transistor digital computer, or TRADIC, was only the first of hundreds of electronic computers which could process data faster and cheaper than ever before.

These early computing machines only worked in machine language, which is composed of 1s and 0s. This made programming tedious and error prone. A team at IBM, headed by John Backus, devised a way of writing a program using mathematical notation and common typewriter symbols, instead of machine language. Introduced in 1957 and called formula translation, or FORTRAN, it is still in use. By 1960, COBOL, or common business-oriented language, was added to the growing list of high-level languages.

Cray produced the first special-purpose second-generation computer in 1956, and IBM produced the first general-purpose transistorized computer in 1959, the IBM 7090/7094. In 1959, the integrated circuit was independently invented by Robert Noice at Fairchild Semiconductor and by Jack Kilby at Texas Instruments, and this period from roughly 1948 to 1959 covers the era of the second generation of digital computers.

By 1961, Fairchild and Texas Instruments were producing commercial integrated circuits. This invention, combined with high-level language capability, introduced the third generation of integrated digital computers.

In 1960, Atalla and Kahn of Bell Labs reported the first silicon metal-oxide-semiconductor field-effect transistor, or MOSFET. In 1962, Hopstein and Heiman of RCA were awarded a patent for their development of an integrated-circuit MOSFET fabrication process. In 1964, IBM introduced its System 360 Mainframe computer, said to cover the entire 360 degrees of scientific and business applications. This was a third-generation computer, using hybrid integrated circuits consisting of many discrete transistors on a single substrate.

By 1970, manufacturing processes had advanced to the point where over 15,000 transistors could be put on a single chip. This was due to the lower power requirements of first PMOS, then NMOS, and finally CMOS transistor technology, which opened the door to large-scale integrated circuits. The period from about 1960 to the end of the 1970s became the era of the third generation of digital computers.

In 1978, Lynn Conway of Xerox, in Palo Alto, taught the first course in VLSI or very large scale integration at MIT, using notes developed by Carver Mead of Caltech and herself. In 1980, the text, *Introduction to VLSI Systems,* by Mead and Conway, was published by Addison-Wesley Publishing Company, and the era of

VLSI ushered in the fourth generation of digital computers, with single chips holding millions of transistors. VLSI opened the door to microprocessors constructed on a single chip and led to the fast reduced-instruction-set computers, or RISCs.

CHAPTER 1

Number Bases, Codes, and Binary Arithmetic

1.1 INTRODUCTION

A digital system processes discrete information using electronic circuits that respond to and operate in a finite number of states, usually two states, on and off, which can be represented by the binary digits 1 and 0. The information so processed may represent anything from arithmetic integers and letters of the alphabet to values of physical quantities such as mass, velocity, and acceleration, or current, voltage, and impedance. A digital system accepts digital input representing numbers, symbols, or physical quantities, processes this input information in some specific manner, and produces a digital output.

Digital systems are used in all types of control systems due to their flexibility, accuracy, reliability, and low cost. Flexibility is due to the ease with which digital systems can be reprogrammed. Accuracy is limited only by the number of bits (BInary digiTs, consisting of 1s and 0s) one wishes to use in representing the digital quantities being processed. Reliability is due to the ability of digital circuits to correctly interpret logical 1s and 0s. For example, in *transistor-transistor logic* or *TTL* technology, a logical 1 is represented by a voltage in the range of roughly 2.5 to 5.0 V, and a logical 0 is represented by a voltage of from 0 to about 1 V, and minor fluctuations in voltage levels are not misinterpreted by the hardware.

The cost of all digital chips has dropped dramatically in the past three decades. This is primarily due to the number of transistors that can be put on a single chip. This number has been doubling almost every year for three decades, from a single transistor on a chip in 1960 to several million transistors per chip in 1990.

This chapter starts with a discussion of number bases and how to convert from one number base to any other number base. Next, the topics of binary addition and subtraction and then multiplication and division are covered. Following this, binary codes are discussed, specifically the binary-coded decimal, and the excess-three, the Gray, and error-detecting codes are covered. This leads into the concept of Hamming distance and the requirements for detecting and/or correcting codes.

The American standard code for information interchange (ASCII) alphanumeric code is also introduced.

Boolean cubes are defined and discussed as a means of graphically portraying Hamming distances. One's and two's complements and nine's and ten's complements are covered next, followed by an introduction to modulo arithmetic. Complementary arithmetic converts subtraction to addition. Binary subtraction by means of one's and two's complement arithmetic is covered.

1.2 NUMBER BASES

The number system most often used in everyday figuring is the decimal, or base 10, system, which uses ten characters, 0 through 9. This system is not convenient for computer applications, which are normally carried out in binary, or base 2, arithmetic. For this reason, one must be able to convert from base 10 to base 2 at computer/operator interfaces. This is easy to do using a weighted positional notation, and positional notation has been used since the discovery of zero. The Roman numeral system, for instance, is not a weighted positional system.

In a positional number system any number can be represented by a string of characters, with each character position assigned a weight which is a power of the radix or base. In the familiar decimal system the number 1234.5 represents 1 ¥ $1000 + 2 \times 100 + 3 \times 10 + 4 \times 1 + 5 \times 0.1$. Except for leading and trailing 0s, the representation of any number in positional notation is unique. (i.e., 01234.500 is the same number as 1234.5). This idea will be pursued further when we come to complementing numbers, where the representation of the complemented number will depend upon how many leading and/or trailing 0s are to be complemented.

The base or radix of a number system can be any positive or negative number, either rational or irrational, other than the number 1. Thus Π and e could be used as bases of number systems, as could negative numbers such as -1.5 or -7. In Chapter 1 it will be necessary to specify the radix or base of the number system being used, and this will be done with a subscript following the number to indicate its radix. The number 123 in base 5 will be represented as 123_5, representing 1 times 5 squared (1 times 5 to the second power) plus 2 times 5 to the first power plus 3 times 5 to the zeroth power.

In the following discussion, bases will be restricted to positive integers greater than 1. Let $R > 1$ be defined as the *radix* or *base* of a number system. Then any number, N, can be represented as a polynomial in R, viz

$$N = a_n R^n + a_{n-1} R^{n-1} + \cdots + a_1 R + a_0 + a_{-1} R^{-1} + a_{-2} R^{-2} + \cdots + a_{-k} R^{-k}$$

$$= \sum_{i=-k}^{n} a_i R^i \tag{1.1}$$

Break the number into an *integral* part, I, and a *fractional* part, F, such that

$$I = \sum_{i=0}^{n} a_i R^i \tag{1.2}$$

is the integral part of the mixed number N, and

$$F = \sum_{i=-1}^{-k} a_i R^i \tag{1.3}$$

is the fractional part of the mixed number N. For instance,

$$432.4_{10} = 4\left(10^2\right) + 3(10) + 2 + 4/10, \text{ and}$$
$$432.4_{8} = 4\left(8^2\right) + 3(8) + 2 + 4/8 = 282.5_{10}$$

where the subscripts 8 and 10 signify bases 8 and 10, respectively.

The integral part of the number lies to the left of the radix point, and the fractional part of the number lies to the right of the radix point. In the first example above, the radix point is the familiar decimal point, whereas in the second example it is the octal point.

In the number 123.45_{10}, 1 is referred to as the *high-order* or *most significant digit* and 5 is referred to as the *low-order* or *least significant digit*. In the number 100.11_2, the leftmost 1 is called the *high-order* or *most significant bit (MSB)*, and the rightmost 1 is called the *low-order* or *least significant bit (LSB)*.

For any base other than base 10, 10 (read one zero) is used in the positional location of the *radix* or *base*. Table 1.1 gives a partial listing of several number bases.

For any base greater than base 10, new characters are required. One can devise number systems up to and including base 36, simply by utilizing the 10 decimal characters plus the 26 alphabetical characters of the English language. Such a system is referred to as an alphanumerical system. In base 36 one would count from 1 to 9, and continue counting from A to Z, A having the value 10 and Z the value 35 in base 10. Thus, $Z.Z_{36} = 35 + 35/36 = 35.9722_{10}$.

To convert an integer from any original base, A, to any new base, B, consider the following:

If:
$$I = a_n R^n + a_{n-1} R^{n-1} + \cdots + a_1 R + a_0 \tag{1.4}$$

Then:
$$I/R = a_n R^{n-1} + a_{n-1} R^{n-2} + \cdots + a_1 + a_0/R \tag{1.5}$$

TABLE 1.1   Number Bases

Base 10 (decimal)	Base 2 (binary)	Base 3 (trinary)	Base 4 (quaternary)	Base 16 (hexadecimal)
0	0	0	0	0
1	1	1	1	1
2	10	2	2	2
3	11	10	3	3
4	100	11	10	4
5	101	12	11	5
6	110	20	12	6
7	111	21	13	7
8	1000	22	20	8
9	1001	100	21	9
10	1010	101	22	A
11	1011	102	23	B
12	1100	110	30	C
13	1101	111	31	D
14	1110	112	32	E
15	1111	120	33	F
16	10000	121	100	10

Define $I' = (I-a_0)/R$, and I' is seen to be the integral part of I/R.

Now: $$I'/R = a_n R^{n-2} + a_{n-1} R^{n-3} + \cdots + a_2 + a_1/R \qquad (1.6)$$

Proceeding in the above manner, one can obtain all the coefficients of the integral part of the original number, N (base A), in the new base, B.

EXAMPLE 1.1:   Convert 50_{10} to base 2.

SOLUTION:
```
2)50 | Remainder
2)25 | 0 = a₀ = LSB
2)12 | 1 = a₁
 2)6 | 0 = a₂
 2)3 | 0 = a₃
 2)1 | 1 = a₄
   0 | 1 = a₅ = MSB
```

$$\begin{array}{r|l} 2)50 & \text{Remainder} \\ 2)25 & 0 = a_0 = \text{LSB} \\ 2)12 & 1 = a_1 \\ 2)6 & 0 = a_2 \\ 2)3 & 0 = a_3 \\ 2)1 & 1 = a_4 \\ 0 & 1 = a_5 = \text{MSB} \end{array}$$

Read the remainders from the bottom up to obtain the integer in base 2, and 50_{10} = 110010_2. To check the answer: $2^5 + 2^4 + 2^1 = 32 + 16 + 2 = 50_{10}$.

To convert a fraction, consider the following:

If: $$F = a_{-1} R^{-1} + a_{-2} R^{-2} + \cdots + a_{-k} R^{-k} \qquad (1.7)$$

Then: $$R \cdot F = a_{-1} + a_{-2} R^{-1} + \cdots + a_{-k} R^{-k+1} \qquad (1.8)$$

Define $F' = (R \cdot F - a_{-1})$ and F' is seen to be the fractional part of $R \cdot F$.

Now:
$$R \cdot F' = a_{-2} + a_{-3}R^{-1} + \cdots + a_{-k}R^{-k+2} \tag{1.9}$$

By proceeding in the above manner, one can obtain all the coefficients of the fractional part of the original number, N (base A), in the new base, B.

EXAMPLE 1.2: Convert 0.62_{10} to base 2.

SOLUTION:

Carry	
	.62
	×2
$a_{-1} = 1$	.24
	×2
$a_{-2} = 0$	.48
	×2
$a_{-3} = 0$	.96
	×2
$a_{-4} = 1$	.92
	×2
$a_{-5} = 1$	.84
	×2(etc.)

The fraction in the new base is obtained by reading the carries from top down, viz.: $0.62_{10} = 0.10011_2$, truncated to 5 places, or 0.10100_2 if rounded off.

A mixed number is converted by combining the integral and fractional parts of the number after they have been separately converted. From the previous two examples, the number 50.62_{10} converted to base 2 is found to be approximately 110010.10011_2.

> To convert any number N from the original base A to any new base B, repeatedly *multiply the fractional part,* F (*divide the integral part,* I) by the *new base* B, using the arithmetic of the *old base,* A.

To convert a number from a base other than base 10 to a new base, it is usually easier to convert the number directly to base 10, and then to the new base. This confines the multiplication and division to base 10, a base with which we are all familiar.

The adventurous person who is willing to learn multiplication and division in other bases can do the conversion directly, since the above derivation is valid for any base, R. (See Problems 1.54 and 1.55.)

1.3 CONVERSION OF BINARY, QUARTERNARY, OCTAL, AND HEXADECIMAL NUMBERS

As can be seen from Table 1.1, numbers in base 4 can be converted directly to binary by replacing each character in base 4 by its binary equivalent. Thus, to convert 13_4 to binary, replace the 1 by 01 and the 3 by 11 to obtain $13_4 = 0111_2$, and to convert 32_4 to binary, replace the 3 by 11 and the 2 by 10 to obtain 1110_2. The same procedure works for any base 2^n.

Binary, quaternary, octal, and hexadecimal numbers all represent the same string of bits and can be interconverted very easily. The first step is to convert the number to base 2, after which it can be converted to any base which is a multiple of two by correctly grouping bits. Starting from the binary point, group the bits by twos in both directions to convert the number to base 4. Group the bits by threes in both directions from the binary point to convert the number to base 8, group by fours to convert to base 16, etc.

EXAMPLE 1.3: Convert 231.3_4 to base 7.

SOLUTION: $231.3_4 = 2(4^2) + 3(4) + 1 + 3/4 = 45.75_{10}$.

Proceed as before to convert the integer and the fraction separately.

		and			
7)45				.75	
7)6	3			×7	
0	6		5	.25	
				×7	
			1	.75	
				×7	
			5	.25	(repeating)

The fraction repeats, and, to four places: $231.3_4 = 63.\overline{51}_7 \approx 63.5151_7$.

EXAMPLE 1.4: Convert $1011\ 1001\ 0111\ 1101\ .\ 0101\ 10_2$ to bases 4, 8, and 16.

SOLUTION: Group by twos, starting from the binary point to obtain

$$10\ 11\ 10\ 01\ 01\ 11\ 11\ 01\ .\ 01\ 01\ 10 = 23211331.112_4$$

Next, group by threes from the binary point to obtain

$$001\ 011\ 100\ 101\ 111\ 101\ .\ 010\ 110 = 134575.26_8$$

And, lastly, group by fours, from the binary point to obtain

$$1011\ 1001\ 0111\ 1101\ .\ 0101\ 1000 = B97D.58_{16}$$

EXAMPLE 1.5: Convert 123.45_8 to bases 4, 16, and 32.

SOLUTION: First convert the number to binary:

$$123.45_8 = 001\ 010\ 011\ .\ 100\ 101_2$$

Then group by twos, starting from the binary point, to obtain the quaternary equivalent:

$$123.45_8 = 01\ 01\ 00\ 11\ .\ 10\ 01\ 01 = 1103.211_4$$

Group the bits by fours, from the binary point, to obtain the hexadecimal equivalent:

$$123.45_8 = 0101\ 0011\ .\ 1001\ 0100 = 53.94_{16}$$

Group the bits by fives, from the binary point, to obtain the equivalent number base 32:

$$123.45_8 = 00010\ 10011\ .\ 10010\ 10000 = 2J.IG_{32}$$

1.4 BINARY ARITHMETIC

Computers must perform arithmetic operations on binary numbers. The addition of positive numbers or the subtraction of a small number from a larger number generates positive answers, whereas the addition of two negative numbers or the subtraction of a larger number from a smaller number gives a negative answer. To do something as simple as subtract two numbers, the computer must be capable of recognizing and operating upon negative numbers.

The simplest designation of signed numbers is called *sign magnitude representation,* wherein a leading 0 is used to designate a positive number and a leading 1 is used to designate a negative number. In sign magnitude representation, the number +127 can be written as 0111 1111, and the number –127 as 1111 1111. It takes one extra bit to indicate the sign. The decimal numbers +5 and –5 can be represented by 0101 and 1101, respectively, using four bits, or by 0000 0101 and 1000 0101, respectively, using eight bits.

The penalty for using sign magnitude representation is one additional bit, used to represent the sign. Negative numbers can also be represented in complement form (covered in Section 1.7), which requires no additional sign bits.

1.4.1 Binary Addition and Subtraction

Binary arithmetic is simpler than decimal since the addition and multiplication tables are so simple, viz

$$0 + 0 = 0 \qquad\qquad 0 \times 0 = 0$$
$$0 + 1 = 1 \qquad\qquad 0 \times 1 = 0$$
$$1 + 1 = 10 \qquad\qquad 1 \times 1 = 1$$

Binary addition and subtraction can be specified using truth tables. (A truth table is a listing of all possible combinations of inputs, with a column specifying each output. The inputs are customarily written by counting in binary. This practice systematizes the procedure and avoids duplication or omission.) The required truth tables for addition and subtracton are shown in Table 1.2.

EXAMPLE 1.6: Add 01010 plus 00111 (in base 10, 10 + 7 = 17).

SOLUTION:

$$
\begin{array}{r}
111-- \\
01010 \\
+00111 \\
\hline
10001
\end{array}
\qquad
\begin{array}{ll}
\text{A} & 01010 \\
\text{B} & 00111 \\
\hline
\text{Carry} & 111-- \\
\text{Sum} & 10001
\end{array}
$$

EXAMPLE 1.7: Subtract 01010 from 10001 (i.e., 17 – 10 = 7).

SOLUTION:

$$
\begin{array}{r}
111-- \\
10001 \\
-01010 \\
\hline
00111
\end{array}
\qquad
\begin{array}{ll}
\text{A} & 10001 \\
\text{B} & 01010 \\
\hline
\text{Borrow} & 111-- \\
\text{Diff.} & 00111
\end{array}
$$

1.4.2 Binary Multiplication and Division

Binary multiplication and division require a shift operation as well as addition and subtraction operations, the same as in base 10 arithmetic.

EXAMPLE 1.8: Multiply 1100 by 1011 (in base 10, 12 × 11 = 132).

SOLUTION:

$$
\begin{array}{rl}
1100 & \\
\times 1011 & \\
\hline
1100 & \text{(replicate, or copy)} \\
1100 & \text{(shift and replicate)} \\
11000 & \text{(shift, shift, and replicate)} \\
\hline
10000100 & \text{(add)}
\end{array}
$$

TABLE 1.2 Truth Tables for Addition and Subtraction

A	B	Sum	Carry	A	B	Diff.[a]	Borrow
0	0	0	0	0	0	0	0
0	1	1	0	0	1	1	1
1	0	1	0	1	0	1	0
1	1	0	1	1	1	0	0

[a] Diff. is the difference, A − B.

EXAMPLE 1.9: Divide 10110 by 11 (in base 10, 22/3 = 7 + 1/3).

SOLUTION:

$$
\begin{array}{r}
111 \\
11\overline{)10110} \\
\end{array}
$$

11	(shift and replicate)
101	(subtract and carry down)
11	(shift and replicate)
100	(subtract and carry down)
11	(shift and replicate)
1	(subtract; remainder is 1)

1.5 BINARY CODES

While binary numbers are the most appropriate for the internal computations of a digital system, people generally are more facile with, and prefer to deal with, decimal numbers. As a result, interfacing between people and machines is generally done using inputs and displays that are read in decimal. The digital systems still require input and output in 1s and 0s, so a string of 1s and 0s is needed to represent each decimal number.

A set of n-bit strings in which different strings represent different characters is called a *code*, and each valid string of n bits is called a *code word*. Every string of n bits may or may not represent a valid code word. At least four bits are required to represent the ten decimal digits. There are 16!/10!6! different ways to choose 10 out of 16 4-bit code words, and 10! ways to assign each different choice to the 10 digits, giving a total possibility of over 29 billion different 4-bit decimal code words. Some of the more commonly used codes are listed herein, and some are referred to in Problems 1.51 through 1.53.

1.5.1 Binary Coded Decimal

A *weighted code* is one for which each bit position has a weight assigned to it. For example, the weight of a four-bit straight binary character is 8-4-2-1 or 2^n, where n is the column location. There are many possible weighted binary codes. In practice, except for one's or two's complement arithmetic, only positive weights are used, which still leaves many alternatives. What is normally referred to as the *binary-coded decimal* or *BCD code* is an 8-4-2-1 weighted binary code, obtained by converting each decimal to a four-bit character string called a *code word*. The BCD is wasteful of space from a binary or machine point of view, because it only utilizes 10 of 16 possible arrangements of 4 bits; but it is an easy-to-read code for humans. BCD circuits are especially useful in arithmetic calculations which are to be displayed in decimal form using light-emitting diode (LED) displays or liquid-crystal displays (LCD). The BCD code words are obtained by converting each decimal character to an equivalent four-bit binary string.

BCD codes are a good compromise between machine requirements and human requirements and are often used in business applications where there is a lot of interaction with humans, whereas straight binary is used in scientific applications where there is little interfacing between people and computers, but much computation, or number crunching. Numbers stored in BCD are still in decimal form and require no conversion of binary to decimal or decimal to binary when reading data into or out of a computer.

A register is a piece of hardware in which binary numbers can be stored in the form of high or low voltages. The largest number that can be stored in a 16-bit BCD register is $9999_{10} = 1001\ 1001\ 1001\ 1001_{BCD}$. By comparison, the largest binary number that can be stored in a 16-bit register is $2^{16} - 1$ or $65,535_{10} = 1111\ 1111\ 1111\ 1111_2$.

There are no BCD equivalents to hexadecimal characters A, B, C, D, E, or F. Sometimes two BCD digits are combined in one eight-bit byte, with BCD numbers of any desired size represented by using one byte for each pair of digits. This is referred to as a *packed-BCD* representation and allows one byte to represent values from 0 to 99, whereas hexadecimal representation allows the eight-bit byte to represent any number from 00 = 0 to FF = 255.

EXAMPLE 1.10:   Convert 926.54_{10} to BCD.

SOLUTION: $926.54_{10} = 1001\ 0010\ 0110\ .\ 0101\ 0100_{BCD}$.

1.5.2   Binary Coded Decimal Addition

Addition in BCD is accomplished by adding the two numbers the same as in straight binary, except that if the sum of any two digits is greater than 9, one must add 6 to the sum in order to obtain the correct answer. The addition of two BCD numbers will produce a carry to the next higher digit position if either the original sum or the corrected sum produces a carry. Many computers and some calculators have the ability to do packed BCD arithmetic directly. Some examples of BCD addition follow.

EXAMPLE 1.11: Add the decimal numbers 4 and 5, 4 and 6, and 7 and 8 in BCD.

SOLUTION: $4 + 5 = 9_{10}$. In BCD, $0100 + 0101 = 1001$ is correct.

$4 + 6 = 10_{10}$. In BCD, $4 + 6 = 0100 + 0110 = 1010$ is an invalid code word. Adding $6_{10} = 0110_{BCD}$, one obtains

$0100 + 0110 + 0110 = 10000 = 0001\ 0000_{BCD} = 10_{10}$

$7 + 8 = 15_{10}$. In BCD, $7 + 8 = 0111 + 1000 = 1111$ is an invalid code word. Adding 6_{10}, one obtains

$0111 + 1000 + 0110 = 10101 = 0001\ 0101_{BCD} = 15_{10}$

The addition of two BCD numbers may yield a sum of 8 or less, which requires no correction. If the sum equals 9 and there is a carry from the next lower position, the correction factor of 6 must still be added to the number to obtain a valid BCD number.

EXAMPLE 1.12: Add the two decimal numbers 34 and 67 in BCD.

SOLUTION: $34_{10} = 0011\ 0100_{BCD}$ and $67_{10} = 0110\ 0111_{BCD}$.

```
0011      0100
0110      0111
─────     ─────
1001      1011
          0110      (correction is required here)
─────     ─────
1001    1 0001
```

Upon adding the carry to the first code word ($9_{10} = 1001_{BCD}$), the answer is greater than 9 and must be corrected to give

```
1001      0001
   1
0110
─────     ─────
1 0000    0001
```

The correct answer is $0001\ 0000\ 0001_{BCD} = 101_{10}$.

EXAMPLE 1.13: Add the decimal numbers 358 and 769 in BCD.

SOLUTION: In BCD, $358 = 0011\ 0101\ 1000$ and $769 = 0111\ 0110\ 1001$. The BCD addition proceeds as follows:

$$
\begin{array}{ccc}
0011 & 0101 & 1000 \\
0111 & 0110 & 1001 \\
\hline
1010 & 1011 & 1\ 0001 \\
0110 & 0110 & 0110 \\
\hline
1\ 0000 & 1\ 0001 & 1\ 0111
\end{array}
$$

    (there is a carry here)
(add correction factors)

The carries must be added to the next higher BCD code word. When this is done the answer is

$$
\begin{array}{ccc}
1\ 0000 & 0001 & 0111 \\
1 & 1 & \\
\hline
1\ 0001 & 0010 & 0111
\end{array}
$$

The correct answer is $0001\ 0001\ 0010\ 0111_{BCD} = 1127_{10}$.

1.5.3 Excess-Three Code

The *excess-three (XS3)* code is an excess-weighted BCD code obtained by adding 3 to each BCD code word. Excess-weighted refers to the fact that the sum of the weights of the bits of the XS3 code exceeds the value of the decimal digit that the code word represents.

One advantage of XS3 over BCD is that XS3 can distinguish the difference between the transmittal of a 0 and no transmission. A more important reason for using XS3 code is that the decimal words which are nine's complements have code words which are one's complements, a fact that makes subtraction in XS3 simple. It was originally designed to facilitate decimal subtraction in binary form. Complements are discussed in Section 1.7 and subtraction is discussed in Section 1.9.

EXAMPLE 1.14: Convert 926.54_{10} to XS3 code.

SOLUTION: $926.54_{10} = 1100\ 0101\ 1001\ .\ 1000\ 0111_{XS3}$.

1.5.4 Error-Detecting Codes

When information is encoded and transmitted from one medium to another, there is always the possibility of a bit being changed due to electrical noise or other transient failure mechanisms. Codes that detect and/or correct errors are used to minimize the damage due to these errors.

The probability of a single error occurring is small for a reasonably reliable system. From probability theory, the probability of a double error occurring is equal to the square of the probability of a single error occurring, provided the single errors are independent. Thus, if the probability of a single error is of the order of 10^{-6}, then the probability of a double error occurring is of the order of 10^{-12}, and a system that detects single errors is much more accurate than one that does not.

A simple code that detects single errors is the 2-out-of-5 code. In a 2-out-of-5 code, two bits must be 1s and three bits must be 0s for each code word. There are exactly ten combinations of 2-out-of-5 code words, and each combination can correspond to a decimal digit. This type of code is useful for single error detection, since any number of 1s other than two indicates that an error in transmission has occurred.

Other simple codes that detect single errors are 1-out-of-10 codes where a single bit is a 1, and the other nine bits are 0s, and the inverted 1-out-of-10 code, which has a single 0 and nine bits that are 1s. This can be generalized to m-out-of-n codes. Any m-out-of-n code has $n!/m!(n-m)!$ valid words. Thus, a 2-out-of-4 code has only $4!/2!(4-2)! = 6$ valid words, namely, 0011, 0101, 0110, 1001, 1010, and 1100, whereas a 4-out-of-10 code has $10!/4!(10-4)! = 210$ valid words.

1.5.5   Unit Distance Codes

In a *unit distance code (UDC)* the next valid code word is obtained by changing only one bit in the present word. These codes are also called *Hamming distance-one codes*. The Hamming distance specifies the minimum number of bits that must change to get from any valid code word to another valid code word. UDCs are useful for error minimization, especially when mechanical wheels are involved. For example, in straight binary, two wheels must move to change from 001 to 010. If there is a slight misalignment of one wheel, it is possible for the number to change from 001 to 011 to 010, or to change from 001 to 000 to 010. If different instructions are sent elsewhere when the wheels read either 011 or 000, then there is an ambiguity in making this change. The worst problem occurs when the counter has reached all 1s and resets to all 0s, at which time all bits change. This problem cannot arise when using a UDC since only one wheel at a time changes.

Some binary codes are *reflected codes*. A reflected word is characterized by the fact that it is imaged about the center entries with only one bit changed. A specific UDC which is reflective in its MSB is referred to as a *Gray code*.

UDCs can be generated with the help of Karnaugh Maps, as will be shown later. A UDC (modulo 10) is shown in Table 1.3, along with BCD, XS3, and 2-out-of-5 codes.

TABLE 1.3   Some Common Binary Codes

Decimal	BCD	XS3	2-out-of-5	UDC
0	0000	0011	00011	0000
1	0001	0100	00101	0001
2	0010	0101	00110	0011
3	0011	0110	01001	0010
4	0100	0111	01010	0110
5	0101	1000	01100	1110
6	0110	1001	10001	1010
7	0111	1010	10010	1011
8	1000	1011	10100	1001
9	1001	1100	11000	1000

1.5.6   Parity Checking

By adding a single bit to any character string, one can devise a code in which all the valid code words have either *even parity* (an even number of 1s), or *odd parity* (an odd number of 1s). Thus, the word 0100 0001 contains two 1s and is said to have even parity, whereas 1100 0001 contains three 1s in it and has odd parity. This simple addition will detect all single errors. By adding more bits, double errors or triple errors can be detected. Since single errors are far more likely to occur than double errors, it is important to be able to detect and correct them. In a "noisy" system or in a situation requiring high accuracy, multiple error detection and/or correction is also important.

A simple technique for correcting single errors consists of using a *parity word* at the end of a block of code words. This parity word checks parity in a dimension orthogonal to that checked by a parity bit and can thus flag the bit that is in error as well as the word in which the bit occurs. When a single error occurs, this scheme detects the row and column with the error, as shown in Figure 1.1(a). The row and the column in which an error occurs are both tagged by an E.

When a double error occurs in the same word, two columns register an error, as shown in Figure 1.1(b), although the row is unknown because no row error is detected. If a double error occurs in one column, the column appears correct, while two rows indicate an error, as shown in Figure 1.1(c). If two errors occur in two different rows and in two different columns, both words and both columns are flagged, but again it is not known which combination of errors occurred. This is shown in Figure 1.1(d), where the errors could be at the dark nodes or at the light nodes. Three errors occurring in one row or column would all be flagged, whereas the three errors shown in Figure 1.1(e) cause the detector to flag a single error in the one correct location.

This gives the code an effective distance of 4, because any pattern with one, two, or three bits in error causes an incorrect parity of a row or a column or both. If four errors occur in a rectangular pattern, as shown in Figure 1.1(f), with two errors in each of two words and two errors in each of two columns, this scheme cannot detect the four errors. With an effective distance of 4, single errors can be

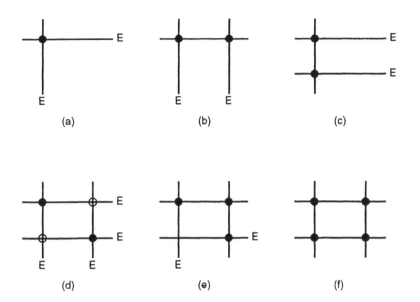

FIGURE 1.1. (a) One, (b), (c), and (d) two, (e) three, and (f) four errors in a cross-parity code.

corrected and double or triple errors can be detected, although it is not known what was sent and they cannot be corrected. If single error correction is being done, triple errors will be treated as false single errors and the wrong bit will be corrected. This approach is also called horizontal-redundancy checking, or cross-parity checking.

Single errors can be detected in a distance-2 code since changing any single bit results in an invalid code word in this code. Since it is not known which bit is the erroneous one, distance-2 codes do not allow one to correct detected errors. A distance-3 code can detect and correct single errors. This is because there are two noncode words between any pair of valid words, and a code word with a single error will be closer to the correct code word than to another valid word. One can determine the correct transmission and correct the erroneous word.

From Table 1.3 it can be seen that the Hamming distance of a BCD code is 1, since only one bit must be changed to go from 0 to 1, from 2 to 3, 4 to 5, 6 to 7, and 8 to 9 in BCD. The Hamming distance in XS3 code is likewise 1, since only one bit has to change to go from 1 to 2, etc. in XS3. The Gray code is a reflective distance-1 code. No errors can be detected in a distance-1 code, since changing any bit might result in another valid code word.

The Hamming distance is 2 for the 2-out-of-5 code shown in Table 1.3, since at least two bits must be changed to get from any valid code word to another valid code word. As another example, a distance-3 code might have the two valid code words 100111 and 101001. If the bit string 101011 is received, and a single error occurred, the correct word transmitted was 101001, and the erroneous bit was the fifth one.

In the rare event that two errors occur in transmission of a distance-3 code it can detect the double error, but if it is correcting single errors, an erroneous correction will be made. If, in the above example, 100111 was transmitted and both bits 3 and 4 are in error, the word 101011 is again received, only now it is "corrected" to 101001, which is not what was transmitted. Usually the probability of three or more errors occurring is very small and can be ignored, but a distance-3 code used to correct single errors will make mistakes in those rare cases when double errors occur.

In general, if a code has a minimum distance of d + 1, it can detect errors in up to d bits, while if it has a minimum distance of 2c + 1, it can be used to correct errors that affect up to c bits. If a code has a minimum distance of 2c + d + 1, it can be used to correct errors in up to c bits, and to detect errors in up to d additional bits. Thus, a distance-2 code can only detect single errors, while a distance-3 code can correct single errors or detect, but not correct, single and double errors. This is because a double error appears to be a single error in a distance-3 code, and the wrong bit would be corrected. A distance-4 code could detect single, double, and triple errors while correcting none, or it could correct single errors and detect double errors. In this case, a three-bit error appears to be a one-bit error and is erroneously corrected. A distance-5 code could be used to detect up to quadruple errors, to correct single errors and detect double or triple errors, or to correct single and double errors. When correcting single and double errors, triple errors would appear to be double errors and quadruple errors would appear to be single errors, and they would both be erroneously corrected.

1.5.7   The ASCII Code

Many digital applications require the handling of data that consist of both numerical and alphabetical characters, as well as special characters such as "+", "-", "?", "*", "%", "$", etc. These codes come under the general heading of alphanumeric codes, and are said to consist of an alphanumeric character set.

Two of the most widely used alphanumeric codes are the ASCII code and the extended binary coded decimal interchange code (EBCDIC).

ASCII is a seven-bit code and includes special control characters used for teletype and other data communication devices. The ASCII code incorporates the 10 decimal numbers, the 26 upper-case characters and the 26 lower-case characters of the English language, as well as a large number of special characters, such as SOH (Start Of Heading), STX (Start of TeXt), ETX (End of TeXt), EOT (End Of Transmission), etc. It must also handle special functions on a computer keyboard, such as ESCape, ConTRoL, Caps Lock, Enter or Carriage Return, and many other functions.

Seven bits gives $2^7 = 128$ code words. The numerals and the upper- and lower-case letters add up to 62 characters, leaving 66 7-bit binary strings available for other words. An eighth bit can be added as a parity bit used to check transmissions for errors.

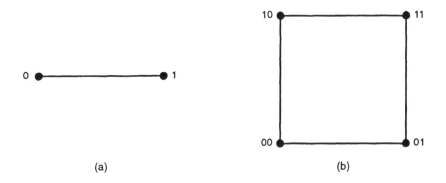

FIGURE 1.2. Boolean cubes in (a) one dimension, (b) two dimensions, (c) three dimensions, and (d) four dimensions.

The ASCII word for upper-case A is 100 0001. This word has even parity which can be maintained by prefixing a 0 before the string, or it can be converted to odd parity by prefixing a 1 to the string. ASCII words can be transmitted in either format. The ASCII word for lower-case a is 110 0001 and has odd parity. Affixing a 0 to this word gives an odd parity string 0110 0001, whereas affixing a 1 to the word changes it to even parity, 1110 0001. By proceeding in this manner, all seven-bit ASCII character strings can be converted to either even or odd parity eight-bit binary strings.

1.6 BOOLEAN CUBES

The concept of *boolean cubes* allows one a geometric interpretation of Hamming distance, making it easy to visualize. Let each bit location of a binary string be associated with a dimension in *boolean space*. One can then construct boolean cubes as follows.

Each boolean variable can take on the value 1 or 0, and these values can be represented as the endpoints of a single line in boolean space. Two variables can be represented by a plane, with the four boolean values of these two variables forming the vertices of a square. Three variables can be represented by a cube in three dimensions with the vertices corresponding to the possible values of the three variables. Four variables require four dimensions and are harder to draw.

Dimensions greater than four are impossible to visualize, although computer programs can be written that operate on these *hypercubes*. One-, two-, three-, and four-dimensional "cubes" in boolean space are shown in Figure 1.2.

The problem of designing an n-bit UDC is now reduced to the problem of finding a path along the edges of an n cube that visits each vertex exactly once. Paths for three- and four-bit unit distance reflected codes are shown in Figure 1.3. These are three- and four-bit Gray codes, respectively.

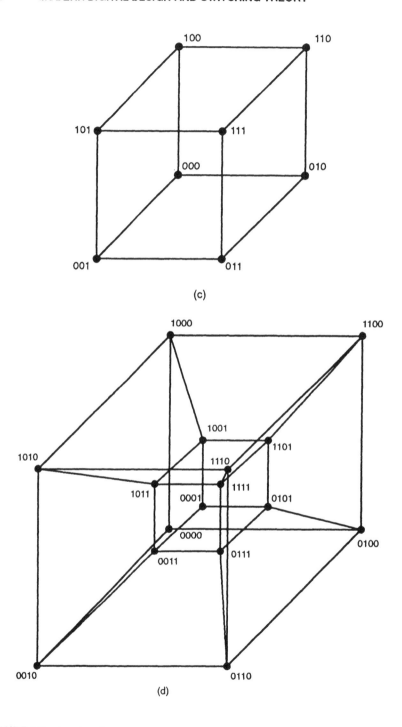

FIGURE 1.2. (continued).

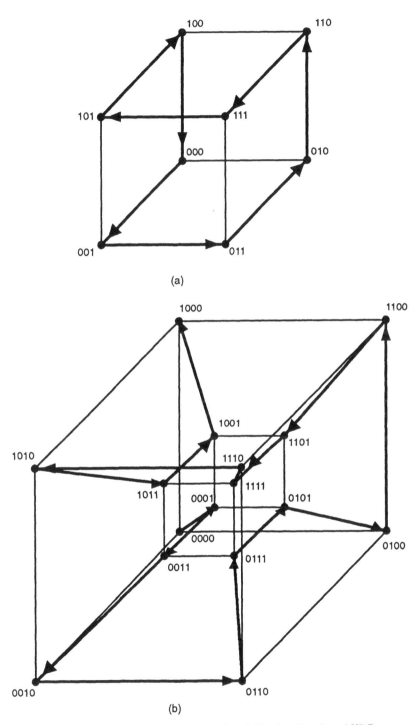

FIGURE 1.3. Cubic representation of (a) a three-bit and (b) a four-bit reflected UDC.

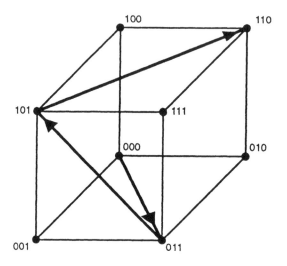

FIGURE 1.4. Cubic representation of a three-bit distance-2 code.

Boolean cubes provide a geometrical interpretation for the concept of Hamming distance. In terms of n cubes, the distance between any two words of a code is the length of the path between the two corresponding vertices. Two adjacent vertices have a distance of 1, whereas vertices 000 and 101 of a 3 cube have a distance of 2. Figure 1.4 shows a three-bit distance-2 code consisting of the characters 000, 011, 101, 110.

1.7 COMPLEMENTS

To avoid ambiguity when complementing numbers, one must specify the modulus of the register which contains the number, i.e., the number of storage elements available in that register. This will indicate how many bits are involved, each of which must be complemented. Since $2^5 = 32$, a modulo-32 counter or register requires 5 storage locations or flip-flops which can output 32 numbers, 00000 through 11111. The complementary outputs (one's complements) are also available from many registers with no extra circuitry.

1.7.1 Logical or One's Complement

The logical or one's complement (from the Latin complementum, that which completes) of 1 is 0, and the logical complement of 0 is 1. The logical complement can be written with a tilde, ˜, or with a bar over the character, such as $1˜ = \overline{1} = 0$, $0˜ = \overline{0} = 1$ and $(011.01)˜ = \overline{011.01} = 100.10$.

TABLE 1.4   Examples of Nine's and Ten's Complements

Number	Nine's complement	Ten's complement
147.0	852.9	853.0
0.53	9.46	9.47
147.53	852.46	852.47
0147.530	9852.469	9852.470

TABLE 1.5   Examples of One's and Two's Complements

Number	One's complement	Two's complement
1.011	0.100	0.101
1010.0	0101.1	0110.0
0.101	1.010	1.011
1010.101	0101.010	0101.011

1.7.2   Two's Complement

The two's complement of a number is obtained by taking the one's complement and adding 1 to the rightmost bit.

EXAMPLE 1.15:   Find the two's complement of 011.01

SOLUTION: The two's complement of $011.01 = 100.10 + 0.01 = 100.11$

1.7.3   Nine's and Ten's Complement

The nine's complement of a number is obtained by subtracting each digit from 9, while the ten's complement is obtained by adding 1 to the rightmost digit of the nine's complement of the number.

EXAMPLE 1.16:   Obtain the nine's and ten's complement of 54.3.

SOLUTION: The nine's complement of 54.3 is $99.9 - 54.3 = 45.6$ and the ten's complement of 54.3 is $99.9 - 54.3 + 0.1 = 45.7$. The ten's complement is also $100.0 - 54.3 = 45.7$

Several more examples are given in Tables 1.4 and 1.5 below.

XS3 codes are self-complementing in that the nine's complement of a decimal number is the one's complement of the binary equivalent of the decimal number, viz

$$0_{10} = 0011_{XS3} \text{ and } 9_{10} = 1100_{XS3}$$

$$4_{10} = 0111_{XS3} \text{ and } 5_{10} = 1000_{XS3}$$

The UDC of Table 1.3 is a reflective code. This means that the nine's complement of a decimal number is formed by changing the most significant bit of the binary equivalent of the decimal number. Thus,

$$0_{10} = 0000_{UDC} \text{ and } 9_{10} = 1000_{UDC}$$
$$1_{10} = 0001_{UDC} \text{ and } 8_{10} = 1001_{UDC}$$

The difference between straight binary and two's complement binary is the weight attached to the MSB. This weight of the MSB is negative in two's complement arithmetic, whereas all the bits have positive weight in straight binary code.

1.8 MODULO ARITHMETIC

Two numbers, A and B, are said to be *congruent, modulo* C, if the difference of the two numbers divided by C is an integer. This is written as follows: $A \equiv B$ MOD $C'(A - B)/C$ is an integer, and is read as "A is congruent to B modulo C, if and only if $A - B$ divided by C is an integer."

EXAMPLE 1.17: Show that 11 is congruent to 3 modulo 8, 15 is congruent to –1 modulo 8, and 5 is congruent to –11 modulo 8.

SOLUTION: $+11 \equiv +03$ MOD 8 because $(11 - 3)/8 = 1$
$+15 \equiv -01$ MOD 8 because $(15 + 1)/8 = 2$
$+05 \equiv -11$ MOD 8 because $(5 + 11)/8 = 2$

Two common examples of modulo systems are clocks and automobile odometers. The normal clock is a modulo-12 counter, while the military clock counts modulo 2400. If the time is now 2 p.m., the time 43 hours from now is obtained by first dividing 43 hours by 12, obtaining the remainder 7. (i.e., 43/12 = 3, with remainder 7.) In 43 hours the clock hour hand will make three complete revolutions from 2 o'clock to 2 o'clock and will then advance 7 hours to read 9 o'clock. The correct time will then be 9 a.m. In modulo arithmetic, $43 \equiv 7$ MOD 12. The time 43 hours from 2 p.m. is also 48 hours minus 5 hours from 2 p.m., since 43 $\equiv -5$ MOD 12 and $43 \equiv -5$ MOD 24. This just means that 43 hours from 2 p.m. is 5 hours short of being exactly 2 days from that time.

An automobile odometer is a modulo-100,000 counter. If the odometer reads 99,994 miles at the start of a 12-mile trip, it will read 00,006 miles at the end of the trip; $100,006 \equiv 00,006$ MOD 100,000. If a new automobile with 0 miles on the odometer is driven backwards 5 miles, the odometer would read 99,995 since – 5 $\equiv 99,995$ MOD 100,000.

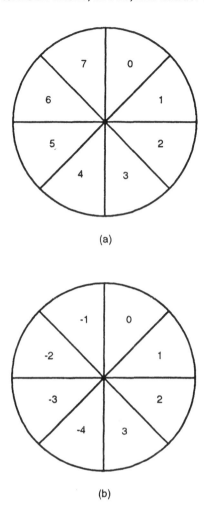

(a)

(b)

FIGURE 1.5. Representation of modulo-8 arithmetic, for (a) positive numbers only, and (b) positive and negative numbers.

Because $95 \equiv -5$ MOD 100, $18 - 5 = 13$ is congruent to $18 + 95 = 113$, both modulo 100. Thus, in modulo-100 arithmetic, subtracting 5 from a number is equivalent to adding 95 to the number. In other words, one can perform subtraction simply by obtaining the ten's complement of the number and adding, being careful to drop the overflow.

Computers do modulo arithmetic. A modulo-N (divide-by-N) counter can unambiguously count from 0 to $N - 1$, which is a total of N events. The next (N + 1) event resets it. The modulo-N counter requires B storage latches, where $2^B > N$ and each latch stores one bit.

TABLE 1.6   Negative Numbers in Two's Complement Form

Base 10	Base 2	Base 10	Base 2
0	0000	-0	0000
1	0001	-1	1111
2	0010	-2	1110
3	0011	-3	1101
4	0100	-4	1100
5	0101	-5	1011
6	0110	-6	1010
7	0111	-7	1001
		-8	1000

Parity checking, discussed in Section 1.5.6 above, relies on modulo-2 addition of bits. If the sum of two bits is 0 modulo 2, then there were either zero 1s or two 1s. If the sum of two bits is 1 modulo 2, then a single 1 occurred.

Modulo-8 arithmetic may be visualized by considering Figure 1.5. Starting from any location in Figure 1.5, a number can be added modulo 8 by rotating in a clockwise direction, while a number can be subtracted modulo 8 by proceeding in a counterclockwise direction.

1.9   BINARY SUBTRACTION

The standard method of subtracting two numbers is easy for humans, but it is difficult to program machines to determine when to borrow, and a more machine-oriented method of subtraction is desired. If this method can be realized with the same hardware as that used for addition, a great convenience in designing hardware and a great savings in equipment costs and complexity can also be realized.

The logical complement of a number is easy to obtain and transforms the operation of subtraction into the operation of addition. This allows adders to perform subtraction also. To see why this is so, consider subtracting number B from number A. In base 10, $A - B = A + (10 - B) - 10$; $(10 - B)$ is the ten's complement of B. Adding this to A gives an answer that is larger than $A - B$ by 10. Subtract the 10 by dropping the carry. Thus, $9 - 4 = 9 + (10 - 4) - 10 = 9 + 6 - 10 = 15 - 10 = 5$.

1.9.1   Subtraction Using One's Complements

Binary subtraction can be done using one's complement arithmetic as follows:

1.   Find the one's complement of the subtrahend.
2.   Add this number to the minuend.
3.   (a) If an overflow occurs, add it to the least significant bit; (b) if an overflow does not occur, the result is negative and is in one's complement form.

EXAMPLE 1.18: Subtract 7_{10} from 10_{10} and 10_{10} from 7_{10} in one's complement arithmetic.

SOLUTION: $7_{10} = 0111_2 = 1000_{1C}$ where 1C represents the one's complement form of the number. $10_{10} = 1010_2$. Thus, $10 - 7$ is found to be $1010 + 1000 = 1\ 0010$. Since a carry occurs, add it to the LSB and the answer is $+0011_2 = +3_{10}$.
$10_{10} = 1010_2 = 0101_{1C}$ and $7 - 10$ is found to be $0111 + 0101 = 1100$. There is no carry; hence the answer is negative and is in one's complement form. The correct answer is $-0011_2 = -3_{10}$.

1.9.2   Subtraction Using Two's Complements

Binary subtraction can be done using two's complements also. Two's complement subtraction proceeds in four steps, as follows:

1.   Find the one's complement of the subtrahend.
2.   Add 1 to obtain the two's complement of the subtrahend.
3.   Add this complemented number to the minuend.
4.   (a) If an overflow occurs, discard it, and the positive answer remains; (b) if there is no overflow, the answer is negative, and is in two's complement form.
5.   If one is subtracting a number from a negative number, drop the overflow and the answer is in two's complement form.

To see how subtraction in two's complement arithmetic works, consider the representation of negative numbers base 10 in the two's complement form, as shown in Table 1.6, and note the following:

1.   Negative 0 and positive 0 are represented by the same binary string. This is very convenient and is not true if one's complements are used for negative numbers, in which case $+0 = 0000$ and $-0 = 1111$.
2.   With a modulo-16 counter, one can only represent numbers between $+7$ and -8 inclusive.
3.   Negative numbers have a 1 in the MSB position, while positive numbers have a 0 in the MSB position. The MSB is thus a weighted sign bit.

A modulo-16 counter has available three bits to represent the number and a fourth bit (the MSB) to represent the sign of the number.

EXAMPLE 1.19: Add $+5$ or -5 to $+6$ or -6, using two's complement arithmetic, modulo 16.

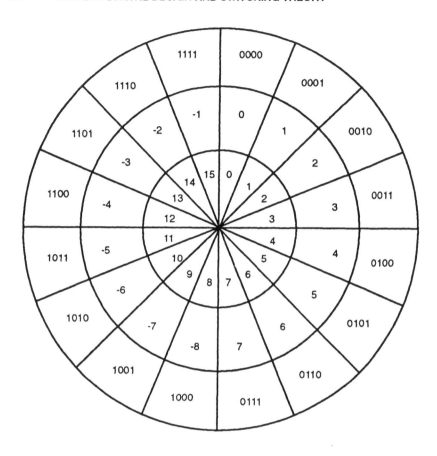

FIGURE 1.6.   A modulo-16 counter can be interpreted to represent straight binary (inner ring) or two's complement binary (middle ring).

SOLUTION: $+5 + (+6) = 0101 + 0110 = 1011_2 = +11_{10}$ in straight binary. However, in two's complement binary $1011_2 = -5_{10}$. Note that $-5 \equiv 11$ MOD 16.

$+5 + (-6) = 0101 + 1010 = 1111_2 = 15_{10}$ in straight binary, but $1111_2 = -1_{10}$ in two's-complement binary; $-1 \equiv 15$ MOD 16.

$-5 + (+6) = 1011 + 0110 = 10001_2 = 17_{10}$. By dropping the carry, the correct answer of $+1_{10}$ is obtained. Again, $+1 \equiv 17$ MOD 16.

$-5 + (-6) = 1011 + 1010 = 10101_2 = 21_{10}$ in straight binary or -11_{10} in two's complement binary, but this requires a five-bit counter.

Note that $21 \equiv -11$ MOD 16. Since the carry is lost, the counter returns the number $0101 = 5_{10}$, which is also congruent to -11, modulo 16.

TABLE 1.7   A Modulo-256 Two's Complement Weight Table

-128	64	32	16	8	4	2	1	Base 10
0	0	0	1	1	0	1	0	26
1	0	0	1	1	0	1	0	-102

The two wrong answers ($5 + 6 = -5$, and $-5 + (-6) = 5$) arose due to overflow errors, because the answers are too big for a modulo-16 counter, which can only count from -8 to $+7$ in two's complement arithmetic. A modulo-32 counter has five bits, of which four can be used to represent the number, leaving one bit for the sign of the number. It can count from -16 to $+15$ and will return the correct answers for $5 + 6$ and $-5 + (-6)$, as shown in Example 1.20. An example of adding negative numbers in base 10 is shown in Example 1.21.

EXAMPLE 1.20:   Add +5 or –5 to +6 or –6, using two's complement arithmetic, modulo 32.

SOLUTION: $+5 + (+6) = 00101 + 00110 = 01011_2 = 11_{10}$
$+5 + (-6) = 00101 + 11010 = 11111_2 = -1_{10}$
$-5 + (+6) = 11011 + 00110 = 100001$. The 1 is an overflow and is discarded, and the number $00001_2 = +1_{10}$ is returned.
$-5 + (-6) = 11011 + 11010 = 110101_2$. Again, the overflow is discarded, and the answer returned is $10101_2 = -11_{10}$.

EXAMPLE 1.21:   Add ±75 to ±56 in base 10.

SOLUTION: Observe that adding both positive or both negative numbers gives a number greater than 100; thus, at least three significant figures are required. Let the two numbers be represented as 075 and 056, and there will be no overflow errors. Then, $075 + 056 = 131$ and the answer is correct.
$075 - 056 = 075 + (1000 - 056) - 1000 = 075 + 944 - 1000 = 1019 - 1000$. Add the 10's complement of 056 to 075, drop the overflow, and the correct answer is 019.
$-075 + 056 = (1000 - 075) + 056 = 925 + 056 = 981$, and 981 $\equiv -19$ MOD 1000. The 10's complement of $981 = 1000 - 981 = 019$, and the correct answer is -19.
$-075 - 056 = (1000 - 075) + (1000 - 056) = 925 + 944 = 1869$. $1869 \equiv 869 \equiv -131$ MOD 1000. Drop the overflow and the answer is 869. The 10's complement of 869 is $1000 - 869 = 131$, and the correct answer is -131.

Most microprocessor instruction sets provide a subtraction capability using two's complement arithmetic. Most instruction sets also include the capability of independently complementing a number stored in the accumulator, without using the subtraction routine.

"Two's complement" is a procedure for converting a negative decimal number to a weighted-sign binary number, also called a signed binary number. The expression "weighted sign" means that the leftmost bit of the binary number represents both a binary value and a sign, with 0 representing a positive number and 1 representing a negative number. A very simple way to view two's complement arithmetic is to consider the MSB to have a negative-weighted binary value, while the rest of the bit positions maintain their normal positive bit values.

A modulo-16 counter using two's complement arithmetic may be considered to contain bits of weights –8, 4, 2, and 1, respectively. Thus, the largest positive number this counter can display is $0111_2 = 7_{10}$, and the most negative number the counter can display is $1000_2 = -8_{10}$. The counter can handle numbers from –8 to +7, as indicated in Table 1.6, while a modulo-32 counter can handle numbers from $10000_2 = -16$ to $01111_2 = +15_{10}$.

A modulo-16 counter requires four bits and can count from either 0 to 15 in straight binary or from –8 to +7 in two's complement binary, as shown in Figure 1.6. The inner and middle arrays of numbers in Figure 1.6 are congruent MOD 16, the inner array representing the decimal equivalent if the bits are interpreted in straight binary, and the middle array if they are interpreted in two's complement binary.

An eight-position two's complement number can be represented symbolically as shown in Table 1.7. To determine the value of any number between +127 and –128, simply add the positional weight values above each box which contains a 1 in it. This is a modulo-256 system which can count from –128 to +127.

In Table 1.7, row 1 represents $16 + 8 + 2 = 26_{10}$, while row 2 represents $-128 + 26 = -102_{10}$.

1.10   SUMMARY

Numbers can be converted from any original base A to any new base B by repeatedly multiplying the fractional part, F, and dividing the integral part, I, by the new base, B, using the arithmetic of the old base, A. If the old base is anything other than base 10, it is simpler to convert the original number to base 10 first. The arithmetic is then done in base 10.

Bases 2, 4, 8, 16, 32, etc. can be easily converted from one base to another. If the number is given in a base other than base 2, it is first converted to base 2. If one then groups the bits two at a time, the number can be converted to base 4. Grouping the bits three at a time gives the number in base 8, grouping 4 bits at a time converts the number to base 16, and grouping the bits five at a time would convert the number to base 32.

BCD is a very convenient weighted binary code when interfacing with human operators, since numbers can be converted from decimal to BCD or vice versa by inspection. BCD addition requires correction factors since there is no BCD equivalent to the hexadecimal characters A, B, C, D, E, or F.

The XS3 code is a type of BCD that was devised to facilitate subtraction. In XS3 code decimal numbers which are nine's complements of each other have binary representations which are one's complements of each other.

Parity refers to the number of 1s in a code word. Even-parity words contain an even number of 1s, while odd-parity words contain an odd number of 1s. Parity is used to increase the Hamming distance of codes in order to detect or correct errors. The Hamming distance of a code specifies the minimum number of bits that must be changed to go from any valid code word to any other valid code word.

UDCs cannot detect errors. A distance-2 code can only detect single errors. A distance-3 code can either correct single errors or detect single and double errors. If the code corrects single errors it will erroneously correct a double error by treating it as a single error. A code with a minimum distance of $2c + d + 1$ can be used to correct up to c erroneous bits and detect errors in up to d additional bits.

Boolean cubes offer a geometrical interpretation of binary numbers. A binary number with n bits can take on 2^n unique values. These values can be represented by the vertices of an n-dimensional cube.

The logical or one's complement of a binary number is obtained by changing all its 0s to 1s and all its 1s to 0s. The two's complement of a binary number is obtained by adding 1 to the least significant bit in the one's-complemented number. Negative numbers start with a 1 in two's complement representations, whereas positive numbers start with a 0. The first bit doubles as a sign bit, telling whether the number is positive or negative.

The number of bits that are to be complemented depends upon the modulus of the register. A modulo-2^n register can store n bits, and can count up to 2^n distinct binary words.

One's and two's complement numbers can be used to convert binary subtraction to addition. This allows one to design a binary adder which will perform subtraction by adding one's or two's complement numbers. The adder must be capable of recognizing whether the answer is a positive or a negative number. If the answer to an addition using two's complement arithmetic is a number with a 0 sign bit, it represents a positive number. If a 1 sign bit occurs, the answer is a negative number and is in two's complement form.

With subtraction capability, one can perform all four basic operations: addition, subtraction, multiplication, and division. Care is required to warn of overflow. Multiplication in base 2 requires a shift operation. When multiplying by 1, the number being multiplied is added to the product, followed by a shift left. When multiplying by 0, only a shift left occurs. Long division in base 2 is essentially the same as long division in base 10, the divisor being subtracted either once or zero times, followed by a shift right operation.

PROBLEMS

1.1. Convert 25.62_{10} to base 7.

1.2. Convert 33.45_{10} to base 6.

1.3. Convert 34567_8 to base 10.

1.4. Convert 143112_5 to base 10.

1.5. Convert 72.25_{10} to bases 3, 5, and 7.

1.6. Convert 65.43_{10} to bases 4, 6, and 9.

1.7. Convert 22.34_5 to base 7.

1.8. Convert 53.2_7 to base 6.

1.9. Convert 110011.0011_2 to bases 4, 8, and 16.

1.10. Convert 101010.1010_2 to bases 4, 8, and 16.

1.11. Convert 1111010.0011_2 to bases 4, 8, and 16.

1.12. Convert 10111001.111101011_2 to bases 4, 8, and 16.

1.13. Convert 122.17_{10} to bases 2, 4, 8, and 16.

1.14. Convert 421.6095_{10} to bases 2, 4, 8, and 16.

1.15. Convert $AB.CD_{16}$ to bases 2, 4, and 8.

1.16. Convert $BA.DC_{16}$ to bases 2, 4, and 8.

1.17. Convert $CB.A9_{16}$ to bases 2, 4, and 8.

1.18. Convert $DC.BA_{16}$ to bases 2, 4, and 8.

1.19. Convert $A98B_{12}$ to base 3.

1.20. Convert $EF64H_{20}$ to base 6.

1.21. Find 1111.0011_2 plus 1011.1111_2.

1.22. Find 1010.1010_2 plus 1001.0101_2.

1.23. Find 1111.0011_2 divided by 101_2.

1.24. Find 1010.1100_2 divided by 110_2.

1.25. Find 1111.0011_2 times 101_2.

1.26. Find 1010.0110_2 times 110_2.

1.27. Convert 97.53_{10} to BCD and XS3 codes.

1.28. Convert 12.34_{10} to BCD and XS3 codes.

1.29. Convert 6820_{10} to BCD and XS3 codes.

1.30. Convert 32.1_{10} to BCD and XS3 codes.

1.31. Convert 123.45_{10} to BCD and XS3 codes.

1.32. Convert 541.23_{10} to BCD and XS3 codes.

1.33. Find 1001_{XS3} plus 0100_{XS3}. Give the answer in BCD and XS3 codes.

1.34. Find 0111_{XS3} plus 1000_{XS3}. Give the answer in BCD and XS3 codes.

1.35. Find 1010_{XS3} minus 0110_{XS3}. Give the answer in BCD and XS3 codes.

1.36. Find 1100_{XS3} minus 0111_{XS3}. Give the answer in BCD and XS3 codes.

1.37. Find the one's and two's complements of 1010.0101_2.

1.38. Find the one's and two's complements of 1011.1010_2.

1.39. Find the one's and two's complements of 11010.10101_2.

1.40. Find the one's and two's complements of 11011.10101_2.

1.41. Find 01011.10_2 minus 00110.11_2 in one's and in two's complement arithmetic.

1.42. Find 01010.10_2 minus 01110.11_2 in one's and in two's complement arithmetic.

1.43. Find 01111.0011_2 minus 01011.1111_2 in one's and two's complement arithmetic.

1.44. Find 01011.1111_2 minus 01111.0011_2 in one's and two's complement arithmetic.

1.45. Find the nine's and ten's complements of 123.45_{10}.

1.46. Find the nine's and ten's complements of 453.67_{10}.

1.47. Find the nine's and ten's complements of 954.67_{10}.

1.48. Find the nine's and ten's complements of 056.92_{10}.

1.49. Add 65_{10} and -57_{10} in ten's complement arithmetic.

1.50. Add 57_{10} and -65_{10} in ten's complement arithmetic.

1.51. Devise two possible 5321 codes. A 5321 code is a binary code whose decimal weights are 5, 3, 2, and 1, respectively.

1.52. Devise a 6311 code.

1.53. Why is a 5411 code useless?

1.54. Convert 231.3_4 to base 7, using arithmetic in base 4. To do this, use the tables for addition and multiplication base 4, given below.

1.55. Convert 132.1_4 to base 5, using arithmetic in base 4. To do this, use the tables for addition and multiplication base 4, given below.

+	0	1	2	3
0	0	1	2	3
1	1	2	3	10
2	2	3	10	11
3	3	10	11	12

×	0	1	2	3
0	0	0	0	0
1	0	1	2	3
2	0	2	10	12
3	0	3	12	21

1.56. The Hamming distance between two valid code characters A and B can be represented by vertices and edges of the n cube as shown below. Consider an eight-bit distance-4 code with vertices of its eight-cube A = 11000101, and B = 11001010.

(a) If code word A is sent, show the vertices V_i corresponding to errors in bit position 4, in bit positions 1 and 4, and bit positions 1, 3, and 4. Bit position 1 is the LSB.

(b) What words are received when as many errors as possible are corrected.

(c) Prove that in this case the Hamming distance is $2c + d + 1$.

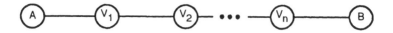

1.57. Repeat problem 1.56 for A = 11001010 and B = 11000101.

1.58. Convert 123_{10} to the smallest positive number, modulo 8, 16, and 20.

1.59. Convert 321_{10} to the smallest positive number, modulo 7, 9, and 11.

1.60. Convert 123_{10} to the number of smallest absolute value, modulo 8, 16, and 20.

1.61. Convert 321_{10} to the number of smallest absolute value, modulo 7, 9, and 11.

1.62. Find all N between 1 and 10 for which $15 \equiv 9$ MOD N.

1.63. Find all N between 10 and 30 for which $N \equiv 5$ MOD 7.

SPECIAL PROBLEMS

1.64. Write a flow chart to multiply two eight-bit binary words together. Let the words be labeled X and Y, and the result be labeled Z. Use I for an increment counter, and let N be the decimal value of the most significant 1 in Y. Then the loop must be repeated N times. Show the results of each pass through the loop for $X = 0C_{16}$, $Y = 0B_{16}$, and $N = 04_{16}$.

1.65. Listed below is a BASIC subroutine which converts any decimal or alphabetical character string first into an ASCII character, and then into a decimal number. Use this program to convert a number from any base between 2 and 36 inclusive to any other base between 2 and 36 inclusive. To do this, separate the number into an integral part and a fractional part. The integral part must then be repeatedly divided by the new base, and the fractional part multiplied by the new base. The remainders resulting from the division of the integral part must be reversed in order, combined with the carries, reconverted to ASCII, and printed out. N$ is the number to be converted, and NN is the length of N$. B.OLD is the original base, and B.NEW is the new base. ASCII character 46 is the decimal point. Inputs are N$, B.OLD, and B.NEW, and the output is the value of N$ in the new base, B.NEW. Test the program by converting $FF.FF_{16}$ to base2, base 2 to base 4, base 4 to base 8, base 8 to base 10, base 10 to base 32, then back to base 16.

```
10  REM SUBROUTINE TO CONVERT ORIGINAL NUMBER TO ASCII
    EQUIVALENT
20  NN = LEN(N$)
30  FOR I = 1 TO NN
40  A(I) = ASC(MID$(N$,I))
```

50 IF A(I) = 46 THEN FLAG = 1: GOTO 110: REM SET FLAG FOR RADIX POINT
60 REM CONVERT ASCII NUMBER TO DECIMAL NUMBER
70 IF A(I) > 47 AND A(I) < 58 THEN A(I) = A(I) − 48
80 IF A(I) > 64 AND A(I) < 91 THEN A(I) = A(I) − 55
90 IF FLAG THEN R = R + 1: REM TO LOCATE RADIX POINT
100 A = A * B.OLD +A(I)
110 NEXT I
120 A = A/B.OLD^R: REM TO INSERT RADIX POINT

A printout of some ASCII characters follows:

A(I) value	Alphanumeric character	A(I) value	Alphanumeric character
46	. (RADIX POINT)	69	E
48	0	70	F
49	1	71	G
50	2	72	H
51	3	73	I
52	4	74	J
—	—	75	K
56	8	76	L
57	9	77	M
65	A	—	—
66	B	89	Y
67	C	90	Z

CHAPTER 2

Boolean Algebra and Implementation

2.1 INTRODUCTION

The mathematical basis for the design of digital systems is *boolean algebra,* also known as the *algebra of propositions* or the *algebra of sets.* George Boole was one of the first people to develop a mathematical approach to the way humans reason in 1849.[1] Using his system, one can formulate propositions that are true or false, combine them to create new propositions, and determine the truth or falsehood of these new propositions.

Augustus DeMorgan, who lived from 1806 to 1871, was a contemporary of George Boole who published works on mathematics and symbolic logic for many years. His most famous work on logic was entitled *Formal Logic, or The Calculus of Inference, Necessary and Probable* in 1847.

The next major step in the development of a mathematics of logic and reasoning was due to E. V. Huntington, first in 1904, and later in 1933.[2,3] The work of Boole and Huntington paved the way for Claude E. Shannon, a Bell Labs researcher, who in 1938 showed how to apply boolean algebra to the analysis of networks of relays in order to describe and predict the behavior of circuits built from relays.[4]

In Shannon's *switching algebra,* the condition of a relay contact, open or closed, is represented by a variable that can have one of two possible values, 0 or 1. Relay circuits were the most commonly used digital logic elements of that time. Today, switching algebra is applied to a wide variety of physical conditions; a voltage level being high or low, a light being on or off, a capacitor charged or discharged, a fuse blown or not, etc.

Switching algebra is a term often used interchangeably with boolean algebra. As shown in Chapter 3, Huntington's postulates of boolean algebra apply to sets having more than two elements. Switching algebra is a subset of boolean algebra called binary boolean algebra in which the only elements are 0 and 1, corresponding to a switch or relay being open or closed.

Transistors have replaced relays in most digital applications, and they, like mechanical switches, are operated in one of two states also. Strictly speaking, relays and most mechanical switches are bilateral devices and allow the flow of information in either direction. Transistors are often operated as unilateral devices which only allow information to flow in one direction. Switching functions can be implemented either way, but the most common devices in use today are unilateral and are called gates. In this chapter the basic gates used in modern digital design will be studied.

Binary or switching logic is a process of classifying information into one of two valid states. Traditionally, these states have been referred to as *true* and *false*. In digital circuit applications "true" and "false" are meaningless appellations. A switch can be either open or closed, and a binary voltage level can be either high or low. Neither case is more "true" than the other. It makes more sense to speak of two assertion levels; a variable or a function can be asserted high or asserted low. One of these two levels can be given the value 1 and the other can be given the value 0. A high voltage level can be defined to be either a logical 1 *(positive logic)* or a logical 0 *(negative logic),* with the low voltage level assuming the other logical value.

This chapter introduces the fundamental boolean operations of symbolic logic, AND, OR, and NEGATION; and the corresponding electronic building blocks, AND, OR, and INVERT gates. The relationship between logical operations and electronic circuits is investigated for both positive and negative logic, and conversion from one type of logic to the other is discussed.

NAND gates, NOR gates, exclusive-OR gates, and exclusive-NOR gates are additional electronic building blocks whose logical operations are defined by means of truth tables. Each row of a truth table corresponds to a product of all the literals that make up the logical expression. This product is called a minterm.

Functionally complete sets of logic gates are investigated, DeMorgan's theorem is introduced, and applications of it are given. The chapter concludes with logical operations applied to binary strings or vectors.

2.2 FUNDAMENTAL BOOLEAN OPERATIONS

There are three fundamental boolean operations: complementing, ANDing, and ORing. The symbols normally used to represent these operations are shown in Table 2.1. These three operations were originally defined grammatically, AND being the logical intersection, OR being the logical union, and NEGATE complementing the value of the function.

The connection between grammar and circuitry is of importance in formulating a grammatical logic statement which can be implemented directly in hardware. This concept that logical functions performed and the circuits used to implement them can both be described in terms of boolean algebra is due to Claude Shannon.[4]

The order in which the operations are performed is complement, followed by

TABLE 2.1   Boolean Symbols

Operation	Digital circuit	Boolean logic
Complement	$\overline{}$ or B AR	(Negate) $\tilde{}$
AND	·	(Intersection) ∩ or ∧
OR	+	(Union) ∪ or ∨

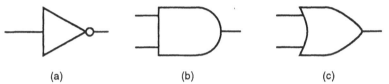

(a)   (b)   (c)

FIGURE 2.1.   The (a) inverter symbol, (b) AND symbol, and (c) OR symbol.

AND, followed by OR. For example, to evaluate the expression $A + B \cdot \overline{C}$, one would first complement C, then AND B with $\overline{C}$, and lastly OR A to the product of $B \cdot \overline{C}$.

There are electronic building blocks called "gates" which perform each of the three basic operations, although in practice, inverting gates merely add time delays and power dissipation and are to be avoided whenever possible. A *gate* is a *unilateral* electronic or mechanical switching device which implements a logical operation by allowing data to flow from input to output. The usual symbols for these three basic gates are shown in Figure 2.1. The "bubble" at the output of the inverter indicates an output which is the logical complement of the input. This is discussed further in Section 2.5.3.

There is no necessary one-to-one correspondence between logic gates and logic functions, as will be shown shortly. An AND gate can perform the OR operation, and an OR gate can perform the AND operation under the proper circumstances (see Figures 2.4 and 2.5, for example).

Logical or one's complements have already been discussed. The truth tables which define the logical AND and OR operations are given below and may be considered the definitions of these boolean operations (see Table 2.2). Truth tables make more sense if 1s (0s) are considered to represent the function or variable being *asserted (not asserted)*, rather than referring to 1s (0s) as *true (false)*. Some synonyms for asserted are *excited, invoked,* and *active*.

EXAMPLE 2.1:   Realize the function $F = A + B \cdot \overline{C}$.

SOLUTION: First C must be complemented, then $\overline{C}$ is ANDed with B. Last the output of the AND gate is ORed with A. The circuit is shown in Figure 2.2.

TABLE 2.2   AND and OR Truth Tables

A	B	A·B	A + B
0	0	0	0
0	1	0	1
1	0	0	1
1	1	1	1

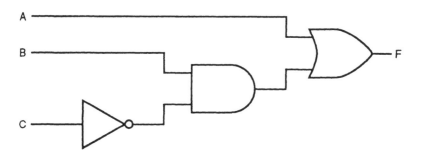

FIGURE 2.2.   Circuit realization of $F = A + B\overline{C}$.

2.3   LITERALS AND MINTERMS

Each appearance of a variable or its complement is referred to as a *literal,* and a product which contains every variable either in the complemented or in the uncomplemented form is called a *minterm.* For example, if $F(A,B,C,D) = ABCD + \overline{A}BC\overline{D} + AB\overline{C}\,\overline{D} + \overline{A}\,\overline{B}\,C$, then F consists of 4 variables, 15 literals, and 3 minterms. $\overline{A}\,\overline{B}\,C$ is not a minterm because the variable D is missing in it. The function $G(A,B,C,D) = A\overline{B}\,\overline{C}\,\overline{D} + A\overline{B}CD + \overline{A}\,\overline{C}D + \overline{A}B$ contains 4 variables, 13 literals, and 2 minterms, $A\overline{B}\,\overline{C}\,\overline{D}$ and $A\overline{B}CD$.

Each literal is an input to a logic gate. Each minterm is a vertex of an n cube in the n variables of the function. Each minterm is also a product term which can be the output of an AND gate provided all the variables are present as inputs, as shown in Figure 2.3.

2.4   ASSERTED AND NOT-ASSERTED CONDITIONS
AND TRUTH TABLES

Table 2.3 shows a truth table in terms of the voltages associated with the variables. The truth table can be interpreted in either of two ways, depending upon whether the variables are asserted high or low. These choices are referred to as *positive* and *negative* logic, respectively, and gates are normally referred to by their positive logic function.

FIGURE 2.3.   A three-input AND gate. For a function of three variables, the output is a minterm.

TABLE 2.3   AND/OR Gate Truth Table

A[a]	B	C	A	B	C	A	B	C
LV	LV	LV	0	0	0	1	1	1
LV	HV	LV	0	1	0	1	0	1
HV	LV	LV	1	0	0	0	1	1
HV	HV	HV	1	1	1	0	0	0

[a] LV — low voltage; HV — high voltage.

The first and second set of columns of 1s and 0s both represent the same physical situation and must be logically equivalent. The gate representing the first set of columns is obtained by replacing a high-voltage condition by a 1, and a low-voltage condition by a 0. This is a *positive-logic AND* gate. The output, C, is *asserted high* only if the inputs A and B are both *asserted high*. This is indicated in Figure 2.4(a) by the letter H in parentheses following the name of the variable.

The second set of columns represents a logical OR function. These columns were created by replacing a low-voltage condition by a 1, and a high-voltage condition by a 0; hence, the ones in the truth table correspond to the *asserted-low* condition. The output is asserted low when either A or B or both are asserted low. A gate drawn with bubbles at those inputs (and the output) which are asserted low will symbolize this interpretation of the original truth table. This is shown in Figure 2.4(b), and the letter L in parentheses designates the correct assertion level of the inputs and the output.

Another truth table is shown in Table 2.4, this time for a *positive-logic OR* gate, or a *negative-logic AND* gate, as indicated in the second and third sets of columns shown. The corresponding gates are shown in Figure 2.5.

A-BAR means A-NOT (the complement of A) is *asserted,* and A-BAR also means A is *not-asserted.* Therefore, read A-BAR as *"A not asserted".* Thus,

$$\overline{F(H)} = F(L) \quad \text{and} \quad \overline{F(H)} = \overline{F}(H)$$

$$\overline{F(L)} = F(H) \quad \text{and} \quad \overline{F(L)} = \overline{F}(L)$$

$$\overline{F(H)} = \overline{F}(H) = F(L) \quad \text{and} \quad \overline{F(L)} = \overline{F}(L) = F(H) \quad\quad (2.1)$$

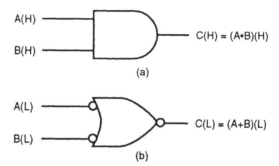

$$C(H) = (A \cdot B)(H)$$

(a)

$$C(L) = (A + B)(L)$$

(b)

FIGURE 2.4.   (a) A positive-logic AND gate and (b) a negative-logic OR gate.

TABLE 2.4   OR/AND Gate Truth Table

A[a]	B	C	A	B	C	A	B	C
LV	LV	LV	0	0	0	1	1	1
LV	HV	HV	0	1	1	1	0	0
HV	LV	HV	1	0	1	0	1	0
HV	HV	HV	1	1	1	0	0	0

[a] LV — low voltage; HV — high voltage.

2.5   CONVERSION OF ASSERTION LEVELS

It is often necessary to interface positive and negative logic. There is no problem mixing both positive- and negative-logic inputs, provided a systematic procedure is followed. If the inputs are labeled according to whether they are asserted high or low, both assertion levels can be shown on the same circuit diagram. The term *ambipolar notation* will be used to describe this situation, since the asserted-high and the asserted-low polarities both coexist. In a complex system, it may be necessary to convert the positive or negative logic to ambipolar notation first, after which the circuit can be simplified without regard to whether a given symbol had its origin in the positive- or negative-logic system. After simplification, the result can be changed to either positive or negative logic, as desired.

One must answer the question of what to do when there is a mismatch between an ambipolar assertion level and the logic assertion level at a gate input. A mismatch occurs when the literal is asserted high and the input to the gate indicates a low asserted state due to the presence of a "bubble", or when the literal is asserted low and the input to the gate indicates a high asserted state by the absence of a "bubble".

Whenever a mismatch of assertion levels occurs at a gate input, the mismatched variable will appear in the answer in complemented form. Whenever a

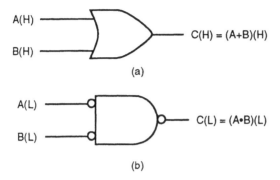

A(H) ───╲
          ╲
           ╲──── C(H) = (A+B)(H)
          ╱
B(H) ───╱

(a)

A(L) ───o╲
          ╲
           │──o── C(L) = (A•B)(L)
          ╱
B(L) ───o╱

(b)

FIGURE 2.5. (a) A positive-logic OR gate and (b) a negative-logic AND gate.

mismatch of assertion levels occurs at the output of a gate, the entire output expression is complemented. It is often easier to trace signals through a circuit when the polarities are matched at each interface between the output of one gate and the input to the succeeding gate.

2.5.1   Ambipolar Notation and Positive Logic

To convert from positive-logic notation to ambipolar notation, replace all *uncomplemented* variables by *asserted-high* variables and all *complemented* variables by *asserted-low* variables. To convert from ambipolar notation to a positive-logic system, replace all *asserted-high* variables by *uncomplemented variables* and all *asserted-low* variables by *complemented variables*.

EXAMPLE 2.2(a):   Convert A, $\overline{B}$, and C from positive-logic notation to ambipolar notation.

SOLUTION: If A, $\overline{B}$, and C are high, then the answers are A(H), B(L), and C(H).

EXAMPLE 2.2(b):   Convert A(L), B(H), and C(L) to positive logic.

SOLUTION: The answers are $\overline{A}$, B, and $\overline{C}$, since these literals are high.

2.5.2   Ambipolar Notation and Negative Logic

To convert from negative-logic notation to ambipolar notation, replace all *uncomplemented* variables by *asserted-low* variables, and all *complemented* variables by *asserted-high* variables. To convert from ambipolar notation to a negative-logic system, replace all *asserted-low* variables by *uncomplemented* variables, and all *asserted-high* variables by *complemented* variables.

EXAMPLE 2.3(a): Convert D, $\overline{E}$, and F from negative-logic notation to ambipolar notation.

SOLUTION: If D, $\overline{E}$, and F are low, then the answers are D(L), E(H), and F(L).

EXAMPLE 2.3(b): Convert D(H), E(L), and F(H) to negative logic.

SOLUTION: The answers are $\overline{D}$, E, and $\overline{F}$, since these literals are asserted low.

2.5.3 The Inverter and Assertion Levels

The inverter is a gate that changes assertion voltage levels. Thus, if A is asserted high, then $\overline{A}$ is asserted low, and if A is asserted low, then $\overline{A}$ is asserted high. In ambipolar notation, if the input is asserted high, the output is asserted low and vice versa. This is shown in Figure 2.6 with the symbols for matched assertion levels.

2.5.4 Mixed-Input Assertion Levels

The output of a postive-logic AND gate is asserted high if, and only if, both inputs are asserted high. In Figure 2.7, input A is asserted low, which means that $\overline{A}$ is asserted high. Thus, the output, asserted high, is $C(H) = (\overline{A} \cdot B)(H)$, as shown.

In positive-logic notation, A(L) can be replaced by $\overline{A}$(H), as shown in Figure 2.8(a), or an invert bubble can be shown at the A input to the gate, as shown in Figure 2.8(b). Since input A is asserted high while the gate input is asserted low due to the presence of an invert bubble, there is a mismatch of assertion levels, and A is complemented in the output expression. Since inverters change assertion levels, one can visualize the bubble as representing an inverter, as seen in Figure 2.8(c). All three circuits in Figure 2.8 are equivalent and they all realize the same output function.

By the same token, an invert bubble at the output of a gate indicates that the output is asserted low, as shown in Figure 2.9(a). In positive logic notation there is a mismatch in assertion levels at the output, and the output is asserted low or the function which is asserted high is complemented. If the output is C(L), it is also $\overline{C}$(H). The invert bubble at the output can also be thought of as an inverter, as shown in the circuit of Figure 2.9(b).

FIGURE 2.6.   The INVERTER changes assertion voltage levels (a) from high to low or (b) from low to high.

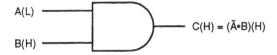

FIGURE 2.7.   Mixed-input assertion levels for a positive-logic AND gate.

Whether discussing an input or an output variable in positive logic, if the variable is asserted low, a bubble is required to indicate this. Then, if a bubble appears at an input to the gate, that variable will be complemented in the positive-logic output expression. If there is a bubble at the output of a gate, the output itself is active low and appears complemented in positive-logic notation.

In negative-logic notation, an input that is asserted high will be complemented in the output expression, and the output will be asserted low. In Figure 2.10(a), input B is asserted high while the input to the gate is asserted low. This is a mismatch in assertion levels between input B and the gate input, and B is complemented in the output expression. The circuit can also be drawn with a missing bubble at the B input, as shown in Figure 2.10(b), or an inverter and a bubble can be visualized as in Figure 2.10(c). All three representations of Figure 2.10 are equivalent, and the output is the same function in all three cases.

EXAMPLE 2.4:   Realize $F = A\overline{B} + \overline{C}D$ with AND and OR gates. The inputs A, B, C, and D are asserted high, and F is asserted low.

SOLUTION: Since F is asserted low, the output must be from a negative-logic OR gate as shown in Figure 2.11. In Figure 2.11, the inputs are asserted low. Therefore, we need negative-logic AND gates to realize $(A\cdot\overline{B})(L)$ and $(\overline{C}\cdot D)(L)$, as shown in Figure 2.12. The polarity of the inputs is to be asserted high as specified in the statement of the problem. This requires inverters at two inputs, giving the final circuit shown in Figure 2.13.

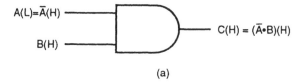

(a)

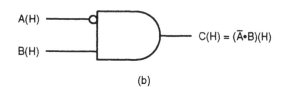

(b)

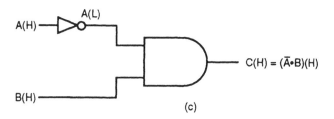

(c)

FIGURE 2.8.   A low input, A(L), in positive logic can be represented by (a) $\overline{A}$ high, or (b) an invert bubble, or (c) an inverter at the input.

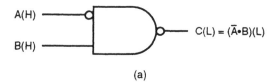

(a)

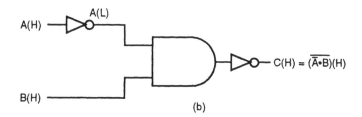

(b)

FIGURE 2.9.   A positive-logic AND gate with (a) input A and output C asserted low and (b) inverters replacing the invert bubbles.

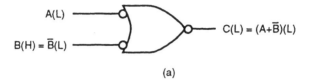

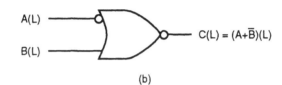

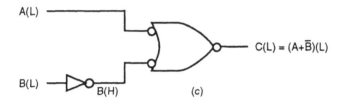

FIGURE 2.10. A high input, B(H), in negative logic can be represented by (a) $\overline{B}$ low, or (b) an active high input (no bubble), or (c) an inverter at the input.

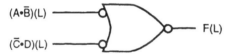

FIGURE 2.11. A negative-logic OR gate to realize $F(L) = A\overline{B} + \overline{C}D$.

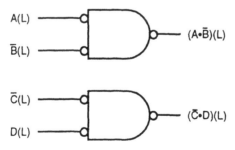

FIGURE 2.12. Two negative-logic AND gates to realize $A\overline{B}$ (L) and $\overline{C}D(L)$.

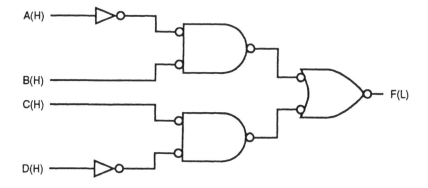

FIGURE 2.13.   Realization of the circuit for F(L) = A$\overline{\text{B}}$ + $\overline{\text{C}}$ D.

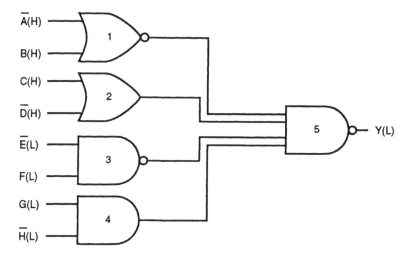

FIGURE 2.14.   A circuit composed of mixed inputs and mismatched assertion levels.

As a final example, consider the mixed-logic system shown in Figure 2.14. The first four inputs, $\overline{A}$, B, C, and $\overline{D}$, are obtained from a positive-logic system and are asserted high, as shown; while the next four inputs, $\overline{E}$, F, G, and $\overline{H}$, are derived from a negative-logic system and are asserted low, as shown. A solution is desired in which all the inputs match the corresponding gate-input assertion levels, the output is asserted high, and the number of bubbles used is minimized, consistent with these requirements.

The first step is to change gate 5 to have an asserted-high output. The new gate 5 will have to have all its inputs asserted low. To match these input assertion levels, change gate 2 from a positive-logic OR gate to a negative-logic AND gate, and change gate 4 from a positive-logic AND gate to a negative-logic OR gate.

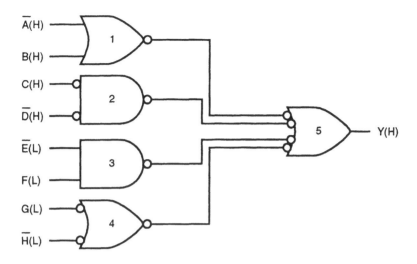

FIGURE 2.15. The partially modified circuit of Figure 2.14.

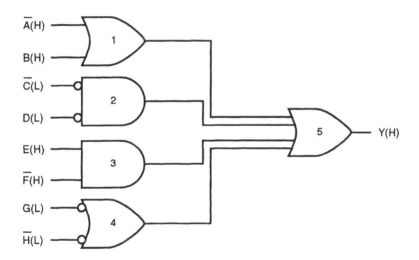

FIGURE 2.16. Circuit of Figure 2.14 redrawn as specified.

The result of these steps is shown in Figure 2.15.

Next, remove the double bubbles between the first-level gates and the second-level gate; adjust the inputs of gates 1 through 4 to match the assertion levels of the respective gates. This requires changing C(H) to $\overline{C}$(L), and $\overline{D}$(H) to D(L) at the input to gate 2, and changing $\overline{E}$(L) to E(H) and F(L) to $\overline{F}$(H) at the input to gate 3. The desired circuit is shown in Figure 2.16. Since all assertion levels are matched, the output can be read directly as Y(H) = ($\overline{A}$ + B + $\overline{C}$D + E$\overline{F}$ + G + $\overline{H}$)(H).

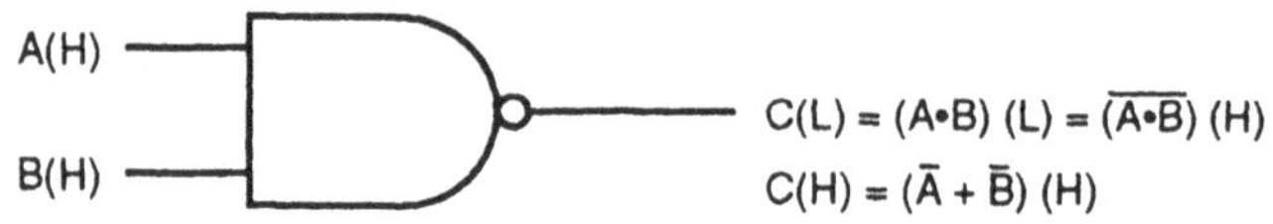

A(H) ─────────┐
              ⟩o──────── $C(L) = (A \cdot B)\ (L) = \overline{(A \cdot B)}\ (H)$
B(H) ─────────┘ $C(H) = (\bar{A} + \bar{B})\ (H)$

FIGURE 2.17. A positive-logic NAND gate.

A(H) ─────────┐
              ⟩⟩o──────── $C(L) = (A+B)\ (L) = \overline{(A+B)}\ (H)$
B(H) ─────────┘ $C(H) = (\bar{A} \cdot \bar{B})\ (H)$

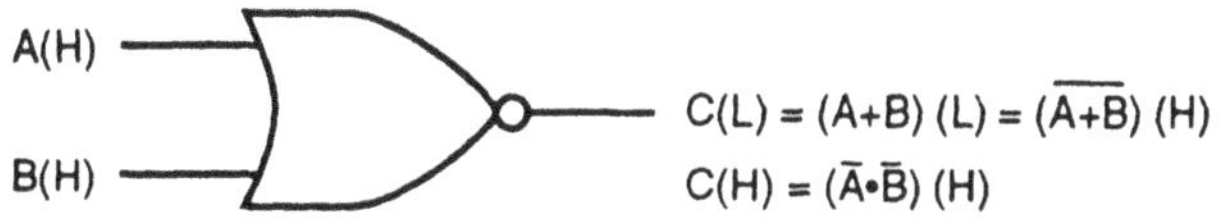

FIGURE 2.18. A positive-logic NOR gate.

2.6 NAND AND NOR GATES

If the output of an AND gate is inverted, one has a NOT-AND or NAND gate. If the output of an OR gate is inverted, one has a NOT-OR or NOR gate. These are very popular gates in all major logic families.

2.6.1 The NAND Gate

The NAND or NOT-AND gate outputs an *asserted-low* AND of the *asserted-high* inputs, as shown in Figure 2.17. Since a positive-logic AND gate is the same as a negative-logic OR gate, the output can be read as either $(A \cdot B)(L)$ or $(\bar{A} + \bar{B})(H)$. The fact that $(A \cdot B)(L)$ is equal to $(\bar{A} + \bar{B})(H)$ is shown in Section 2.8 to be one form of DeMorgan's law.

2.6.2 The NOR Gate

The NOR or NOT-OR gate outputs an *asserted-low* OR of the *asserted-high* inputs, as shown in Figure 2.18. Since a positive-logic OR gate is the same as a negative-logic AND gate, the output can be read as either $(A + B)(L)$ or $(\bar{A} \cdot \bar{B})(H)$. The fact that $(A + B)(L)$ is equal to $(\bar{A} \cdot \bar{B})(H)$ is shown in Section 2.8 to be the second form of DeMorgan's law.

The positive-logic NAND gate is a negative-logic NOR gate, and the positive-logic NOR gate is a negative-logic NAND gate. This is shown in terms of assertion levels in Figures 2.19 and 2.20.

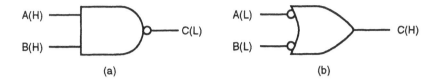

FIGURE 2.19.   The equivalence of (a) a positive-logic NAND gate and (b) a negative logic NOR gate.

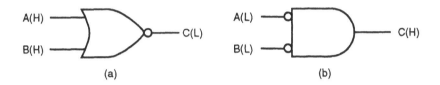

FIGURE 2.20.   The equivalence of (a) a positive-logic NOR gate and (b) a negative-logic NAND gate.

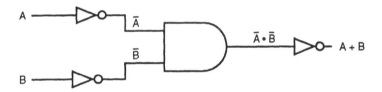

FIGURE 2.21.   Realization of an OR operation with AND and INVERT gates only.

2.7   FUNCTIONALLY COMPLETE SETS OF LOGIC GATES

AND, OR, and INVERT gates are sufficient to realize any function, because all three boolean operations are covered by these three gates. The question that naturally arises is whether one can get by with less than three kinds of gates. AND gates and OR gates cannot invert; hence, they will not suffice alone. If AND and INVERT gates can perform the logical OR operation, they will form a functionally complete set of logic gates, as will OR and INVERT gates if they can perform the logical AND operation.

2.7.1   AND and INVERT Gates Only

If the OR operation can be synthesized with AND and INVERT gates alone, then they will form a functionally complete set of gates. The realization of an OR gate with AND gates and INVERTERS is shown in Figure 2.21. Positive logic is assumed.

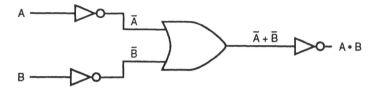

FIGURE 2.22. Realization of an AND operation with OR and INVERT gates only.

2.7.2 OR and INVERT Gates Only

If the AND operation can be synthesized with OR and INVERT gates alone, then they will also form a functionally complete set of gates. The realization of an AND gate with OR gates and INVERTERS is shown in Figure 2.22. Positive logic is again assumed.

2.7.3 NAND Gates and NOR Gates

NAND gates are functionally complete of themselves, as are NOR gates, since both can invert signals without additional help. The realization of all three boolean operations using only NAND gates, and the realization of all three boolean operations using only NOR gates is shown in Figure 2.23.

2.8 DEMORGAN'S THEOREM

DeMorgan's theorem is basic to the simplification of boolean expressions. For two variables, DeMorgan's theorem can be written in two ways, which are duals of each other.

$$\overline{A \cdot B} = \overline{A} + \overline{B} \qquad \text{or} \qquad \overline{A + B} = \overline{A} \cdot \overline{B} \qquad (2.2)$$

DeMorgan's theorem can be proven by using a truth table (Table 2.5).

The proof consists of observing that the sixth and seventh columns of the truth table are identical, as are the ninth and tenth columns. Since we have tried every combination of values for the variables A and B in the truth table, this consists of complete mathematical induction. Because $\overline{A \cdot B} = \overline{A} + \overline{B}$ for all four pairs of values of A and B, and $\overline{A + B} = \overline{A} \cdot \overline{B}$ for all four pairs of values, the expressions are always interchangeable.

Actually, DeMorgan's theorem is a statement of the fact that a *positive-logic* NAND gate is equivalent to a *negative-logic* NOR gate, and a *positive-logic* NOR gate is equivalent to a *negative-logic* NAND gate. This is shown in Figure 2.24.

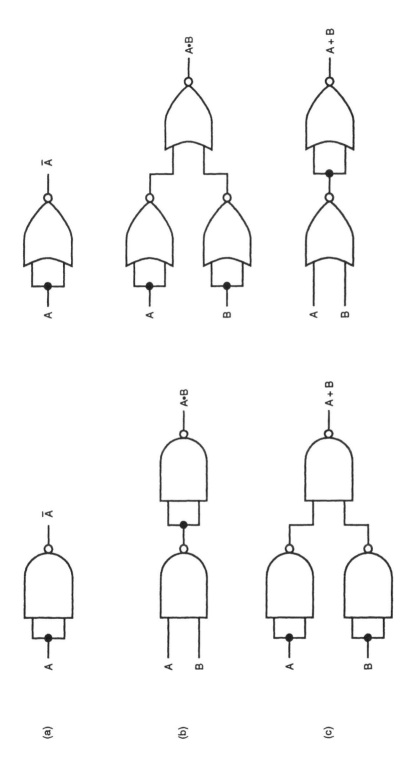

FIGURE 2.23. NAND and NOR gates alone can perform (a) the INVERT operation, (b) the AND operation, and (c) the OR operation.

TABLE 2.5 Proof of DeMorgan's Theorem

A	B	$\overline{A}$	$\overline{B}$	A·B	$\overline{A \cdot B}$	$\overline{A} + \overline{B}$	A+B	$\overline{A+B}$	$\overline{A} \cdot \overline{B}$
0	0	1	1	0	1	1	0	1	1
0	1	1	0	0	1	1	1	0	0
1	0	0	1	0	1	1	1	0	0
1	1	0	0	1	0	0	1	0	0

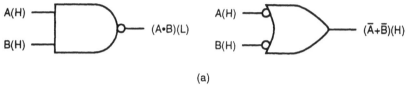

(a)

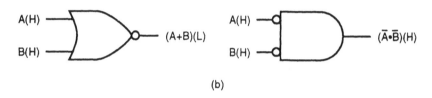

(b)

FIGURE 2.24. Circuit applications of DeMorgan's theorem. (a) A positive-logic NAND/negative-logic NOR gate and (b) a positive-logic NOR/negative-logic NAND gate.

DeMorgan's theorem can easily be generalized to n variables by applying mathematical induction, as follows:

$$\overline{A \cdot B \cdot C} = \overline{A} + \left(\overline{B \cdot C}\right) = \overline{A} + \overline{B} + \overline{C}$$
$$\overline{A \cdot B \cdot C \cdot D} = \overline{A} + \overline{B} + \left(\overline{C \cdot D}\right) = \overline{A} + \overline{B} + \overline{C} + \overline{D}$$
$$\overline{A + B + C} = \overline{A} \cdot \left(\overline{B + C}\right) = \overline{A} \cdot \overline{B} \cdot \overline{C}$$
$$\overline{A + B + C + D} = \overline{A} \cdot \overline{B} \cdot \left(\overline{C + D}\right) = \overline{A} \cdot \overline{B} \cdot \overline{C} \cdot \overline{D} \qquad (2.3)$$

The above can be generalized to obtain the complement of any boolean expression by

1. Complementing each variable ($\overline{0} = 1$ and $\overline{1} = 0$).
2. Complementing each operation ($\overline{\text{AND}}$ = OR and $\overline{\text{OR}}$ = AND).

TABLE 2.6   Truth Tables for EOR and ENOR Gates

A	B	A ⊕ B	A ⊙ B
0	0	0	1
0	1	1	0
1	0	1	0
1	1	0	1

EXAMPLE 2.5:   Apply DeMorgan's theorem to find the complement of $AB + \overline{C}D$.

SOLUTION: $\overline{A \cdot B + \overline{C} \cdot D} = \left(\overline{A} + \overline{B}\right) \cdot \left(C + \overline{D}\right)$.

EXAMPLE 2.6:   Find the complement of $A + B\overline{C} + \left(\overline{D} + E\right)F$.

SOLUTION: $\overline{A + B \cdot \overline{C} + \left(\overline{D} + E\right) \cdot F} = \overline{A} \cdot \left(\overline{B} + C\right) \cdot \left(D \cdot \overline{E} + \overline{F}\right)$.

EXAMPLE 2.7:   Find the complement of $\overline{A + B \cdot \overline{C} + \overline{\overline{D}}}$.

SOLUTION: $\overline{\overline{A + B \cdot \overline{C} + \overline{\overline{D}}}} = A + B \cdot \overline{C} + \overline{D}$.

2.9   EXCLUSIVE-OR AND EXCLUSIVE-NOR GATES

Two more very common gates are the *exclusive-OR (EOR* or *XOR)* gate, and the *exclusive-NOR (ENOR* or *XNOR)* gate. The ENOR gate is sometimes called an *equivalence* or *coincidence* gate.

The EOR function is defined in Table 2.6. Examination of the table reveals the EOR function to be *asserted high* when one input is *asserted high* and the other input is *asserted low*. It excludes the case where both inputs are asserted high, therefore $A + B = A \oplus B + AB$.

The complement of this is the ENOR gate, whose output is *asserted high* only if both inputs are identical or if there is a "co-incidence" of both inputs. There are several symbols for these two mathematical operations. The most commonly used symbols are ⊕ and ⊙, sometimes referred to as *ring-sum* and *ring-dot*.

The most common circuit symbols for the EOR and the ENOR gates are shown in Figure 2.25.

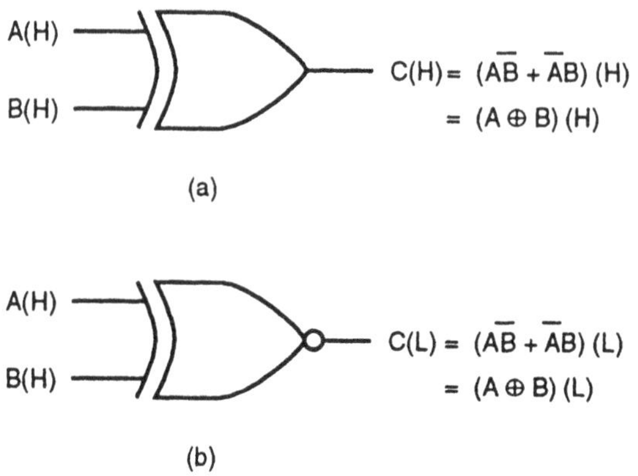

(a)

(b)

FIGURE 2.25. (a) The EOR gate and (b) the ENOR gate.

2.10 LOGICAL OPERATIONS ON BINARY WORDS

Binary words can be operated upon the same as single bits. Operating on words is done by performing the operation bitwise over the entire binary word or vector. The AND of two vectors is the vector-wide (bitwise) AND of the two vectors. The inclusive and exclusive OR of two vectors is also defined as the vector-wide OR or EOR operation. Some examples of word operations follow.

EXAMPLE 2.8: Find the vector complement of A, and the vector-wide AND, OR, and EOR of A = 1011 and B = 1100.

SOLUTION: $\overline{A} = \overline{1011} = 0100$
$A \cdot B = (1011) \cdot (1100) = 1000$
$A + B = (1011) + (1100) = 1111$
$A \oplus B = (1011) \oplus (1100) = 0111$

EXAMPLE 2.9: Find $(33_{16}) \cdot (F0_{16})$, $(33_{16}) + (F0_{16})$ and $(33_{16}) \oplus (F0_{16})$ in hexidecimal.

SOLUTION: $(33_{16}) \cdot (F0_{16}) = (0011\ 0011) \cdot (1111\ 0000) = 0011\ 0000 = 30_{16}$.
$(33_{16}) + (F0_{16}) = (0011\ 0011) + (1111\ 0000) = 1111\ 0011 = F3_{16}$. $(33_{16}) \oplus (F0_{16}) = (0011\ 0011) \oplus (1111\ 0000) = 1100\ 0011 = C3_{16}$.

EXAMPLE 2.10: The AND of $0F_{16}$ with any ASCII number will convert it to its BCD equivalent. Show that this is true for ASCII numbers 2 and 7.

SOLUTION: $2_{ASCII} = 32_{16} = 0011\ 0010_{BCD}$, and $7_{ASCII} = 37_{16} = 0011$
0111_{BCD}. $32_{16} \cdot 0F_{16} = (0011\ 0010) \cdot (0000\ 1111) = 0000\ 0010_{BCD}$
$= 02_{10}$. $37_{16} \cdot 0F_{16} = (0011\ 0111) \cdot (0000\ 1111) = 0000\ 0111_{BCD} =$
07_{10}.

2.11   SUMMARY

Binary boolean algebra or switching theory is a mathematical system for dealing with two-valued statements that are either true or false. Precise formal meanings are assigned to the three fundamental or primitive grammatical terms AND, OR, and NOT. The three basic boolean operations, COMPLEMENT, AND and OR, are done in that sequence.

A gate is a unilateral piece of hardware, usually electronic or mechanical, which implements a logical operation and outputs one of two voltage levels. Three circuit elements, called INVERTERS, AND gates, and OR gates, perform the three basic boolean operations. One must not confuse the logic operators with specific logic gates. A positive-logic AND gate is a negative-logic OR gate and vice versa. Hence, the logical operation performed depends upon the point of view.

An appropriate combination of AND, OR, and COMPLEMENT steps can represent any given boolean expression, and all logic operations can be reduced to these three operations. Logic circuits that can realize arbitrarily complex functions are built out of these elementary gates that perform logical operations on binary variables. Electronic gates are commercially available that perform the operations of NAND, NOR, EOR, and ENOR.

Sometimes it is necessary to distinguish between logical and arithmetic operations. In this event, it is necessary to use $+$ and $\cdot$ for arithmetic operations and use other symbols for boolean operations. The most common symbols used by mathematicians and logicians are the logical union, $\cup$ or $\vee$, sometimes called "cup", and the logical intersection, $\cap$ or $\wedge$, sometimes called "cap".

Logic functions can be described by truth tables. A truth table is a systematic listing of all the input states or values, with the corresponding output state or value. The output value corresponding to a given input is called a minterm, and is numerically designated by the decimal equivalent of the input variables corresponding to that row of the truth table. The truth table is a valuable tool for proving theorems, as will be shown in Chapter 3.

From the engineering hardware point of view, the INVERTER, the NAND gate, and the NOR gate are the basic electronic circuit building blocks. NAND gates alone or NOR gates alone can realize any boolean expression, and circuits of arbitrary complexity can be constructed with combinations of these two gates.

EOR and ENOR gates are easy to build and have many applications. The EOR symbol is $\oplus$, and the ENOR symbol is $\odot$. $A \oplus B = A\overline{B} + \overline{A}B$, and $A \odot B = \overline{A \oplus B} = \overline{A}\ \overline{B} + AB$. The EOR and ENOR logical operations each consist of two

AND and one OR operation, and each can be constructed from two AND gates and one OR gate. EOR and ENOR gates find application in circuits for generating and detecting even or odd parity.

Assertion levels have more meaning in digital circuitry than true and false. Low assertion levels are associated with the presence of the symbol "_", often referred to as a "bubble", and high assertion levels are associated with the absence of this symbol. For example, a gate might have a CLEAR input which forces the output state of that gate to 0. If the CLEAR input is asserted high, then it must normally be kept at a low or 0 level. When it is raised to a high level it is asserted and the output of the gate is forced to 0. If the CLEAR input is asserted low, this is indicated by the presence of a "bubble", and the input is normally kept in the high or 1 position. Whenever the CLEAR input goes low, the clear operation is asserted and the output of the gate is forced to the 0 state.

High voltages and low voltages can represent 1s and 0s, respectively (positive logic), or 0s and 1s, respectively (negative logic).

DeMorgan's theorem is basic to simplifying boolean expressions, so that the hardware can be implemented with a minimum number of gates and inputs, leading to minimum cost and chip area.

PROBLEMS

In Problems 2.1 through 2.6, give the number of variables, literals, and minterms present in the function.

2.1. $F(A,B,C) = \overline{A}\,\overline{B}\,\overline{C} + \overline{A}BC + A\overline{B}C + AB\overline{C}$.

2.2. $F(A,B,C) = \overline{A}\,\overline{B}C + \overline{A}B\overline{C} + A\overline{B}\,\overline{C} + ABC$.

2.3. $F(A,B,C,D) = \overline{A}\,\overline{B}\,\overline{C}\,\overline{D} + \overline{A}BCD + \overline{A}B\overline{C}D + AC\overline{D}$.

2.4. $F(A,B,C,D) = \overline{A}\,\overline{B}C\overline{D} + A\overline{B}\,\overline{C}D + ABC\overline{D} + AB\overline{C}$.

2.5. $F(A,B,C,D) = \overline{A}B\overline{C}D + AB\overline{C}\,\overline{D} + \overline{A}\,\overline{B}C + AC$.

2.6. $F(A,B,C,D) = \overline{A}BCD + ABC\overline{D} + \overline{A}\,\overline{B}\,\overline{C} + A\overline{C}$.

In Problems 2.7 through 2.10, specify the positive-logic output of each gate below if the inputs are A(H), B(H), and C(H).

2.7. A positive-logic NAND gate.
2.8. A positive-logic NOR gate.
2.9. A negative-logic NAND gate.
2.10. A negative-logic NOR gate.

In Problems 2.11 through 2.14, specify the negative-logic output of each gate below if the inputs are A(L), B(L), and C(L).

2.11. A positive-logic NAND gate.
2.12. A positive-logic NOR gate.
2.13. A negative-logic NAND gate.
2.14. A negative-logic NOR gate.

In Problems 2.15 through 2.18, specify the ambipolar logic output of a positive-logic gate having the following inputs:

2.15. A NAND gate with inputs A(H), B(H), and C(H).
2.16. A NOR gate with inputs A(H), B(H), and C(H).
2.17. An AND gate with inputs $\overline{A}$ (L), $\overline{B}$ (L), and $\overline{C}$ (L).
2.18. An OR gate with inputs $\overline{A}$ (L), $\overline{B}$ (L), and $\overline{C}$ (L).

In Problems 2.19 through 2.24, draw the logic diagrams to realize the function Y. Use the positive-logic gates specified and match all assertion levels.

2.19. Y = (AB + CD)(H) with all the inputs asserted high. Use AND and OR gates.
2.20. Y = (AB + CD)(H) with all the inputs asserted low. Use NOR and OR gates.
2.21. Y = (AB + CD)(L) with all the inputs asserted high. Use AND and NOR gates.
2.22. Y = (AB + CD)(L) with all the inputs asserted low. Use NOR gates only.
2.23. Y = (AB + CD)(L), with all inputs asserted high. Use NOR gates only.
2.24. Y = (AB + CD)(H), with all inputs asserted low. Use NAND gates only.
2.25. Simplify Figure 2.13 to a circuit with only three invert bubbles.
2.26. Redraw Figure 2.23, matching all assertion levels.
2.27. Evaluate Z for the truth table shown below.
       (a) For positive logic, and
       (b) for negative logic.

A	B	Z
LV	LV	LV
LV	HV	HV
HV	LV	HV
HV	HV	LV

2.28. (a) Interpret the truth table shown below for positive logic and write

an expression for Y as a sum of minterms.
(b) Repeat part (a) for negative logic.

A	B	C	Y
LV	LV	LV	LV
LV	LV	HV	HV
LV	HV	LV	LV
LV	HV	HV	LV
HV	LV	LV	LV
HV	LV	HV	HV
HV	HV	LV	HV
HV	HV	HV	HV

2.29. Find ABC and $A + B + C$, if $A = 011011$, $B = 011100$, and $C = 001110$.

2.30. Find ABC and $A + B + C$, if $A = 010101$, $B = 101101$, and $C = 101011$.

2.31. Use DeMorgan's law to find $\overline{AB}$ and $\overline{A + B}$, if $A = 101010$ and $B = 100111$.

2.32. Use DeMorgan's law to find $\overline{AB}$ and $\overline{A + B}$, if $A = 010101$ and $B = 011000$.

In problems 2.33 through 2.36, use DeMorgan's law to evaluate the expressions.

2.33. $Y = \overline{A + B + CD}$.

2.34. $Y = \overline{AB + C + D}$.

2.35. $Y = \overline{\overline{A + B} + \overline{CD}}$.

2.36. $Y = \overline{\overline{AB} + \overline{C} + D}$.

2.37. Show algebraically that $\overline{A} \oplus B = A \oplus \overline{B} = A \odot B$.

2.38. Show that $X \oplus 0 = X$, $X \oplus 1 = \overline{X}$, and $X \oplus \overline{X} = 1$.

2.39. Implement the function $G = (A \oplus B) \odot C$ with the least number of two-input EOR and/or ENOR gates.

2.40. Implement the function $H = (A \odot B) \oplus C$ with the least number of two-input EOR and/or ENOR gates.

2.41. Use a truth table to show that $(A \oplus B) \oplus C = A \oplus (B \oplus C)$.

2.42. Use a truth table to show that $(A \oplus B) \odot C = A \odot (B \oplus C)$.

2.43. Use DeMorgan's law to find the complement of $(A + \overline{B})\overline{C}D + A\overline{C}$.

2.44. Use DeMorgan's law to find the complement of $(\overline{A}B + C + \overline{D})(\overline{A} + C)$.

2.45. Implement the function $(AB + C)(\overline{D} + E)$ in positive logic, using AND, OR, and INVERT gates.

2.46. Implement the function $(A + B\overline{C})(D + E)$ in positive logic, using AND, OR, and INVERT gates.

SPECIAL PROBLEMS

2.47. Construct a circuit that will output B when A is asserted and output C when A is not asserted. (Hint: three gates are needed.)

2.48. Design a circuit to realize the function $Y = (A + B)(C + D)$, using
  (a) two positive-logic OR gates and one positive-logic AND gate;
  (b) two positive-logic NOR gates and a negative-logic NAND gate; and
  (c) three positive-logic NOR gates.

2.49. (a) Draw a truth table for a two-bit output, Y and Z, which is the binary equivalent of the sum of the 1s in the three-bit input function A, B, and C (i.e., when the input is ABC = 010, the output is $YZ = 01_2 = 1_{10}$, and when the input is ABC = 101, the output is $YZ = 10_2 = 2_{10}$). The sum of the logical 1s in a boolean vector is called the vector magnitude.
  (b) Write the boolean expression for Y and Z as a sum of minterms.

2.50. (a) Make a truth table for inputs A, B, C, and D. The output Y is to be asserted or true if two 0s are adjacent on the input table. A and D are not adjacent (i.e., for ABCD = 0011, Y = 1, and for ABCD = 0101, Y = 0).
  (b) Write the boolean expression for Y as a sum of minterms.

REFERENCES

1. **Boole, G.,** *An Investigation of the Laws of Thought, on Which are Founded the Mathematical Theories of Logic and Probability,* 1849 (Reprinted by Dover, 1954).
2. **Huntington, E. V.,** Sets of independent postulates for the algebra of logic, *Trans. Am. Math. Soc.,* 5, 288–309, 1904.
3. **Huntington, E. V.,** New sets of independent postulates for the algebra of logic, with special references to Whitehead and Russell's principia mathematica, *Trans. Am. Math. Soc.,* 35, 274–304, 1933.
4. **Shannon, C. E.,** Symbolic analysis of relay and switching circuits, *AIEE Trans.,* 57, 713–723, 1938.

CHAPTER 3

Boolean Algebra and Circuit Realizations

3.1 INTRODUCTION

Because of the importance of switching or binary boolean algebra to the design of calculators, computers, communications systems, control systems, and the burgeoning digital applications of the last few decades, it is important that the student of electrical or computer engineering have a good understanding of the system of switching algebra. With this system one can formulate propositions that are true or false, combine them to create new propositions, and determine the truth or falsehood of these new propositions.

Every scientific field is based upon fundamental *axioms* or *postulates,* which are *a priori assumptions* upon which the field of knowledge is based. These are the basic building blocks of the theory behind a given field and form the basis upon which theorems and proofs of theorems are derived. These postulates are statements about a set of undefined objects called *primitives.* The statements must be both internally consistent and independent. They should also be simple as well as useful. In general, more simple assumptions are usually considered more elegant.

The first set of *a priori* assumptions a mathematics student comes in contact with are usually Euclid's postulates of geometry. From a small set of axioms, all of geometry follows. Euclid postulated that there is only one line through a point parallel to a given line. While this contradicted no other postulates, it was long thought possible to derive this postulate from other postulates. One proof of its dependence would be to assume the postulate is false and show that a contradiction arises. Attempts along this line eventually lead to the discovery of Riemann's and Lobachevsky's non-euclidian geometries.

The basic postulates of boolean algebra are due to Huntington[1,2] and can be used to evaluate or derive other theorems. Huntington's postulates concern several operators defined over a set of primitive (undefined) elements. A *set* is any collection of elements having some common property or properties. The set of undefined elements of boolean algebra can consist of the numerals 0, 1, 2, 3, etc. A general boolean algebra must have 2^n elements.[3] In particular, in switching theory, or binary boolean algebra, the primitive set consists of 0 and 1, about which the operations of ANDing, ORing, and complementing are defined. Huntington found that all the results and implications of Boole's algebra could be derived from six basic postulates.

Starting from the Huntington postulates, various basic theorems can be derived. A listing of the more important theorems will be given, proven, and studied in this chapter. Two-level logic representations of functions lead to *sum-of-products (SOP)* solutions and *product-of-sums (POS)* solutions. These representations can be realized in what are called *minterm expansions* or *canonical SOP* forms and *maxterm expansions*, or *canonical POS* forms. The term *canonical* as used here refers to a form of defining a boolean function such that other functions can be compared to determine equivalence. There is a one-to-one correspondence between a canonical SOP or POS and a truth table which uniquely defines any function.

Any SOP solution to a boolean function can be realized or implemented with two levels of logic consisting of AND gates driving OR gates, called AND-OR logic. An SOP solution can also be realized with two levels of NAND gates only, called NAND-NAND logic, or with OR gates driving NAND gates, called OR-NAND or OR-AND-INVERT logic, or with NOR gates driving OR gates, called NOR-OR logic.

Any POS solution can be realized with two levels of logic consisting of OR gates driving AND gates, called OR-AND logic. A POS solution can also be realized with two levels of NOR gates, called NOR-NOR logic, or with AND gates driving NOR gates, called AND-NOR or AND-OR-INVERT logic, or with NAND gates driving AND gates, called NAND-AND logic. All of these two-level representations will be covered in this chapter.

Shannon's expansion theorem is a method of obtaining a factored form of any boolean expression. This gives solutions of three or more levels of logic. Factoring may reduce the number of gates needed or the number of literals required. Factoring increases the maximum path length of a signal from the input to the output of the circuit, and this may produce a slower circuit. Factoring also reduces the maximum fan-in of a circuit which may offset the added path length. When this is true, the factored form is superior to a two-level realization.

3.2 HUNTINGTON'S POSTULATES

The basic theorems of boolean algebra rest upon a system of axioms and postulates formulated by E. V. Huntington in 1904. Axioms and postulates themselves are statements about undefined objects that are accepted as useful *a priori* concepts. Thus, any proposed boolean algebra system can be tested by application of Huntington's postulates. The binary system is by no means the only possible boolean system. Other boolean systems are considered in the problems at the end of this chapter.

Huntington's postulates can be stated as follows:

1. The set of elements, S, must be closed with respect to all operators. The two fundamental operators in a binary boolean system are AND and OR, written "·" and "+". The element A·B and the element A + B must both be elements of S, provided that A and B are elements of S. In general, A does not equal B. In set-theory notation, S is a set such that if A∈ S and B∈ S then (A·B)∈ S and (A + B)∈ S.

2. There exists an element 1 in S, such that for every A in S, A·1 = A; and there exists an element 0 in S such that for every A in S, A + 0 = A.

3. The elements of S must commute: A·B = B·A and A + B = B + A.

4. The elements of S must obey the distributive laws:
    A + (B·C) = (A + B)·(A + C), and A·(B + C) = (A·B) + (A·C).

5. The elements of S must obey the associative laws:
    (A·B)·C = A·(B·C) = A·B·C, and (A + B) + C = A + (B + C) = A + B + C.

6. Every element in S posesses a unique complement, such that $(\overline{\overline{A}})$ = A, A·$\overline{A}$ = 0 and A + $\overline{A}$ = 1, where 0 and 1 are complements. Thus,

$$S = \{0, 1, A, B, \overline{A}, \overline{B}, A·B, A + B\}$$

The basic postulates and theorems of a binary boolean system are summarized in Table 3.1.

For a more detailed investigation of Huntington's postulates, the interested reader is referred to the original papers of E. V. Huntington.[1,2] Once the basic postulates are agreed upon, other and more powerful theorems can be proven from these *a priori* axioms.

The theorems and postulates listed in Table 3.1 are paired because the theorems and postulates labeled a and b are duals of each other. The dual of an axiom or theorem is obtained by interchanging 0 and 1, · and + throughout the original axiom or theorem. The principle of duality is that if a boolean statement is proven true, the dual of that statement is also true. This reduces the work of proving theorems by half.

The principle of duality is called a metatheorem, a theorem about theorems. It is true because the duals of all the axioms are true, and the duals of all the theorems can be proven using axioms which are the duals of those used to prove the original theorem.

Physically, the commutative and associative laws tell us that it is immaterial in what order gate inputs are connected. Also, one n-input gate can be replaced by (n – 1) two-input gates. Propagation delays and cost are probably higher for the two-input gates, but they perform the same logic as the single gate. The distributive law of logical multiplication over logical addition allows one to "multiply out" an expression to obtain an answer in the form of a sum of products. The dual

TABLE 3.1   Boolean Postulates and Theorems

Idempotent Laws

P1a $A \cdot A = A$ P1b $A + A = A$

Operations with 1 and 0

P2a $1 \cdot 0 = 0 \cdot 1 = 0$ P2b $0 + 1 = 1 + 0 = 1$
P3a $A \cdot 0 = 0$ P3b $A + 1 = 1$
P4a $A \cdot 1 = A$ P4b $A + 0 = A$

Commutative Laws

P5a $A \cdot B = B \cdot A$ P5b $A + B = B + A$

Distributive Laws

P6a $A + (B \cdot C) = (A + B) \cdot (A + C)$
P6b $A \cdot (B + C) = (A \cdot B) + (A \cdot C)$

Associative Laws

P7a $(A \cdot B) \cdot C = A \cdot (B \cdot C) = A \cdot B \cdot C$
P7b $(A + B) + C = A + (B + C) = A + B + C$

Complementarity Laws

P8a $\bar{1} = 0$ P8b $\bar{0} = 1$
P9a $A \cdot \bar{A} = 0$ P9b $A + \bar{A} = 1$
                P10 $\overline{\left(\bar{A}\right)} = A$

Logical Adjacency (Uniting) Theorems

T11a $(A + B) \cdot (A + \bar{B}) = A$ T11b $A \cdot B + A \cdot \bar{B} = A$

Absorption Theorems

T12a $A \cdot (A + B) = A$ T12b $A + A \cdot B = A$
T13a $A \cdot \bar{B} + B = A + B$ T13b $(A + \bar{B}) \cdot B = A \cdot B$

Consensus Theorems

T14a $A \cdot B + B \cdot C + \bar{A} \cdot C = A \cdot B + \bar{A} \cdot C$
T14b $(A + B) \cdot (B + C) \cdot (\bar{A} + C) = (A + B) \cdot (\bar{A} + C)$

Demorgan's Laws

T15a $\overline{A \cdot B} = \bar{A} + \bar{B}$ T15b $\overline{A + B} = \bar{A} \cdot \bar{B}$

TABLE 3.1   Boolean Postulates and Theorems (continued).

Shannon's Expansion Theorems

T16a $F(X_1, X_2, \ldots, X_n) = X_1 \cdot F(1, X_2, \ldots, X_n) + \overline{X}_1 \cdot F(0, X_2, \ldots, X_n)$

T16b $F(X_1, X_2, \ldots, X_n) = [X_1 + F(0, X_2, \ldots, X_n)] \cdot [\overline{X}_1 + F(1, X_2, \ldots, X_n)]$

distributive law of logical addition over multiplication allows one to "add out" an expression to obtain an answer in the form of a product of sums.

The absorption theorems allow simplification of boolean expressions by deleting redundant terms. The consensus theorem also allows the removal of redundant terms. In T14a in Table 3.1, BC is the consensus term. If BC is true then B and C separately must be true. Thus, either AB is true or $\overline{A}C$ is true, and BC is redundant. The same arguement applies to the dual theorem, T14b.

DeMorgan's laws state that any n-variable expression can be complemented by interchanging the operations of · and + and complementing all the variables. Shannon's expansion theorems allow any function to be split into two subfunctions, one of which is multiplied by a variable called the splitting variable, and the other of which is multiplied by the complement of the splitting variable. The subfunctions are independent of the splitting variable, which takes on the value 1 in one of the subfunctions and the value 0 in the other subfunction.

3.3   PROVING BINARY BOOLEAN THEOREMS

Binary boolean theorems can be proven by direct application of the basic postulates, by Venn diagrams, by analysis of switching theory, by analysis of logic gates, and by application of truth tables. Each of these methods has its advantages, and an example of each follows, starting with manipulation of the basic axioms.

EXAMPLE 3.1:   Use basic axioms to prove T12b, $X + X \cdot Y = X$.

SOLUTION: $X + X \cdot Y = X \cdot 1 + X \cdot Y$ (P4a)

$= X \cdot (1 + Y)$ (P6b)

$= X \cdot 1$ (P3b)

$= X$ (P4a)

Switching theorems can also be proven by means of a graph called a Venn diagram. In this approach, it is customary to draw a rectangular box representing all the possible elements of the set or sets in question, referred to as the universe under discourse. If X and Y represent subsets of this universe, then the set of all elements that belong to X and the set of all elements that belong to Y can be represented by smaller boxes, circles, or other convenient shapes within the rectangle.

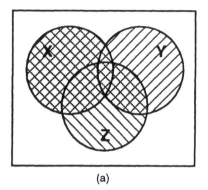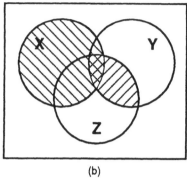

(a)   (b)

FIGURE 3.1.   Venn diagram of (a) (X + Y)(X + Z), and (b) X + YZ.

If X and Y have elements common to both sets, there is a logical intersection X·Y or X∩Y represented by the overlap of X and Y on the diagram. Whether the intersection of X and Y is the null set or not, the union of X and Y, X + Y = X∪Y, consists of all the elements in either X or Y or both X and Y.

The logical AND operation is represented by the *intersection* or overlap of the two areas, and the logical OR operation is represented by the *union* or sum of the two areas. By shading or coloring different areas on the Venn diagram, one can visually interpret the answer to the function so represented. In Example 3.2, the Venn diagram is used to prove the theorem (X + Y)(X + Z) = X + YZ.

EXAMPLE 3.2:   Use a Venn diagram to prove that (X + Y)(X + Z) = X + YZ.

SOLUTION: The Venn diagram for variables X, Y, and Z is shown in Figure 3.1. (X + Y)(X + Z) is the area shaded twice in Figure 3.1(a), and X + YZ is the total shaded area in Figure 3.1(b). They are both the same areas, which completes the proof.

There are two types of logic circuits, those composed of switches and those composed of gates. Both circuits can be used to prove theorems of boolean algebra.

The application of switching theory to prove boolean theorems rests upon the basic behavior of switching circuits. Ideal switches will be assumed in which no current, and therefore no signal, can pass through an open switch (it has infinite resistance to electrical current), and there is no voltage drop across a closed switch (it has zero resistance), and it transmits any signal through it.

There are two categories of switches, *toggle* and *push-button.* Toggle switches, such as light switches on a wall, remain in one state until "toggled" to the other state. This type switch will be discussed with respect to flip-flops and memory devices. The push-button switch is a momentary switch such as those found on a calculator or computer keyboard. These switches have a *normal* position and an *asserted* position and their behavior can be simulated by transistors in a logic switching circuit.

Variables are represented as *closed* or as *open* switches, where *"closed"* can be represented by the value 0 or by the value 1, and *"open"* can be represented by the complementary value. A signal will be transmitted if there is a closed path from the input to the output and no signal will be transmitted if there is an open switch in the path. The transmission of a signal from the input to the output of a switching circuit can be represented by a 1 when there is a closed path and by a 0 when there is no closed path.

A push-button switch that is normally open has no continuity across it until action is taken to close it, whereas a push-button switch that is normally closed has continuity until action is taken to open it. If the variable 1 is associated with action, and the variable 0 is associated with the absence of action, then a normally open switch transmits a signal when an action takes place. In this case, the variable controlling the switch takes on the value 1 when transmission takes on the value 1. This can be called a *positive switch*.

By the same reasoning, a normally closed switch transmits a signal when no action is taken or when the variable controlling it is 0. The switch produces an open circuit when it is activated or when its controlling variable takes the value 1, and transmission ceases when this occurs. Call this push-button switch a *negative switch*. In the following discussion positive switches are assumed.

Since a signal can only be transmitted through two or more switches in series if they are all closed, the logical AND operation is performed by switches in *series*. A signal can be transmitted through either of two paths in parallel, and the logical OR operation is performed by switches in *parallel*.

There is one limitation to switching circuits. To realize a boolean function with a switching circuit, there must be no requirement to complement other than the primary inputs since there is no switch that performs the complement operation. For more information on switching circuits, see R. S. Sandige.[4]

EXAMPLE 3.3:   Use switching theory to prove that $(X + Y)(X + Z) = X + YZ$.

SOLUTION: The switching circuit is shown in Figure 3.2. The outputs of both circuits are the same when the inputs are identical.

One can prove theorems by network analysis of logic gate structures as defined and used in Chapter 2. This requires associating mathematical symbols with specific logic gates. The logical operations of $\cdot$, +, and complement are associated with AND, OR, and INVERT gates, respectively. Example 3.4 gives a switching network proof that $(X + Y)(X + Z) = X + YZ$.

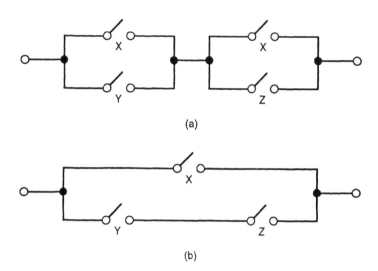

(a)

(b)

FIGURE 3.2. Switching circuit diagram of (a) $(X + Y)(X + Z)$, and (b) $X + YZ$.

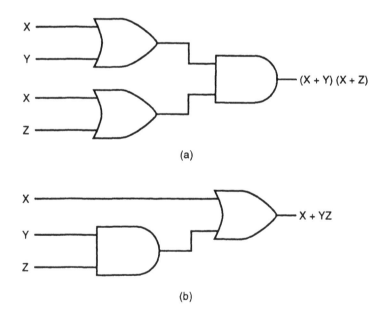

(a)

(b)

FIGURE 3.3. Logic gate reduction of (a) $(X + Y)(X + Z)$, to (b) $X + YZ$.

EXAMPLE 3.4: Use logic gates to prove that $(X + Y)(X + Z) = X + YZ$.

SOLUTION: The logic gate solution is shown in Figure 3.3. Again, the outputs of both circuits are the same when the inputs are identical.

Lastly, one can solve any boolean problem by the application of truth tables. The truth table constitutes complete mathematical induction, and in general we will use a truth table proof for any given problem. In Chapters 4 and 5, advanced methods of simplification based upon truth tables will be covered which will make this the preferred method when very complex circuits are involved.

EXAMPLE 3.5: Use a truth table to prove that $X + X \cdot Y = X$.

SOLUTION:
TABLE 3.2 $X + XY = X$

X	Y	X·Y	X + X·Y
0	0	0	0
0	1	0	0
1	0	0	1
1	1	1	1

The first and last columns are identical, which completes the proof.

EXAMPLE 3.6: Use a truth table to prove that $X + YZ = (X + Y)(X + Z)$.

SOLUTION:
TABLE 3.3 $X + YZ = (X + Y)(X + Z)$

X	Y	Z	YZ	X + YZ	X + Y	X + Z	(X + Y)(X + Z)
0	0	0	0	0	0	0	0
0	0	1	0	0	0	1	0
0	1	0	0	0	1	0	0
0	1	1	1	1	1	1	1
1	0	0	0	1	1	1	1
1	0	1	0	1	1	1	1
1	1	0	0	1	1	1	1
1	1	1	1	1	1	1	1

The fifth and the eighth columns are identical, which proves the theorem.

3.4   TWO-LEVEL LOGIC

One of the major applications of boolean algebra is in the simplification of logic or switching circuits, and the starting point for this is the minimization of boolean expressions. Many techniques for simplifying a given function require a standard form of representation of general functions. Two standard forms are the sum-of-products form and the product-of-sums form.

In logic terminology, the SOP form is called the *disjunctive normal* form since it is a disjunction of conjunctions of literals. As such it can be written as a union of terms which are intersections of more than one literal. In switching terminology, the SOP form is an ORing of ANDed terms. This is often the preferred form of writing boolean expressions.

In logic terminology, the POS form is called the *conjunctive normal* form since it is a conjunction of disjunctions of literals. As such it can be written as the intersection of terms which are unions of more than one literal. In switching terminology, the POS form is an ANDing of ORed terms.

3.4.1   Sum-of-Products Form

A boolean expression is said to be in the SOP form when the expression is multiplied out such that it consists of an SOP of single variables only. The key theorem needed to achieve an SOP form is the distributive law, P6a, namely, $(X + Y)(X + Z) = X + YZ$. In Examples 3.7 through 3.10 inclusive, the commutative laws (P5a and P5b) are assumed.

EXAMPLE 3.7: Reduce the expression $(A + BC)(A + \overline{B}E)$ to an SOP form.

SOLUTION: $(A + BC)(A + \overline{B}E) = A + B\overline{B}CE$ (P6a)

$\qquad\qquad\qquad\qquad\qquad\quad = A$ (P9a)

EXAMPLE 3.8: Obtain an SOP form of the function $(A + B)(A + C)(B + C)$.

SOLUTION: $(A + B)(A + C)(B + C) = (A + BC)(B + C)$ (P6a)

$\qquad\qquad\qquad\qquad\qquad\qquad = A(B + C) + BC(B + C)$ (P6b)

$\qquad\qquad\qquad\qquad\qquad\qquad = AB + BBC + AC + BCC$ (P6b)

$\qquad\qquad\qquad\qquad\qquad\qquad = AB + AC + BC$ (P1a, P1b)

3.4.2   Product-of-Sums Form

A boolean expression is said to be in the *POS* form, when the expression is factored out such that it consists of a product of sums of single variables only. The key theorem needed to achieve a POS form is the distributive law, P6a, namely, $X + YZ = (X + Y)(X + Z)$.

EXAMPLE 3.9: Reduce the expression $AB + CDE$ to a POS form.

SOLUTION: $AB + CDE = (AB + CD)(AB + E)$ (P6a)

$\qquad\qquad\qquad\quad = (AB + C)(AB + D)(A + E)(B + E)$ (P6a)

$\qquad\qquad\quad = (A + C)(B + C)(A + D)(B + D)(A + E)(B + E)$ (P6a)

EXAMPLE 3.10: Obtain a POS form of the function AB + AC + BC.

SOLUTION:
$$
\begin{aligned}
AB + AC + BC &= A(B + C) + BC & \text{(P6b)} \\
&= (BC + A)(BC + B + C) & \text{(P6a)} \\
&= (BC + A)(B + C) & \text{(T12b)} \\
&= (A + B)(A + C)(B + C) & \text{(P6a)}
\end{aligned}
$$

3.4.3 Realization of SOP and POS Forms

The SOP realization of a switching function is also called the *disjunctive* form of the function, while the POS realization of a function is called the *conjunctive* form. The SOP form of a boolean expression yields itself directly to a two-level circuit realization referred to as an AND-OR realization, where the first level consists of AND gates, and the second level consists of OR gates. An AND-OR representation of the function AB + AC + BC is shown in Figure 3.4.

Likewise, a POS form of a boolean expression yields itself directly to a two-level circuit realization referred to as an OR-AND realization, where the first level consists of OR gates and the second level consists of AND gates. An OR-AND representation of the function (A + B)(A + C)(B + C) is shown in Figure 3.5. Note that Figures 3.4 and 3.5 both represent the same function.

3.5 MINTERMS AND MAXTERMS

A boolean expression that is represented by a truth table is uniquely specified and any other expression having the same truth table is equivalent to it. Because of this, the output values of an expression which occupy rows of the truth table are labeled according to the row occupied and are given the name minterms. There are 2^n distinct possible combinations of an n-variable input function, and thus there are 2^n minterms which uniquely represent this function. This is, therefore, a most useful starting point for simplifying a switching expression.

A *minterm* was defined in Section 2.3 as a product of literals in which each variable appears exactly once, either in complemented or in uncomplemented form. A *maxterm* is defined as a sum of literals in which each variable appears exactly once, either in complemented or in uncomplemented form. Table 3.4 lists the minterms and maxterms of a function of the three variables A, B, and C. It is seen in Table 3.4 that minterm m_i is the complement of maxterm M_i and vice versa.

Any function can be written as a sum of minterms or as a product of maxterms. A *minterm expansion* is also referred to as a *standard* or *canonical SOP* as well as a *disjunctive normal* form, and a *maxterm expansion* is called a *standard* or *canonical* product of sums as well as a *conjunctive normal* form. Canonical, from

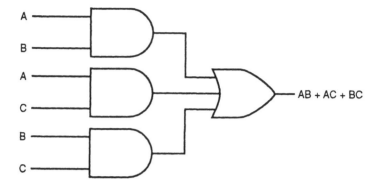

FIGURE 3.4. AND-OR realization of the function AB + AC + BC.

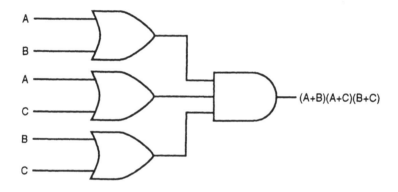

FIGURE 3.5. OR-AND realization of (A + B)(A + C)(B + C).

the Latin word *canon*, a rule, means standard and refers to an equation in its basic or most elementary form. The minterm and maxterm expansions of any function are unique. In general, other representations of a function are not unique. The minterm and maxterm expansions are the largest nonredundant forms of representation of a given function.

TABLE 3.4 Minterms and Maxterms of F(A,B,C)

A	B	C	Minterms	Maxterms
0	0	0	$\overline{A}\cdot\overline{B}\cdot\overline{C} = m_0$	$A + B + C = M_0$
0	0	1	$\overline{A}\cdot\overline{B}\cdot C = m_1$	$A + B + \overline{C} = M_1$
0	1	0	$\overline{A}\cdot B\cdot\overline{C} = m_2$	$A + \overline{B} + C = M_2$
0	1	1	$\overline{A}\cdot B\cdot C = m_3$	$A + \overline{B} + \overline{C} = M_3$
1	0	0	$A\cdot\overline{B}\cdot\overline{C} = m_4$	$\overline{A} + B + C = M_4$
1	0	1	$A\cdot\overline{B}\cdot C = m_5$	$\overline{A} + B + \overline{C} = M_5$
1	1	0	$A\cdot B\cdot\overline{C} = m_6$	$\overline{A} + \overline{B} + C = M_6$
1	1	1	$A\cdot B\cdot C = m_7$	$\overline{A} + \overline{B} + \overline{C} = M_7$

EXAMPLE 3.11: Find the minterm expansion of $F(A,B,C) = AB + \overline{A}BC + AB\overline{C}$.

SOLUTION: Use P9b, $A + \overline{A} = 1$, to add missing literals to the function F.

$$F(A,B,C) = AB(\overline{C} + C) + \overline{A}BC + AB\overline{C}$$
$$= AB\overline{C} + ABC + \overline{A}BC + AB\overline{C}$$
$$= \overline{A}BC + AB\overline{C} + ABC$$

The function, F, is seen to consist of minterms 3, 6, and 7. The function is completely specified by listing its minterms and can be represented in shorthand as $F = m_3 + m_6 + m_7$. This can be further abbreviated to $F(A,B,C) = \sum m\,(3,6,7)$. Thus,

$$F(A,B,C) = \overline{A}BC + AB\overline{C} + ABC = m_3 + m_6 + m_7 = \sum m\,(3,6,7) \qquad (3.1)$$

The complement of a function consists of all the minterms which are absent in the function. A function, F, is described by all the minterms that are asserted true or 1 in the truth table; therefore, the complementary function, $\overline{F}$, is described by all the minterms that are asserted false or 0 in the truth table of the function F. The minterms of $\overline{F}$ would all be asserted true or 1 in a truth table for $\overline{F}$.

The complement of the function F in Example 3.11 is given by

$$\overline{F}(A,B,C) = m_0 + m_1 + m_2 + m_4 + m_5 = \sum m\,(0,1,2,4,5)$$

If this expression for $\overline{F}$ is complemented, one obtains the function F again, this time written as

$$F(A,B,C) = \overline{\overline{F}(A,B,C)} = M_0 \cdot M_1 \cdot M_2 \cdot M_4 \cdot M_5 = \prod M\,(0,1,2,4,5)$$

To obtain a minterm expansion of a function, write an SOP of all those minterms whose output is 1. To obtain a maxterm expansion of a function, write a POS of all those maxterms whose output is 1. For the above example, the function, F, can be described in a canonical SOP form or in a canonical POS form as follows:

$$F(A,B,C) = m_3 + m_6 + m_7 = \sum m\,(3,6,7) \qquad (3.2)$$

or

$$F(A,B,C) = M_0 \cdot M_1 \cdot M_2 \cdot M_4 \cdot M_5 = \prod M\,(0,1,2,4,5) \qquad (3.3)$$

It is seen that if minterm m_i is present in the minterm expansion of F, then maxterm M_i is not present in the maxterm expansion of F, and vice versa.

EXAMPLE 3.12:   Find the minterm and maxterm expansions of the function
$F(A,B,C) = \overline{A} + B\overline{C}$.

SOLUTION: Use P9b, $A + \overline{A} = 1$ to add missing literals to the function, and
obtain the minterm expansion.

$$F(A,B,C) = \overline{A}(\overline{B} + B)(\overline{C} + C) + (\overline{A} + A)B\overline{C}$$

$$= \overline{A}\overline{B}\overline{C} + \overline{A}B\overline{C} + \overline{A}\overline{B}C + \overline{A}BC + \overline{A}B\overline{C} + AB\overline{C}$$

$$= \overline{A}\overline{B}\overline{C} + \overline{A}B\overline{C} + \overline{A}\overline{B}C + \overline{A}BC + AB\overline{C}$$

$$F(A,B,C) = \sum m(0,1,2,3,6) = \prod M(4,5,7)$$

EXAMPLE 3.13:   Find the maxterm and minterm expansions of the function
$F(A,B,C) = (\overline{A} + B)(\overline{A} + \overline{C})$.

SOLUTION: Use P9a, $A\overline{A} = 0$, to add missing literals to the function and
obtain the maxterm expansion.

$$F(A,B,C) = (\overline{A} + B + \overline{C}C)(\overline{A} + \overline{B}B + \overline{C})$$

$$= (\overline{A} + B + \overline{C})(\overline{A} + B + C)(\overline{A} + \overline{B} + \overline{C})(\overline{A} + B + \overline{C})$$

$$= (\overline{A} + B + C)(\overline{A} + B + \overline{C})(\overline{A} + \overline{B} + \overline{C})$$

$$F(A,B,C) = \prod M(4,5,7) = \sum m(0,1,2,3,6)$$

3.6   TWO OR MORE FUNCTIONS

Logical operations that apply to two or more functions can be performed
upon the minterm or maxterm expansions of the functions also. Minterm and
maxterm expansions can be manipulated directly to perform operations such as
ANDing and ORing, inclusively or exclusively, of two or more functions.

From the definition of the AND operation, when two functions F and G are
ANDed together, both functions F and G must have a given minterm asserted in
order for the product F·G to have that minterm asserted. Thus, the only
minterms present in F·G are those present in F and also in G.

F·G is the sum of all the minterms in both F and G.

From the definition of the OR operation, when two functions F and G are ORed
together, either function F or function G must have a minterm asserted in order

for the sum F + G to have that minterm asserted. Thus, all minterms present in either F or G will be present in the sum F + G.

F + G is the sum of all the minterms in either F or G or both.

When two functions F and G are exclusively ORed together, the answer must contain all the minterms present in F but not present in G, and all the minterms present in G but not in F.

F ⊕ G is the sum of all the minterms in F or in G, but not in both F and G.

EXAMPLE 3.14: $F = \sum m\,(0,2,3,5,9)$ and $G = \sum m\,(0,3,9,11)$. Find the minterm and maxterm expansions of F·G, F + G and F ⊕ G.

SOLUTION: $F{\cdot}G = \sum m\,(0,3,9) = \prod M\,(1,2,4,5,6,7,8,10,11,12,13,14,15)$

$F + G = \sum m\,(0,2,3,5,9,11) = \prod M\,(1,4,6,7,8,10,12,13,14,15)$

$F \oplus G = \sum m\,(2,5,11) = \prod M\,(0,1,3,4,6,7,8,9,10,12,13,14,15)$

The variables must be specified before the answers can be rewritten in terms of literals.

3.7   TWO-LEVEL GATE NETWORKS

Any function can be represented by a minterm expansion as well as by a maxterm expansion. A minterm expansion is in the form of an SOP, and a maxterm expansion is in the form of a POS. A minterm expansion can always be realized with two levels of logic, commonly referred to as AND-OR logic; and a maxterm expansion of a function can always be represented by two levels of logic, commonly referred to as OR-AND logic.

3.7.1   Sum-of-Products Realizations of a Function

Once a function is in an SOP form, the function can be implemented with two-level logic in several ways. The SOP form transforms directly into AND-OR logic. AND-OR logic can be transformed into NAND-NAND logic, OR-NAND logic (OR-AND-INVERT or OAI logic), and NOR-OR logic. This is shown in Example 3.15.

EXAMPLE 3.15:   Implement $F = A + B\overline{C} + CD$ in AND-OR, NAND-NAND, OR-AND-INVERT, and NOR-OR logic.

SOLUTION: The function is in the SOP form and is implemented directly in AND-OR logic, as shown in Figure 3.6. The function can be converted to the NAND-NAND form by complementing it twice, as follows:

$$\overline{\overline{F}} = F = A + BC^\sim + CD = \overline{\overline{A} \cdot \overline{BC^\sim} \cdot \overline{CD}}$$

There is a one-to-one correlation between bars and invert bubbles at gate outputs, as seen in Figure 3.7.

To convert the function to OR-AND-INVERT logic, apply DeMorgan's law to the NAND-NAND representation of the function. The circuit is shown in Figure 3.8.

$$F = \overline{\overline{A} \cdot \overline{\left(BC^\sim\right)} \cdot \overline{\left(CD\right)}} = \overline{A^\sim \cdot \left(B^\sim + C\right) \cdot \left(C^\sim + D^\sim\right)}$$

Apply DeMorgan's law once more to the OR-AND-INVERT form, and

$$F = \overline{A^\sim \cdot \left(B^\sim + C\right) \cdot \left(C^\sim + D^\sim\right)} = A + \overline{B^\sim + C} + \overline{C^\sim + D^\sim}$$

This is the NOR-OR representation of the function and is shown in Figure 3.9.

Applying DeMorgan's law again gives the original expression. There are only four unique representations of any function written in an SOP form.

3.7.2   Product-of-Sums Realizations of a Function

Once a function is reduced to a POS form, the function can be implemented with two-level logic in several ways. The POS form transforms directly into OR-AND logic. OR-AND logic can be transformed into NOR-NOR logic, AND-NOR logic (AND-OR-INVERT or AOI logic) and NAND-AND logic. This is shown in Example 3.16.

EXAMPLE 3.16:   Implement the function $F = A + B\overline{C} + CD$ in OR-AND logic.

SOLUTION: First the function must be converted to a minimum POS form.

$$F = A + B\overline{C} + CD = \left(A + B\overline{C} + C\right)\left(A + B\overline{C} + D\right)$$

$$= (A + B + C)(A + B + D)\left(A + \overline{C} + D\right)$$

$(A + B + D)$ is the consensus of the other two terms, thus the POS form of the function is $F = (A + B + C)(A + \overline{C} + D)$. The circuit in OR-AND form is shown in Figure 3.10.

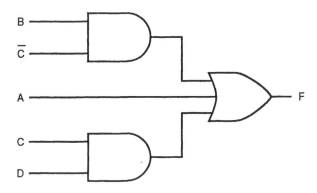

FIGURE 3.6.   AND-OR implementation of $F = A + B\overline{C} + CD$.

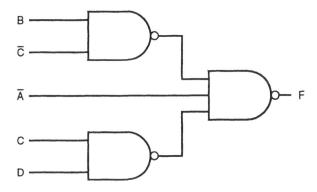

FIGURE 3.7.   NAND-NAND representation of $F = A + B\overline{C} + CD$.

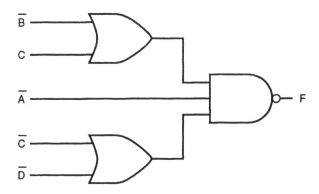

FIGURE 3.8.   OR-NAND or OAI representation of $F = A + B\overline{C} + CD$.

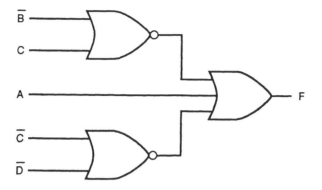

FIGURE 3.9.   NOR-OR representation of $F = A + B\overline{C} + CD$.

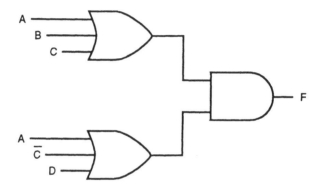

FIGURE 3.10.   OR-AND implementation of $F = (A + B + C)(A + \overline{C} + D)$.

The function can be converted to the NOR-NOR form by complementing it twice, as follows:

$$\overline{\overline{F}} = \overline{\overline{(A + B + C) \cdot (A + C^\sim + D)}} = \overline{\overline{(A + B + C)} + \overline{(A + C^\sim + D)}}$$

Once again, there is a one-to-one correlation between the BAR and the invert bubbles at the gate outputs as seen in Figure 3.11.

Next, apply DeMorgan's law to the NOR-NOR realization to obtain the function in the AND-NOR or AND-OR-INVERT form as shown in Figure 3.12.

$$F = \overline{\overline{(A + B + C)} + \overline{(A + C^\sim + D)}} = \overline{A^\sim \cdot B^\sim \cdot C^\sim + A^\sim \cdot C \cdot D^\sim}$$
$$= \overline{A^\sim B^\sim C^\sim + A^\sim C D^\sim}$$

Apply DeMorgan's law to the AND-NOR representation, and

$$F = \overline{A^\sim B^\sim C^\sim + A^\sim C D^\sim} = \overline{(A^\sim B^\sim C^\sim)} \cdot \overline{(A^\sim C D^\sim)}$$

This is the NAND-AND form of the function as seen in Figure 3.13. Applying DeMorgan's law again gives the original expression. There are only four unique representations of any function written in a POS form.

3.8 SHANNON'S EXPANSION THEOREM AND MULTILEVEL LOGIC

There are times when much simpler circuits can be realized with more than two levels of logic gates. In order to obtain multilevel circuits, the boolean function to be realized must be factored. Factoring can be done using Shannon's expansion theorem, viz:

$$F(X_1, X_2, \ldots, X_n) = X_1 \cdot F(1, X_2, \ldots, X_n) + \overline{X}_1 \cdot F(0, X_2, \ldots, X_n) \qquad (3.4)$$

$$F(X_1, X_2, \ldots, X_n) = [X_1 + F(0, X_2, \ldots, X_n)] \cdot [\overline{X}_1 + F(1, X_2, \ldots, X_n)] \qquad (3.5)$$

Shannon's theorem states that a function can be split into two subfunctions, using any variable which occurs in both the true and complemented form in the function. The variable used to split the function is referred to as the *splitting variable*. Shannon's expansion theorems can be written more concisely as

$$F(X) = X \cdot F(1) + \overline{X} \cdot F(0) = [X + F(0)][\overline{X} + F(1)] \qquad (3.6)$$

where $F(0)$ is the value of $F(X)$ when X, the splitting variable, is replaced by 0 throughout the expression, and $F(1)$ is the value of $F(X)$ when X is replaced by 1 everywhere. The proof of Shannon's theorem is as follows.

$$F(0) = 0 \cdot F(1) + 1 \cdot F(0) = F(0) \text{ and } F(1) = 1 \cdot F(1) + 0 \cdot F(0) = F(1) \qquad (3.7)$$

The dual theorem is similarly proved.

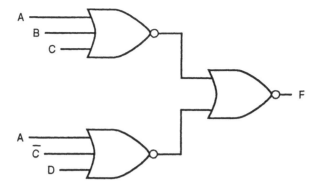

FIGURE 3.11. NOR-NOR representation of $F = (A + B + C)(A + \overline{C} + D)$.

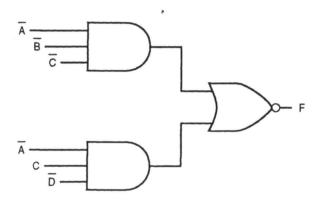

FIGURE 3.12. AND-NOR or AOI representation of $F = (A + B + C)(A + \overline{C} + D)$.

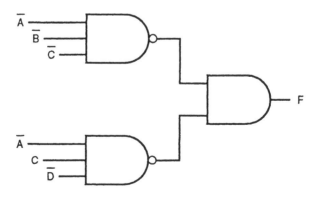

FIGURE 3.13. NAND-AND representation of $F = (A + B + C)(A + \overline{C} + D)$.

Shannon's theorem can be applied to two splitting variables, X_1 and X_2, to obtain

$$F(X_1,X_2) = \overline{X}_1\overline{X}_2F(0,0) + \overline{X}_1X_2F(0,1) + X_1\overline{X}_2F(1,0) + X_1X_2F(1,1) \qquad (3.8)$$

$$\begin{aligned} F(X_1,X_2) = &[\overline{X}_1 + \overline{X}_2 + F(1,1)]\cdot[\overline{X}_1 + X_2 + F(1,0)] \\ &\cdot[X_1 + \overline{X}_2 + F(0,1)]\cdot[X_1 + X_2 + F(0,0)] \end{aligned} \qquad (3.9)$$

One can continue in this manner until the function is completely factored.

EXAMPLE 3.17: Factor the function $F(A,B,C,D) = AB\overline{C} + A\overline{B}D + \overline{A}BD + \overline{A}\,\overline{B}\,\overline{C}$, using A as the splitting variable, and realize the factored form.

SOLUTION: Splitting the function on variable A, one obtains

$$F(A) = B\overline{C} + \overline{B}D, \text{ and } F(\overline{A}) = BD + \overline{B}\,\overline{C}$$
$$F(A,B,C) = A\cdot F(A) + \overline{A}\cdot F(\overline{A}) = A(B\overline{C} + \overline{B}D) + \overline{A}(BD + \overline{B}\,\overline{C})$$

This is realized with four levels of logic in Figure 3.14.

EXAMPLE 3.18: Factor $F(A,B,C,D) = AB\overline{C} + A\overline{B}\,\overline{C}D + ABD + \overline{A}\,\overline{B}\,\overline{C}D + \overline{A}\,\overline{B}C\overline{D}$, using A and B as splitting variables, and realize the factored form.

SOLUTION: $F(A,B,C,D) = \overline{A}\,\overline{B}\cdot F(0,0) + \overline{A}B\cdot F(0,1) + A\overline{B}\cdot F(1,0) + AB\cdot F(1,1)$

$F(0,0) = \overline{C}D + C\overline{D}$, $F(0,1) = 0$, $F(1,0) = \overline{C}D$, and $F(1,1) = \overline{C} + D$,

$F(A,B,C,D) = \overline{A}\,\overline{B}(\overline{C}D + C\overline{D}) + A\overline{B}(\overline{C}D) + AB(\overline{C} + D)$

The function is realized in Figure 3.15 with four levels of logic.

Factored solutions are not always better than SOP and POS solutions. *Propagation delay* is the time for the output of the gate to respond to a change at the input. In general, factoring creates a slower circuit since some signals pass through more than two gates, and each gate the signal passes through adds propagation delay. The maximum path length of any signal is a measure of how slow the circuit is.

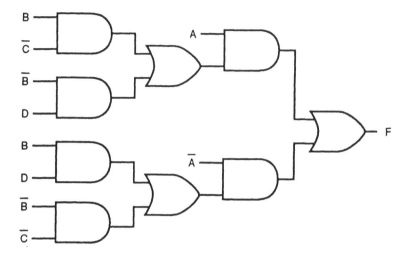

FIGURE 3.14.   Realization of $F(A,B,C,D) = A(B\overline{C} + \overline{B}D) + \overline{A}(BD + \overline{B}\,\overline{C})$.

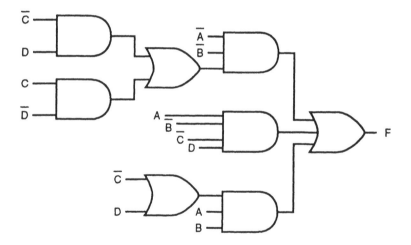

FIGURE 3.15.   Realization of $F(A,B,C,D) = \overline{A}\,\overline{B}(\overline{C}D + C\overline{D}) + A\overline{B}(\overline{C}D) + AB(\overline{C} + D)$.

Each input to a gate has capacitance associated with it, and this capacitance requires a finite time to charge or discharge when the input changes. The more inputs a gate has, the more input capacitance it has and, in general, the longer it takes for its output to respond to input changes. An SOP or POS solution might require a large fan-in to the second level of logic which slows the circuit down. In this case, factoring may give a faster circuit.

Both the above should be considered when comparing different realizations of a given boolean expression. In the following examples, the *maximum path length*

(MPL) and the *maximum fan-in (MFI)* are compared for SOP and factored realizations of the functions. A suitable weight factor, determined by the logic family of gates being used, could be created to ascertain which implementation is superior.

When there are few or no product terms in a factor, the factored form is usually superior to an SOP form. Consider the function

$$W = AB + AC + A\overline{D} + AE + AF = A(B + C + \overline{D} + E + F) \qquad (3.10)$$

The SOP form has an MFI of 5 and an MPL of 2, while the factored form also has an MFI of 5 and an MPL of 2. The SOP realization requires 6 gates, 10 literals, and 15 gate inputs, whereas the factored solution requires only 2 gates, 6 literals, and 7 gate inputs. In this case the factored form is clearly superior to the SOP solution. Had there been product terms in the factored expression, a third level of AND gates would have been required, as in the function

$$X = ABC + ADE + A\overline{F}G = A(BC + DE + \overline{F}G) \qquad (3.11)$$

This time, the SOP form has an MFI of 3 and an MPL of 2, while the factored form has an MFI of 3 and an MPL of 3. The SOP solution is faster and smaller since it requires 4 gates, 9 literals, and 12 gate inputs compared to 5 gates, 7 literals, and 11 gate inputs for the factored form. For

$$Y = A\overline{C}D + AC\overline{D} + B\overline{C}D + BC\overline{D} + C\overline{D}E$$
$$= (A + B)\overline{C}D + (A + B + E)C\overline{D} \qquad (3.12)$$

the SOP form requires 6 gates, 15 literals, and 20 gate inputs, with an MFI of 5 and an MPL of 2, while the factored form requires 5 gates, 9 literals, and 13 gate inputs, with an MFI of 3 and an MPL of 3. The factored solution requires one less gate, seven less gate inputs, and is probably not much slower since it has a smaller MFI.

The function Y can also be factored as below.

$$Y = A\overline{C}D + AC\overline{D} + B\overline{C}D + BC\overline{D} + C\overline{D}E$$
$$= (A + B)(\overline{C}D + C\overline{D}) + C\overline{D}E$$
$$= (A + B)(C \oplus D) + C\overline{D}E \qquad (3.13)$$

This factored form requires 5 gates, 7 literals, 11 gate inputs, and has an MFI of 3 and an MPL of 3 if an EOR gate is used. This is a superior solution. This factored form requires 7 gates, 9 literals, 15 gate inputs, and has an MFI of 3 and an MPL of 4 if only AND and OR gates are used, which is not as good a solution as the first factored form.

Consider a last case, the function Z, where

$$Z = B\overline{C}D + A\overline{C}D + \overline{A}\,\overline{B}C\overline{D} + C\overline{D}E$$
$$= (A + B)\overline{C}D + (\overline{A}\,\overline{B} + E)C\overline{D}$$
$$= (A + B)\overline{C}D + [\,(\overline{A + B}) + E]C\overline{D} \qquad (3.14)$$

The SOP form requires 5 gates, 13 literals, 17 gate inputs, and has an MFI of 4 and an MPL of 2, whereas the factored form requires 6 gates, 9 literals, 14 gate inputs, and has an MFI of 3 and an MPL of 4. If one recognizes that $\overline{A}\,\overline{B} = \overline{A + B}$, the factored form requires 6 gates (including one inverter), 7 literals, 13 gate inputs, and has an MPL of 5 and an MFI of 3. The SOP solution uses one less gate and will be much faster than a circuit with a path length of five gates.

3.9   SUMMARY

Binary boolean algebra is a mathematical system for dealing with statements that take on one of two values. Precise mathematical and physical meanings are assigned to the primitive logic operations AND, OR, and INVERT. These definitions, or logical postulates, are then used to derive other more powerful theorems, which in turn can be used to transform complicated expressions into simpler, more desirable forms which are logically equivalent to the original expression. Once the mathematical expression is in an appropriate form, a circuit can be realized that represents the function.

Switching algebra is based on Huntington's six basic postulates for a boolean set S consisting of two elements, 0 and 1, and two undefined binary operations, + and ·, namely

1. a.   If A and B are in the set S, then $A \cdot B$ is in S.
    b.   If A and B are in the set S, then $A + B$ is in S.
2. a.   There is an element 1 such that $A \cdot 1 = A$ for all A.
    b.   There is an element 0 such that $A + 0 = A$ for all A.
3. a.   The elements of S commute with respect to ·, $A \cdot B = B \cdot A$.
    b.   The elements of S commute with respect to +, $A + B = B + A$.
4. a.   The operation · is distributive over +, $A \cdot (B + C) = A \cdot B + A \cdot C$.
    b.   The operation + is distributive over ·, $A + B \cdot C = (A + B) \cdot (A + C)$.

    Every element A in S has a unique complement $\overline{A}$ such that

5. a. $A \cdot \overline{A} = 0$
    b. $A + \overline{A} = 1$
6. There are at least two distinct elements in S.

The operations of OR (+), AND (·), and complement are defined as follows:

A	B	A + B
0	0	0
0	1	1
1	0	1
1	1	1

A	B	A · B
0	0	0
0	1	0
1	0	0
1	1	1

A	$\overline{A}$
0	1
1	0

In binary boolean algebra, the symbols 0 and 1 are not numerals, but truth values corresponding to conditions such as switches being open or closed. T and F, representing true and false, respectively, can be used in place of 0s and 1s. The boolean operations of ORing and ANDing, represented by the symbols + and ·, refer to the logical operations of union (disjunction), and intersection (conjunction), respectively. They bear some similarity to algebraic operations and obey certain algebraic rules, such as the commutative, distributive, and associative laws. To minimize the need for parentheses, the conventional algebraic ordering of multiplication over addition is also extended to intersection taking precedence over union. Complementation is unique to boolean algebra and takes precedence over both the union and the intersection operations.

The elements of a boolean set under consideration are called variables. A boolean proposition or expression can be proved by verifying that it is true for all possible values of the variables in the proposition. The table listing all possible values of the variables along with the appropriate values of the expression is called a truth table.

Switches can take on one of two values, open or closed, and switching circuits can take on one of two values, transmission of the input to the output or no transmission. Usually, a closed switch is represented by a 1 and an open switch by a 0, with transmission from input to output represented by a 1, and an open circuit (no transmission) by a 0. Binary logic gates have inputs and outputs that can take on one of two voltage values, either high or low voltage. Both switching circuits and logic circuits can be represented by binary boolean algebra.

A minterm is a product term which contains every variable of a function exactly once, either in the true or complemented form, and a maxterm is a sum of literals in which each variable appears exactly once, either in the true or in the complemented form. Each minterm or maxterm can be associated with a specific row of a truth table. Since the rows of a truth table are unique, there is a unique minterm or maxterm expansion for any binary boolean function. For this reason, minterm and maxterm expansions are referred to as standard or canonical expansions.

There are four equivalent two-level logic circuits corresponding to any SOP realization of a function. The function can be realized in the AND-OR, NAND-NAND, OR-AND-INVERT (OAI), and NOR-OR forms. Any POS representation of a function can be realized in four two-level logic circuits also, the OR-AND,

NOR-NOR, AND-OR-INVERT (AOI), and NAND-AND forms. The first level of logic is referred to as the primary level, and the inputs to this level are the primary inputs of the function, as represented by the literals present in the SOP or POS representation.

Sometimes a simpler and/or faster circuit can be obtained by factoring the boolean expression to be realized. In simple cases the factors can be obtained by inspection; in more complex cases Shannon's expansion theorem can be used to factor the function. Shannon's theorem offers a method of programming the factoring of very complicated functions of many variables. Factoring leads to three or more levels of logic and is generally slower than two-level logic realizations; although when factoring greatly reduces fan-in, it can offer a faster solution than two-level logic realizations. Factoring usually reduces the number of gate inputs required. If it also reduces the number of gates needed, the size reduction may offset any increased delay time.

PROBLEMS

In Problems 3.1 through 3.12 find the minterm expansions of the expressions.

3.1. $F(A,B,C) = A\overline{B} + AC$

3.2. $F(A,B,C) = \overline{B} + \overline{A}\,C$

3.3. $F(A,B,C) = A\overline{B} + \overline{A}C + AB\overline{C}$

3.4. $F(A,B,C) = \overline{B}\,\overline{C} + \overline{A}\,BC + AB$

3.5. $F(A,B,C,D) = \overline{B}\,C\overline{D} + AB\overline{C}D + \overline{A}\,BD + \overline{B}\,D$

3.6. $F(A,B,C,D) = \overline{A}\,\overline{B} + B\overline{C}D + ABC\overline{D} + ABC$

3.7. $F(A,B,C,D) = \overline{A}\,C D + \overline{B}\,C\overline{D} + ABC + ACD$

3.8. $F(A,B,C,D) = \overline{A}\,CD + BC\overline{D} + A\overline{B} + A\overline{C}D$

3.9. $F(A,B,C,D) = AD + BC + A(B + C)D$

3.10. $F(A,B,C,D) = (A + B)(A + C) + B\overline{D}$

3.11. $F(A,B,C) = A \oplus B \oplus C$

3.12. $F(A,B,C) = A \odot B \oplus C$

In Problems 3.13 through 3.18 find the maxterm expansions of the expressions.

3.13. $F(A,B,C,D) = (\overline{A} + \overline{B} + D)(\overline{B} + C + \overline{D})$

3.14. $F(A,B,C,D) = (\overline{A} + \overline{C} + D)(\overline{B} + \overline{C} + \overline{D})$

3.15. $F(A,B,C,D) = (\overline{A} + \overline{B} + C)(\overline{A} + C + \overline{D})(A + B + C)$

3.16. $F(A,B,C,D) = (A + \overline{C} + D)(\overline{B} + \overline{C} + D)(\overline{B} + C + \overline{D})$

3.17. $F(A,B,C) = (A\overline{B} + \overline{A}BC) \cdot (\overline{ABC})$

3.18. $F(A,B,C) = (\overline{ABC}) \cdot (\overline{\overline{A}BC + B\overline{C}})$

Find the maxterm expansions of the functions in Problems 3.19 through 3.26.

3.19. $F(A,B,C) = \sum m\,(0,2,4,6)$

3.20. $F(A,B,C) = \sum m\,(1,2,4,5)$

3.21. $F(A,B,C) = \sum m\,(0,3,5,6,7)$

3.22. $F(A,B,C) = \sum m\,(1,3,4,6,7)$

3.23. $F(A,B,C,D) = \sum m\,(0,4,6,7,9,14)$

3.24. $F(A,B,C,D) = \sum m\,(1,2,5,6,11,15)$

3.25. $F(A,B,C,D) = \sum m\,(2,4,10,11,13,14)$

3.26. $F(A,B,C,D) = \sum m\,(1,3,8,9,12,15)$

Find the minterm expansions of the functions in Problems 3.27 through 3.34.

3.27. $F(A,B,C) = \prod M\,(0,2,4,6)$

3.28. $F(A,B,C) = \prod M\,(1,2,4,5)$

3.29. $F(A,B,C) = \prod M\,(0,3,5,6,7)$

3.30. $F(A,B,C) = \prod M\,(1,3,4,6,7)$

3.31. $F(A,B,C,D) = \prod M\,(0,4,6,7,9,14)$

3.32. $F(A,B,C,D) = \prod M\,(1,2,5,6,11,15)$

3.33. $F(A,B,C,D) = \prod M\,(2,4,10,11,13,14)$

3.34. $F(A,B,C,D) = \prod M\,(1,3,8,9,12,15)$

3.35. Given the two expressions $W(A,B,C) = \sum m\,(0,2,4)$ and $X(A,B,C) =$
       $\prod M\,(1,3,6,7)$,
       (a) Find the maxterm expansion of W.
       (b) Find the minterm expansion of X.
       (c) Find the minterm expansion of W·X.
       (d) Find the maxterm expansion of W + X.
       (e) Find the minterm expansion of W ⊕ X.

3.36. Given the two expressions $W(A,B,C) = \prod M\,(0,2,4)$ and $X(A,B,C) =$
       $\sum m\,(0,3,6,7)$,
       (a) Find the minterm expansion of W.
       (b) Find the maxterm expansion of X.
       (c) Find the minterm expansion of W·X.
       (d) Find the maxterm expansion of W + X.
       (e) Find the minterm expansion of W ⊕ X.

3.37. Repeat Problem 3.35 for $W(A,B,C) = \sum m\,(0,1,2,5)$ and $X(A,B,C) = \prod M\,(1,3,5,7)$.

3.38. Repeat Problem 3.36 for $W(A,B,C) = \prod M\,(0,1,2,5)$ and $X(A,B,C) = \sum m\,(1,2,5,7)$.

3.39. Repeat Problem 3.35 for $W(A,B,C,D) = \sum m\,(3,5,6,12,13,15)$ and $X(A,B,C,D) = \prod M\,(0,7,9,11,12)$.

3.40. Repeat Problem 3.36 for $W(A,B,C,D) = \prod M\,(1,6,7,8,10,12,15)$ and $X(A,B,C,D) = \sum m\,(2,6,9,12,14)$.

3.41. $W = \sum m\,(1,2,4,6,7)$, $X = \sum m\,(0,3,4,6,7)$, and $Y = \sum m\,(0,2,3,5,7)$. Show W, X, and Y on a Venn diagram with the minterms located in the correct areas of the diagram.

3.42. Repeat Problem 3.41 for $W = \sum m\,(0,3,4,5,7)$, $X = \sum m\,(0,1,2,4,6)$, and $Y = \sum m\,(1,2,4,5,7)$.

3.43. $Y(A,B,C) = \sum m\,(1,2,7)$.
   (a) Implement Y with AND-OR logic.
   (b) Implement Y with NAND-NAND logic.
   (c) Implement Y with OR-AND-INVERT logic.
   (d) Implement Y with NOR-OR logic.

3.44. $Z(A,B,C) = \prod M\,(0,4,7)$.
   (a) Implement Z with OR-AND logic.
   (b) Implement Z with NOR-NOR logic.
   (c) Implement Z with AND-OR-INVERT logic.
   (d) Implement Z with NAND-AND logic.

3.45. Repeat Problem 3.43 for $Y = ABC + \overline{D}\,\overline{E}\,\overline{F}$.

3.46. Repeat Problem 3.44 for $Z = (A + B + C)(\overline{D} + \overline{E} + \overline{F})$.

3.47. Realize the function $W = AB + AC + A\overline{D} + AE + AF = A(B + C + \overline{D} + E + F)$ in an SOP and a factored form and verify the number of gates, gate inputs, literals, MFIs, and MPLs required.

3.48. Repeat Problem 3.47 for the function $X = ABC + ADE + A\overline{F}G = A(BC + DE + \overline{F}G)$.

3.49. Repeat Problem 3.47 for the function $Y = A\overline{C}D + AC\overline{D} + B\overline{C}D + BC\overline{D} + C\overline{D}E$, with Y factored both as $(A + B)\overline{C}D + (A + B + E)C\overline{D}$ and as $(A + B)(C \oplus D) + C\overline{D}E$.

3.50. Repeat Problem 3.47 for the function $Z = B\overline{C}D + A\overline{C}D + \overline{A}\,\overline{B}C\overline{D} +$ $C\overline{D}E$ with Z factored both as $(A + B)\overline{C}D + (\overline{A}\,\overline{B} + E)C\overline{D}$ and as $(A + B)\overline{C}D + [\,\overline{(A + B)} + E]C\overline{D}$.

3.51. (a) Use Shannon's theorem to factor $W = AB + \overline{A}BC + ACD + \overline{A}D$.
       (b) Implement W with AND and OR gates.
       (c) Compare the SOP solution to the factored solution for number of gates, literal count, maximum fan-in, and maximum path length.

3.52. Repeat Problem 3.51 for $X = \overline{A}BC + \overline{B}\,\overline{C}D + A\overline{B}\,\overline{C} + BCD$.

3.53. Repeat Problem 3.51 for $Y = ABC + \overline{A}BD + A\overline{B}D + \overline{A}\,\overline{B}\,\overline{D}$.

3.54. Repeat Problem 3.51 for $Z = A\overline{C}D + BCD + \overline{A}BC + \overline{A}\,\overline{C}\,\overline{D}$.

SPECIAL PROBLEMS

3.55. (a) How many minterms are there for a function of N variables?
       (b) What is the maximum number of literals possible for a function of N variables?
       (c) How many functions of N variables are there.

3.56. The tables defining "+", "·", and complement for a set of four elements are given below. Show that these are possible boolean systems by proving that $A + BC = (A + B)(A + C)$ and $A(B + C) = AB + AC$ for the special cases $ABC = 012$, 120, and 211.

+	0	1	2	3	Complement
0	0	1	2	3	1
1	1	1	1	1	0
2	2	1	2	1	3
3	3	1	1	3	2

·	0	1	2	3
0	0	0	0	0
1	0	1	2	3
2	0	2	2	0
3	0	3	0	3

(a)   (b)

3.57. (a) Show that for the operation "+" table (a) in the structure below does not meet Huntington's postulates for a boolean-algebra system. (Hint: test $A + \overline{A} = 1$.)

+	0	1	2	Complement
0	0	1	2	1
1	1	1	1	0
2	2	1	2	2

+	0	1	2	Complement
0	0	1	2	1
1	1	1	1	0
2	2	1	1	2

(a)   (b)

(b) Show that table (b) does not yield a boolean algebra system. (Hint: test $A + A = A$.)

3.58. It is desired to use a logic system to automatically control the time during which a sprinkler is in operation. Let W denote the output of the logic operation, with the value 1 denoting that the sprinkler is to be turned on or stay on, and the value 0 to denote that the sprinkler is to be turned off or stay off. The conditions for determining when the sprinkler is to be operated are

1. The sprinkler is to operate automatically during the time interval of 7:00 a.m. to 9:00 a.m. Let the variable T be a 1 during this period of the day, and a 0 otherwise. A timing device will be used to provide these inputs to the system, and a person can override this constraint manually.

2. It is not desired to water the lawn while it is raining. Let the variable R be a 1 if it is raining and a 0 if it is not raining. A suitable rain detector will provide this input.

3. The system is to be capable of manual or automatic operation. Let the variable M be a 1 when the system is in manual operation and a 0 when the system is in automatic operation. If it rains while the system is in manual operation, the sprinkler is to be turned off. A control switch will be provided for this purpose.

4. An ON/OFF power switch is to be provided. Let the variable P be a 1 when the system is operating, either in manual or in automatic, and 0 when the system is turned off.

Procedure:

(a) Obtain the truth table for the above system.

(b) Obtain the boolean expression for the output, W, from the truth table.

(c) Translate the result of b into English suitable for an instruction manual to accompany the system.

(d) Reduce the answer of part b to the simplest form, and again state the result in boolean algebra and in suitable English.

(e) Implement the system with a minimum number of gates.

3.59. Four stockholders in a small corporation hold 100 shares. A has 10 shares, B has 20 shares, C has 30 shares, and D has 40 shares. Each share is worth 1 vote, and 67 votes are required to pass a motion.

Procedure:

(a) Obtain the truth table which represents the various voting combinations which are sufficient to pass a motion.

    (b) Obtain the boolean expression for a motion to pass, X, from the truth table.

    (c) Translate the result of b into English suitable for entry in the bylaws of the company.

    (d) Factor the answer of part b to the simplest form, and again state the result in boolean algebra and in suitable English.

    (e) Implement the function with a minimum number of gates.

3.60. Design a two-way light switch. Let U be the upstairs switch and D be the downstairs switch. Represent a closed switch by a logical 1 and an open switch by a logical 0. A closed path from the input (120 V ac) to the output (light bulb) causes the light to turn on.

    (a) Show a circuit diagram of the switching network.

    (b) Give a truth table for the circuit and specify the boolean function of U and D which results in the light being on.

3.61. A three-input minority gate and a three-input majority gate are defined by the functions MIN and MAJ in the truth table shown below and in Figure 3.16. Use a box with "MIN" for a minority gate, and a box with "MAJ" for a majority gate as shown.

    (a) Why can't you build a NAND or a NOR gate using only majority gates?

    (b) Construct a two-input NAND gate using a minimum number of minority gates. (Hint: let $C = 0$, and the inputs are A, B, and 0. Also, $X + XY = X$.)

    (c) Construct a two-input NOR gate using a minimum number of minority gates. (Hint: let $C = 1$, and the inputs are A, B, and 1.)

    (d) One can combine parts b and c to obtain $Y = \overline{C}\ (\overline{AB}) + C \cdot (\overline{A + B})$. Give a realization of this function. Input C can be thought of as a select input which causes the circuit to output either the NAND or the NOR of inputs A and B.

A	B	C	MIN	MAJ
0	0	0	1	0
0	0	1	1	0
0	1	0	1	0
0	1	1	0	1
1	0	0	1	0
1	0	1	0	1
1	1	0	0	1
1	1	1	0	1

(a)

FIGURE 3.16.

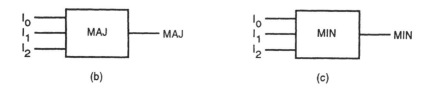

(b)   (c)

FIGURE 3.16. (continued).

REFERENCES

1. **Huntington, E. V.,** Sets of independent postulates for the algebra of logic, *Trans. Am. Math. Soc.,* 5, 288–309, 1904.
2. **Huntington, E. V.,** New sets of independent postulates for the algebra of logic, with special references to Whitehead and Russell's principia mathematica, *Trans. Am. Math. Soc.,* 35, 274–304, 1933.
3. **Mendelson, E.,** *Boolean Algebra and Switching Circuits, Schaum's Outline Series,* McGraw-Hill, New York, 1970, 135.
4. **Sandige, R. S.,** *Modern Digital Design,* McGraw-Hill, New York, 1990, 7–20.

CHAPTER 4

Mapping Boolean Expressions

4.1 INTRODUCTION

The digital designer is bound by constraints on the size, speed, power dissipation, and cost of any digital circuit and, in general, seeks the solution which optimizes all four of these constraints. Generally, all four parameters can be optimized at the same time by minimizing the number of gates involved and the number of inputs. To do this, it is necessary to minimize the switching function prior to implementing it in hardware. In this chapter methods of minimizing logic functions are studied.

If a minimum boolean function can be found, it can be implemented with the least number of small-scale logic gates and represents a best-case design in terms of cost and size. Each logic gate requires at least one connection to the plus and minus supplies also and furnishes an electrical path from the positive to the negative supply. Thus, reducing the total number of gates also reduces the total power dissipation of the circuit. If the number of gates a signal must traverse is also minimized, the fastest circuit will normally result.

In large-scale integrated circuits, the minimum area typically yields the lowest cost and best performance for a given circuit, and the total area required is proportional to the number of gates and the number of inputs needed.

If two or more minimum solutions to a given function are found that have the same number of gates, the one which uses the least number of inputs or literals is generally the best. In s*mall-scale integration (SSI)*, each literal requires an input pin and adds circuit complexity. In modern *large-scale integration (LSI)*, circuit wiring is proportional to the number of literals and is critical. It is essential in SSI and LSI to minimize the number of literals of a large circuit since each literal usually requires a connecting wire.

Inverters introduce delay and dissipate power. If two solutions have the same number of literals, that solution which minimizes the number of inverters is generally the best solution. Thus, the design engineer wants to minimize the number of gates, the number of inputs, and the number of inverters needed.

Various two-level logic realizations of sum-of-products (SOP) and product-of-sums (POS) solutions, as well as multilevel factored solutions, were studied in Chapter 3. The canonical SOP or canonical POS solution is the starting point for many logic minimization techniques.

The Karnaugh map is a very useful tool for simplifying functions of four or less variables and will be studied in this chapter. The Karnaugh map approach requires a knowledge of the minterms and/or maxterms of the function to be simplified. An implicant of a function is any product term or cube which, when true, will cause the function to be asserted true. An implicant which is not a subset of any other implicant is called a prime implicant. The minimum solution in an SOP form consists of a sum of prime implicants of the function.

A Karnaugh map can be used to loop the optimum number of minterms or the optimum number of maxterms that will cover a given function. Often there are inputs which never occur, or inputs which do occur but have no effect on the function. The corresponding minterms are called "don't care" terms and can be considered as either 1s or 0s, whichever helps simplify the function. Variables can also be entered on the Karnaugh map when necessary. For each such variable entered, the size of the map is reduced by one variable.

There is a one-to-one correspondence between cubes and implicants. In cubic terminology a minterm is a zero-dimensional cube (0-cube), and combining 0-cubes yields cubes of higher dimensions. A cube which is not contained in any other cube is a prime cube and is a prime implicant of the function under discussion. Obtaining a minimum set of prime implicants is the same as obtaining a minimum set of prime cubes of a function.

Additional techniques can be applied to mapping. Factoring on a Karnaugh map will be studied, as well as the identification of EOR and ENOR functions on Karnaugh maps. The application of EOR/ENOR gates to parity checking circuits, boolean difference applications, and the realization of half adders and full adders will also be covered.

4.2 ALGEBRAIC SIMPLIFICATION

In general, the lower the gate count, the lower the power dissipation, and the smaller the number of gates a signal must pass through, the faster the circuit. While the minterm and maxterm expansions of any function are unique, the implementation of a minterm or maxterm expansion is in general very wasteful of gates, and simplified SOP and POS solutions are needed.

Direct application of the postulates and theorems listed in Table 3.1 will simplify a boolean function. This approach is purely mathematical and requires a thorough familiarity with the theorems and postulates of Huntington. Examples of this approach are considered in this section. The various approaches used can be categorized as follows:

1. *Combining Terms.* The two basic theorems are P1b and T11b, namely
    $A + A = A$ and $AB + A\overline{B} = A$.

EXAMPLE 4.1: Simplify (a) $F = ABCD + ABC\overline{D}$ and (b) $G = A\overline{B}C +$
    $ABC + \overline{A}BC$.

SOLUTION: (a) $F = ABCD + ABC\overline{D} = ABC(D + \overline{D}) = ABC$
    (b) $G = (A\overline{B}C + ABC + \overline{A}BC) + ABC$
    $= AC(\overline{B} + B) + BC(\overline{A} + A) = AC + BC$

2. *Eliminating Terms.* The two basic theorems are T12b and T14a,
    $A + AB = A$ and $AB + \overline{A}C + BC = AB + \overline{A}C$.

EXAMPLE 4.2: Simplify (a) $H = A\overline{B} + A\overline{B}C$ and (b) $J = AB + ABC +$
    $ABD + ABE$.

SOLUTION: (a) $H = A\overline{B} + (A\overline{B})C = A\overline{B}$
    (b) $J = (AB + ABC) + ABD + ABE = (AB + ABD) + ABE$
    $= AB + ABE = AB$

3. *Eliminating Literals.* The basic theorem is T13a, $A + \overline{A}B = A + B$.

EXAMPLE 4.3: Simplify (a) $K = AB + \overline{AB} \cdot CD$ and (b) $L = \overline{A}B +$
    $\overline{A}\,\overline{B}\,\overline{C}\,\overline{D} + ABC\overline{D}$.

SOLUTION: (a) $K = AB + \overline{AB} \cdot CD = AB + CD$
    (b) $L = \overline{A}B + \overline{A}\,\overline{B}\,\overline{C}\,\overline{D} + ABC\overline{D} = \overline{A}(B + \overline{B} \cdot \overline{C}\,\overline{D}) + ABC\overline{D}$
    $= \overline{A}(B + \overline{C}\,\overline{D}) + ABC\overline{D} = \overline{A}\,\overline{C}\,\overline{D} + B(\overline{A} + AC\overline{D})$
    $= \overline{A}\,\overline{C}\,\overline{D} + B(\overline{A} + C\overline{D}) = \overline{A}\,\overline{C}\,\overline{D} + \overline{A}B + BC\overline{D}$

4. *Adding Redundant Terms.* The basic theorems are P9a, P9b, T12b, and T14a
    or 14b. One can always add $A\overline{A} = 0$, or add AB to A or to B or to both at
    any time, or multiply by $(\overline{A} + A) = 1$, or add a consensus term.

To use the consensus theorems, T14a and T14b, one must be able to spot the
consensus, or redundant, term. Given two variables in a pair of terms, with one
variable also complemented, the consensus term is the product of the original two
terms with the selected variable omitted. Thus, given the two terms $XY + \overline{X}Z$, the

consensus term is $Y \cdot Z$. The consensus term of $AB + \overline{A} C$ is $B \cdot C$, and the consensus term of $ABD + \overline{B} \overline{D} E$ is either $AB \cdot \overline{B} E$ or $AD \cdot \overline{D} E$, both of which are null terms.

EXAMPLE 4.4: Use the consensus theorem to simplify the function

$$F = \overline{A} \, \overline{C} D + \overline{A} BD + BCD + ABC + AC\overline{D}.$$

SOLUTION: $F_1 = (\overline{A} \, \overline{C} D + \overline{ABD} + BCD) + (BCD + \underline{ABC} + AC\overline{D})$.

The term BCD has been used with the first two terms and with the last two terms. The redundant consensus term in each group is underlined and can be eliminated. BCD need only be counted once, giving a function of three terms.

$F_1 = (\overline{A} \, \overline{C} D + BCD) + (BCD + AC\overline{D}) = \overline{A} \, \overline{C} D + BCD + AC\overline{D}$.

The function can be simplified in a different manner by eliminating the term BCD, a redundant consensus term, giving another solution, F_2. F_2 consists of four terms and is a nonminimal, nonredundant solution.

$F_2 = \overline{A} \, \overline{C} D + (\overline{A} BD + \underline{BCD} + ABC) + AC\overline{D} = \overline{A} \, \overline{C} D + \overline{A} BD + ABC + AC\overline{D}$.

In Example 4.4, application of the consensus theorem to eliminate the term BCD leads to a sum of four terms which cannot be further simplified. In fact, given the second expression for F, one must add the redundant term BCD in order to proceed as in the first case to reduce F_2 to F_1. A more systematic approach is required in order to assure that one has arrived at the simplest form of any given boolean expression. The Karnaugh map is such an approach.

4.3 KARNAUGH-MAP SIMPLIFICATION

A pictorial or diagrammatic approach called the Karnaugh map on which the function is plotted is often a better method of simplifying a switching function. From a mapping of the function, groups of minterms which can combine are easily identified. This approach is limited to six variables or less, as will be shown below. Methods suitable for more than six variables will be studied in Chapter 5.

The *Karnaugh map* is a truth table folded into a rectangle with a *cell*, or *box*, in which to place each minterm.[1,2] For two variables, four cells (boxes) are required for minterms 0 through 3 as shown in Figure 4.1(a); while for three variables, eight cells or boxes are required for minterms 0 through 7, as shown in Figure 4.1(b). A Gray code is used for the two variables B and C, and only one variable, B or C, changes in going from one cell to an adjacent cell.

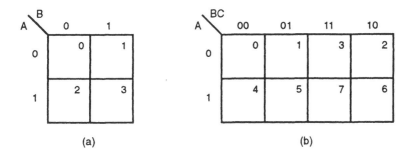

FIGURE 4.1.   Karnaugh maps for (a) two varibles and (b) three variables.

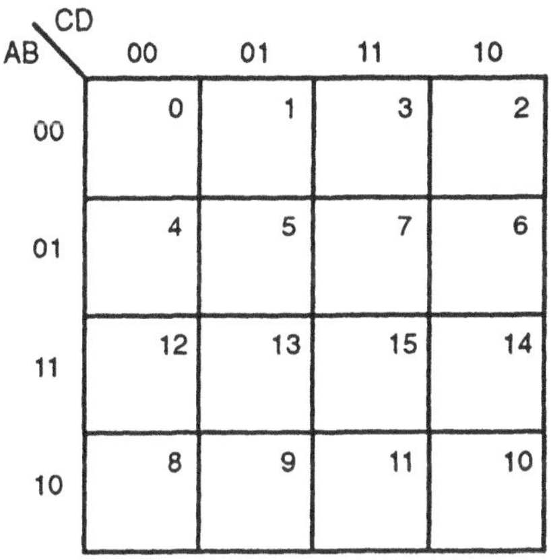

FIGURE 4.2.   The four-variable Karnaugh map.

The two-variable Karnaugh map is trivial, the three-variable Karnaugh map can be thought of as one half of a four-variable Karnaugh map, and maps of more than four variables can be treated as combinations of four-variable maps. Hence, it is sufficient to concentrate on four-variable mappings. The four-variable Karnaugh map is shown in Figure 4.2.

A Gray code is used for both pairs of variables, and it can be seen that as one goes from any cell to an adjacent cell the value of exactly one literal changes. Adjacent cells can be combined into one term. Minterms m_0 and m_1 can be combined into one term since D changes from 0 to 1 while the other three variables remain the same, viz $m_0 + m_1 = \overline{A}\,\overline{B}\,\overline{C}\,\overline{D} + \overline{A}\,\overline{B}\,\overline{C}D = \overline{A}\,\overline{B}\,\overline{C}$. Likewise, in going from m_{15} to m_{11}, only B changes value and $m_{15} + m_{11} = ABCD + A\overline{B}CD = ACD$. (The basic reduction theorem being used is $AB + A\overline{B} = A$.)

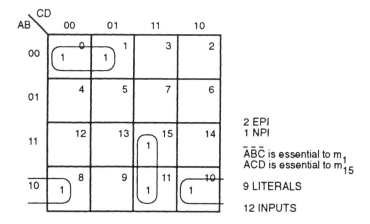

FIGURE 4.3. Karnaugh map representation of $F = \overline{A}\,\overline{B}\,\overline{C} + A\overline{B}\,\overline{D} + ACD$.

When one goes from m_8 to m_{10}, C is the only variable which changes, i.e., $m_8 + m_{10} = A\overline{B}\,\overline{C}\,\overline{D} + A\overline{B}C\overline{D} = A\overline{B}\,\overline{D}$. In each case, the variable which changes in going from one cell to an adjacent cell is redundant and can be omitted, thus combining two minterms into one product term.

Minterms such as m_8 and m_{10}, or m_0 and m_8 are referred to as *end-around* adjacencies. If the map is rolled into a vertical cylinder, the first column and the last column of the map are seen to be adjacent, while if the map is rolled into a horizontal cylinder, the top and bottom rows of the map are seen to be adjacent. If the map is rolled into a toroidal shape, the four corners are seen to be adjacent also.

Loops of more than two variables can also be combined. After two minterms are combined, an appropriate pair of minterms can be combined into a loop of four minterms. For instance minterms m_0, m_4, m_{12}, and m_8 can be combined, i.e.,

$$F(A,B,C,D) = \sum m(0,4,8,12) = \overline{A}\overline{B}\overline{C}\overline{D} + \overline{A}B\overline{C}\overline{D} + A\overline{B}\overline{C}\overline{D} + AB\overline{C}\overline{D}$$
$$= \left(\overline{A}\overline{B}\overline{C}\overline{D} + \overline{A}B\overline{C}\overline{D}\right) + \left(A\overline{B}\overline{C}\overline{D} + AB\overline{C}\overline{D}\right)$$
$$= \overline{A}\,\overline{C}\,\overline{D} + A\overline{C}\,\overline{D} = \overline{C}\,\overline{D} \qquad (4.1)$$

Groups of appropriate 2, 4, 8 or all 16 minterms can be looped.

EXAMPLE 4.5: Simplify the function $F(A,B,C,D) = \sum m(0,1,8,10,11,15)$.

SOLUTION: $F(A,B,C,D) = (\overline{A}\,\overline{B}\,\overline{C}\,\overline{D} + \overline{A}\,\overline{B}\,\overline{C}D) + (A\overline{B}\,\overline{C}\,\overline{D} + A\overline{B}\,\overline{C}\overline{D}) +$
$(A\overline{B}CD + ABCD) = \overline{A}\,\overline{B}\,\overline{C} + A\overline{B}\,\overline{D} + ACD$
On the map of Figure 4.3, the three pairs of minterms
are looped to indicate they combine to give the simplified
function, F.

4.4   PRIME IMPLICANTS

An expression is said to *imply* a function if, whenever that expression is true, the function is true. Such an expression is called an *implicant* of the function. Mathematically, an *implicant* of a function is any minterm, or two or more minterms which have been combined into a larger loop. In other words, any valid loop consisting of one or more minterms is an implicant of the function.

A *prime implicant* is defined as the largest loop in which a given minterm is contained, i.e., a prime implicant is not entirely contained in a larger implicant, and does not imply any other implicant. Both A and AB are implicants of the function F = A + AB, but only A is a prime implicant since it contains AB and is said to "cover" AB.

The expression ABC and the expression $\overline{A}$ BC both imply the term BC since BC = 1 if either ABC = 1 or if $\overline{A}$ BC = 1. Thus, ABC and $\overline{A}$ BC are implicants of BC. If, further, BC is an implicant of a function F, which does not contain the term B or C separately, then BC is a prime implicant of F. ABC and $\overline{A}$ BC are also implicants of the function F, but they are not prime implicants of the function. The symbol for imply is an arrow, →, hence one can write ABC→BC and $\overline{A}$ BC→BC.

For a function of four variables, a loop of one minterm has four literals, a loop of two adjacent minterms has three literals, a correct loop of four minterms has two literals, and a correct loop of eight minterms has one literal in it.

From the Karnaugh map, the prime implicants can be looped by inspection once the correct loops are known. There are five loops of four minterms of the function G(A,B,C,D), represented by the map in Figure 4.4. (Remember that the four corner minterms are adjacent.) By forming the five loops as shown, the function can be reduced to five terms of two literals each.

Only four of the five loops are actually required to define the function G. Of the two optional prime implicants, one is necessary to specify G and the other becomes redundant. There is no need to count any minterms more than once when obtaining a minimum solution, and redundant prime implicants never appear in a minimum solution to any function. Essential prime implicants must be in every minimal solution. The function represented by the Karnaugh map in Figure 4.4 can be simplifed to either

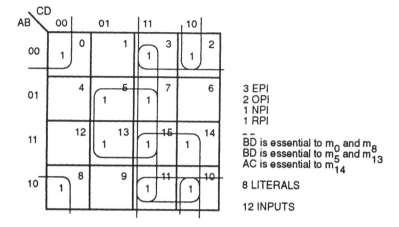

FIGURE 4.4. $G = \sum m\,(0,2,3,5,7,8,10,11,13,14,15) = \overline{B}\,\overline{D} + BD + AC + \overline{B}C + CD$.

$$G(A,B,C,D) = BD + AC + \overline{B}\,\overline{D} + \overline{B}C$$

or

$$G(A,B,C,D) = BD + AC + \overline{B}\,\overline{D} + CD \qquad (4.2)$$

In order to determine the least number of required prime implicants, the basic rule is to form all the largest loops containing minterms whose value is 1, such that each 1 is contained in at least one loop and no loop is completely covered by one or more other loops. In the example in Figure 4.4, the 1s in loop CD are covered by loops $\overline{B}C$, AC, and BD; so CD is a redundant prime implicant. On the other hand, if loop CD is used, loop $\overline{B}C$ is redundant and can be omitted. One has the option of including one prime implicant and omitting the other. In Equation 4.2, both choices require four two-input AND gates, but the second choice may require one less inverter than the first choice, in which case it is the preferred solution.

The valid loops defining implicants of four variables are shown in Figures 4.5 to 4.7.

4.4.1 Categories of Prime Implicants

For any function, a minimum sum is a sum of prime implicants, and no nonprime implicants need be considered in the solution. This can be proved by contradiction. Assume the existence of a product term P in a minimal sum which is not a prime implicant. Then at least one literal can be removed from P to obtain a new product term which still implies the original function. P can be replaced by the new product term in the "minimal" sum, resulting in a sum that still represents the original function but has one less literal. One thus arrives at a logical contradiction.

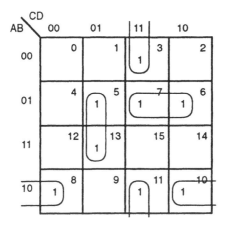

FIGURE 4.5.   The valid loops of two minterms of four variables.

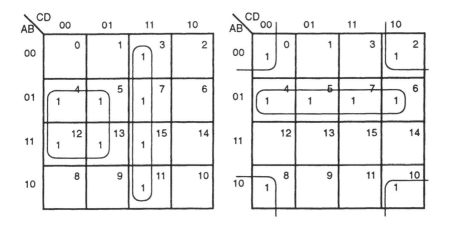

FIGURE 4.6.   The valid loops of four minterms of four variables.

The minimum SOP solution must contain only prime implicants. However, having obtained the prime implicants of a function, it is possible that some of them are unnecessary, and a method of determining the minimum set of prime implicants which represents a given function is needed. To do this, let us classify the prime implicants of a function as follows.

Essential prime implicant (EPI) — If there is only one way to loop a minterm, then that loop is an *essential prime implicant* for the given minterm and must appear in the reduced expression of the function.

Necessary prime implicant (NPI) — If there is only one way to loop two or more minterms together, but they can be separately looped in more than one way,

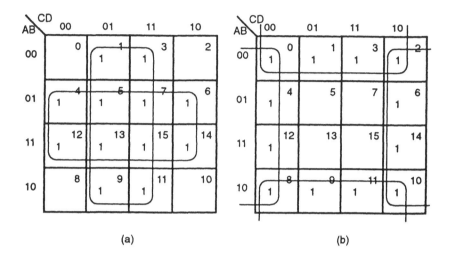

FIGURE 4.7. The valid loops of eight minterms of four variables.

then the loop is referred to as a *necessary prime implicant*. Necessary prime implicants are also referred to as *secondary essential prime implicants*.

Optional prime implicant (OPI) — If there is more than one way to loop a given minterm, the prime implicants are optional. An optional prime implicant is never essential. Choose the prime implicant that minimizes the number of inverters needed, and it becomes necessary.

Redundant prime implicant (RPI) — If all the 1s are covered already, then the loop is redundant. Redundant prime implicants are never included in a minimum SOP solution.

To obtain the EPIs, proceed as follows:

1.   Loop all 1s that cannot combine with (are not adjacent to) any other 1s.
2.   Loop all the 1s that will combine in a loop of two minterms, but not in a loop of four minterms.
3.   Loop all the 1s that will form a loop of four minterms, but not a loop of eight minterms.
4.   Continue until all the 1s are accounted for.
5.   Do not loop any minterm more than necessary.
6.   Make certain that all groups are as large as possible.

All the essential prime implicants are determined first, the necessary prime implicants are determined next, and any additional prime implicants are redundant.

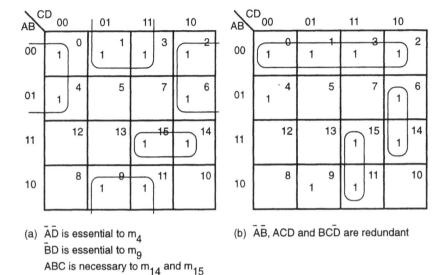

(a) $\overline{A}\overline{D}$ is essential to m_4
   $\overline{B}D$ is essential to m_9
   ABC is necessary to m_{14} and m_{15}

(b) $\overline{A}\overline{B}$, ACD and BCD are redundant

2 EPI, 3 RPI, 4 Gates, 7 Literals, 10 Inputs

FIGURE 4.8.   (a) $F_1 = \overline{A}\,\overline{D} + \overline{B}D + ABC$ is a minimum SOP solution, and (b) RPIs.

Examples of essential, necessary, optional, and redundant prime implicants follow in Figures 4.8 through 4.12 inclusive.

The function, $F_1(A,B,C,D) = \sum m\,(0,1,2,3,4,6,9,11,14,15)$, is mapped in Figure 4.8 and is seen to have two EPIs, one NPI, and three RPIs.

The function, $F_2(A,B,C,D) = \sum m\,(5,7,8,12,13)$, is mapped in Figure 4.9 and is seen to have two EPIs and two OPIs. One OPI is necessary and the other is then redundant. There are two minimum SOP solutions, and each requires at least two inverters, one for $\overline{A}$ and one for $\overline{C}$.

The function, $F_3(A,B,C,D) = \sum(1,3,4,5,6,7,9,11,14,15)$, is mapped in Figure 4.10 and has three EPIs and two RPIs.

Groups of minterms must be as large as possible, but a loop is not necessarily needed just because it is large. The vertical loop $(\overline{C}\,\overline{D})$ of four minterms in Figure 4.11(a) is redundant. It consists of four consensus terms. The square loop $(\overline{A}\,\overline{C})$ of four minterms in Figure 4.11(b) also consists of four consensus terms and is redundant. The four loops shown are essential prime implicants in each case.

Instead of looping the asserted minterms of a function, the minterms that are not asserted (those whose value is 0) can be looped. In this case the answer one obtains is the complement of the function represented by the minterms asserted true. Looping the 0s leads to a minimum POS solution, as shown in Figure 4.12. $\overline{F}_6(A,B,C,D) = AC + B\overline{D}$ has two essential prime implicants.

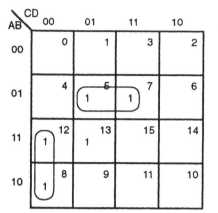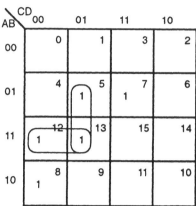

(a) A$\overline{C}\overline{D}$ is essential to m_8
     $\overline{A}$BD is essential to m_7

(b) Optional prime implicants
     (select one to be necessary)

2 EPI, 2 OPI, 1 NPI, 4 Gates, 9 Literals, 12 Inputs

FIGURE 4.9. $F_2 = A\overline{C}\,\overline{D} + \overline{A}\,BD +$ (either $B\overline{C}\overline{D}$ or else $AB\overline{C}$).

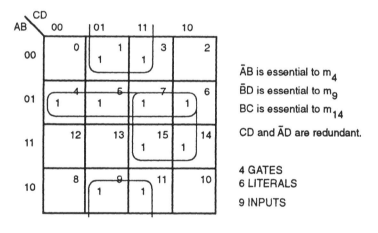

$\overline{A}$B is essential to m_4

$\overline{B}$D is essential to m_9

BC is essential to m_{14}

CD and $\overline{A}$D are redundant.

4 GATES
6 LITERALS
9 INPUTS

FIGURE 4.10. $F_3 = \overline{B}\,D + \overline{A}\,B + BC$. The function F_3 consists of three essential prime implicants, and F_3 is a minimum SOP solution.

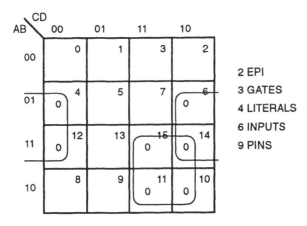

4 EPI, 5 Gates, 12 Literals, 16 Inputs

FIGURE 4.11. (a) $F_4 = \overline{A}\,\overline{B}\,\overline{C} + \overline{A}\,B\overline{D} + AB\overline{C} + A\overline{B}\,\overline{D}$, and (b) $F_5 = \overline{B}\,\overline{C}\,\overline{D} + \overline{A}\,B\overline{D} + \overline{A}\,\overline{B}\,D + B\overline{C}D$.

FIGURE 4.12. $\overline{F}_6 = AC + B\overline{D}$. Hence, $F_6 = \overline{AC + B\overline{D}} = (\overline{A} + \overline{C})(\overline{B} + D)$ is a minimum POS solution.

FIGURE 4.13. $F = \sum m (0,2,5,7,8) + \sum d (10,11,12,13,14,15) = BD + \overline{B}\,\overline{D}$ is in a minimum SOP form.

4.5 DON'T CARE MINTERMS

Often, some minterms do not occur and can be treated as either 1s or 0s, without changing the outcome of the function to be simplified. For instance, if a BCD function is mapped, the inputs corresponding to minterms 10, 11, 12, 13, 14, and 15 do not occur, and it does not matter whether these minterms are considered as 1s or as 0s. The *don't care* minterms can be represented by the letter d and indicated on a map by the letter d, a dash, —, or an X, as shown in Figure 4.13.

Don't care inputs are also generated when they have no effect on the system under consideration. When one input to an AND gate is a 0, the gate is said to be disabled and the output is 0 regardless of the other input values. Likewise, if one input to an OR gate is a 1, the gate is disabled and the output is 1 regardless of the other input values. In both of these cases, minterms generated by varying the remaining inputs are don't care terms.

Since the function is unaffected by don't care terms, they can be treated as 0s or 1s, whichever simplifies the function. In Figure 4.13, minterms 10, 13, and 15 are treated as 1s, and minterms 11, 12, and 14 are treated as 0s.

Don't care terms can be treated as 0s or 1s in looping the zero terms also (the complementary approach), as shown in Figure 4.14.

In Figures 4.13 and 4.14, the don't care minterms were treated the same; don't cares 10, 13, and 15 were all treated as 1s, and don't cares 11, 12, and 14 were all treated as 0s. When this is the case, the minimum SOP solution and the minimum POS solution both give the same answer for the function, and the minimum SOP solution can be complemented twice to obtain the minimum POS solution, and vice versa.

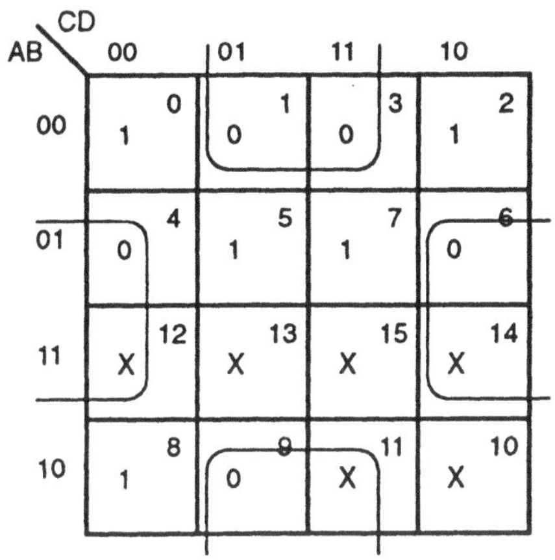

FIGURE 4.14. $F = \sum m\,(0,2,5,7,8) + \sum d\,(10,11,12,13,14,15) = (\overline{B} + D)\cdot(B + \overline{D})$ is in a minimum POS form.

Often one or more optional minterms are treated differently in using the two approaches and the resulting two representations of the function are not the same. An example is shown in Figure 4.15.

In looping the 1s in Figure 4.15, m_0 is treated as a 1 and m_{10} is treated as a 0, while in looping the 0s, m_0 is treated as a 0 and m_{10} is treated as a 1. Hence, two different answers were obtained in simplifying the function.

4.6   FACTORING ON A KARNAUGH MAP

With a little practice, factoring can be done on a Karnaugh Map. An example is shown in Figure 4.16. Looping the 1s as shown in Figure 4.16(a), the standard SOP solution for F is obtained, viz.

$$F(A,B,C,D) = AB + A\overline{C} + \overline{A}\,\overline{B}C + \overline{A}CD \tag{4.3}$$

In Figure 4.16(b), two prime implicants are shown. The loop of four minterms would be the function $\overline{A}C$ if m_6 were a 1. Thus, the loop of four is $\overline{A}C$ but *not* $\overline{A}BC\overline{D}$. This can be written as

$$\overline{A}C\cdot\overline{ABC\overline{D}} = \overline{A}C\cdot(\overline{B} + D) \tag{4.4}$$

Likewise, the loop of eight is missing minterms m_{10} and m_{11} and can be written as A but *not* $A\overline{B}C$, or

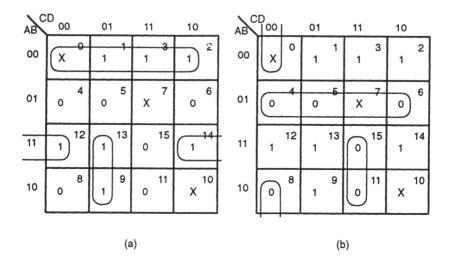

(a)   (b)

FIGURE 4.15. (a) $F_a = \overline{A}\,\overline{B} + A\overline{C}D + AB\overline{D}$ is a minimum SOP, and (b) $F_b = (B + C + D)(\overline{A} + \overline{C} + \overline{D})(A + \overline{B})$ is a minimum POS solution.

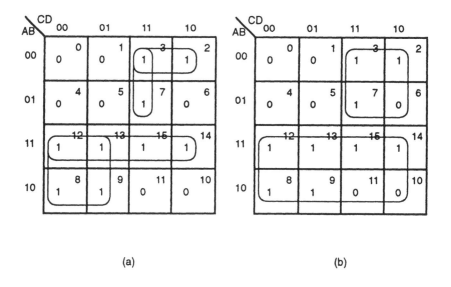

(a)   (b)

FIGURE 4.16. (a) $F_c = AB + A\overline{C} + \overline{A}\,\overline{B}C + \overline{A}CD$ is a minimum SOP solution, and (b) $F_d = \overline{A}C(\overline{B} + D) + A(B + \overline{C})$ is a solution in factored form.

A	B	C	F
0	0	0	1
0	0	1	1
0	1	0	0
0	1	1	0
1	0	0	1
1	0	1	1
1	1	0	0
1	1	1	1

(a)

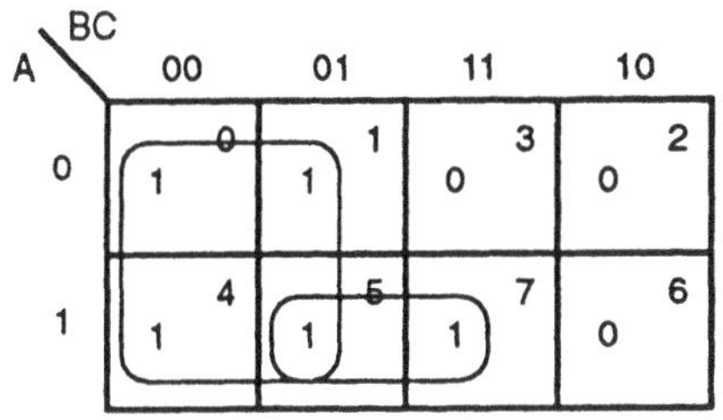

(b)

FIGURE 4.17. (a) Truth-table and (b) Karnaugh-map for the function $F = \overline{B} + AC$.

$$A \cdot \overline{A\overline{B}C} = A(B + \overline{C})$$ (4.5)

The function mapped is thus obtained in a factored form as

$$F(A,B,C,D) = \overline{A}C \cdot \overline{\overline{ABCD}} + A \cdot \overline{A\overline{B}C} = \overline{A}C(\overline{B} + D) + A(B + \overline{C})$$ (4.6)

4.7 VARIABLE-ENTERED MAPS

In Section 4.5, don't care terms were added to 1s and 0s as valid entries on a Karnaugh map. Nothing precludes entering boolean variables directly on a map also. When this is done, it is referred to as *variable-entered mapping,* or *VEM.*

VEM can often reduce the work required in plotting and reading maps by reducing the map size by at least one variable. Thus, a five-variable (or six-variable) map might be reduced to a four-variable (or five-variable) map.

Map entries can now consist of 1s, 0s, don't cares, and boolean variables. The process will be explained with an example. Consider the function, F(A,B,C), represented with the truth table and map shown in Figure 4.17. The truth table is repeated in Figure 4.18 and partitioned in pairs of values of C (C = 0 and C = 1).

The function F(A,B,C) is plotted on the Karnaugh map of Figure 4.17(b), and F is read from the map as $F(A,B,C) = \overline{B} + AC$. The function represented in the truth table of Figure 4.18(a) can be represented on a two-variable Karnaugh map as shown in Figure 4.18(b). Regardless of the value of C, F(A,B,C) is asserted for AB = 00 and AB = 10, while it is not asserted for AB = 01. For AB = 11, however, F(A,B,C) is asserted only if C is asserted. Hence, $F(A,B,C) = \overline{A}\,\overline{B} \cdot 1 + \overline{A}B \cdot 0 + A\overline{B} \cdot 1 + AB \cdot C$, which can also be written as $F(A,B,C) = \sum m\,(0,2,3 \cdot C)$, where $m_i = m_i(A,B)$. The truth table of Figure 4.17 can be rewritten to partition the variable

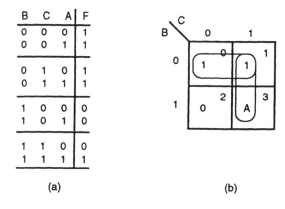

A	B	C	F
0	0	0	1
0	0	1	1
0	1	0	0
0	1	1	0
1	0	0	1
1	0	1	1
1	1	0	0
1	1	1	1

(a)   (b)

FIGURE 4.18.   (a) Truth table of Figure 4.17 partitioned to eliminate variable C and (b) mapped as a function of A and B.

B	C	A	F
0	0	0	1
0	0	1	1
0	1	0	1
0	1	1	1
1	0	0	0
1	0	1	0
1	1	0	0
1	1	1	1

(a)   (b)

FIGURE 4.19.   (a) Truth table partitioned to eliminate variable A and (b) mapped as a function of B and C.

A, as shown in Figure 4.19. Now, $F(A,B,C) = \overline{B}\,\overline{C}\cdot 1 + \overline{B}C\cdot 1 + B\overline{C}\cdot 0 + BC\cdot A$, which can also be written as $\sum m\,(0,1,3\cdot A)$, where $m_i = m_i(B,C)$.

Two or more variables can be entered on a truth table, but this might require ANDing or ORing the variables, which may or may not be a simplification. The method of VEM will be used in the future to reduce maps used in designing multiplexers. The reader is referred to Sandige[3] for a thorough discussion of VEMs.

4.8   CUBES IN n DIMENSIONS

Cubes and cubic representations of binary strings were discussed in Chapter 1. In boolean cartesian geometry, a switching variable can be represented by two points at opposite ends of a single line. Two variables can be represented by a plane, with the four boolean values of the two variables (the minterms) forming the vertices of a square. Three variables can be represented by a cube, with the eight minterms of the three variables forming the vertices of the cube. Cubic representations of a boolean function of one variable, two variables, and three variables are shown in Figure 4.20. These are the same cubes as shown in Figure 1.2, except that there are now coordinate axes associated with the cubes.

A function of n variables consists of a subset of 2^n potential minterms of the function. In an n-dimensional space, 2^n points corresponding to the 2^n potential minterms form the vertices of an *n-dimensional cube (n-cube)* or *boolean hypercube*. There is a one-to-one correspondence between the minterms of n variables and the vertices of an n-cube.[4] A three-dimensional cube (3-cube) is shown in Figure 4.21, with the vertices representing the function $F(A,B,C)$, whose minterm expansion is $\sum m\,(0,2,3,4,7)$.

A function of n variables consists of the set of vertices of the n-cube corresponding to the minterms (0 cubes) of the function. The function $G(A,B,C) = \sum m\,(0,2,4,7)$ is shown in Figure 4.22, with the vertices corresponding to the minterms of G marked.

Two 0-cubes form a 1-cube if they differ in only one coordinate (variable). Two 1-cubes of the function G are shown in Figure 4.22. They are obtained by combining the 0-cubes at the vertices of an edge of the 3-cube and form two edges of the 3-cube. Minterms 0 and 2 combine to give $000 + 010 = 0x0$ where x represents the missing variable B, which is now a don't care. Minterms 0 and 4 also combine to give $000 + 100 = x00$. These two edges of the 3-cube are represented by bold lines in Figure 4.22. Minterm 7 does not combine and is a prime implicant of G. The function G has been simplified to $G(A,B,C) = \overline{A}\,\overline{C} + \overline{B}\,\overline{C} + ABC$.

A 2-cube is obtained by combining a set of four 0-cubes whose coordinate values are the same in all but two variables. The 2-cube forms a face of a 3-cube. Consider the function $H(A,B,C) = \sum m\,(2,3,6,7)$. Minterms 2, 3, 6, and 7 combine to form the 2-cube x1x, shown by the surface or face of the 3-cube which is shaded in Figure 4.23. This corresponds to reducing the function H to $H = B$.

A 4-cube can be represented as shown in Figure 4.24. Minterms (0-cubes) form the 16 vertices of the 4-cube, 1-cubes obtained by combining two 0-cubes form the edges of the hypercube, 2-cubes obtained by combining two 1-cubes or four 0-cubes form the faces of the 3-cubes, and 3-cubes form subvolumes of the 4-cube. Representations of hypercubes of higher order than 4-cubes cannot be drawn conveniently in a two-dimensional plane.

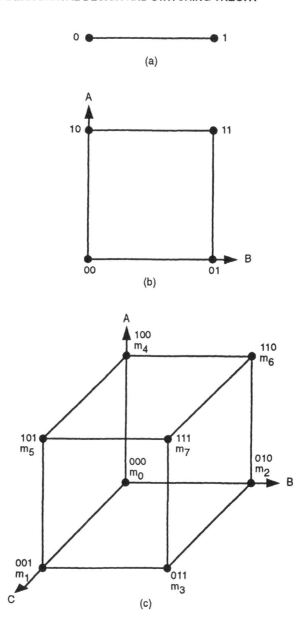

FIGURE 4.20. Cubic representations of a boolean function of (a) one variable, (b) two variables, and (c) three variables.

Figure 4.24 shows the function $J(A,B,C,D) = \sum m (2,4,5,6,7,8,10,12)$. Minterms 8 and 12 combine to give edge $A\overline{C}\,\overline{D}$ ($1000 + 1100 = 1x00$), minterms 2 and 10 combine to give edge $\overline{B}C\overline{D}$ ($0010 + 1010 = x010$), and minterms 4, 5, 6, and 7

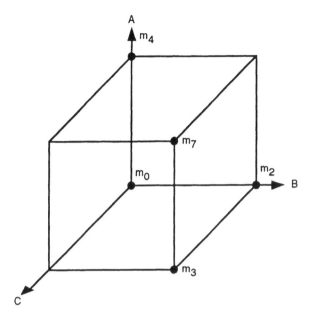

FIGURE 4.21. A cubic representation of the function $F(A,B,C) = \sum m\,(0,2,3,4,7)$.

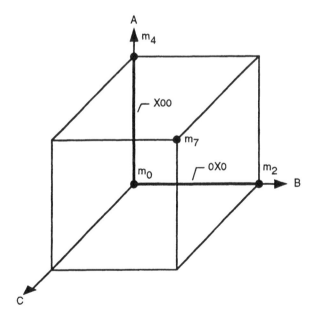

FIGURE 4.22. A cubic representation of the function $G(A,B,C) = \overline{A}\,\overline{C} + \overline{B}\,\overline{C} + ABC$.

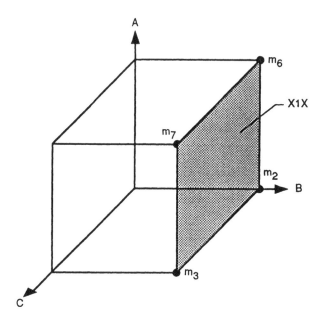

FIGURE 4.23. A cubic representation of the function $H(A,B,C) = B$.

combine to form surface $\overline{A} B$ $(0100 + 0101 + 0110 + 0111 = 01xx)$. The function J is reduced to $J(A,B,C,D) = A\overline{C}\,\overline{D} + \overline{B}C\overline{D} + \overline{A}B$.

In Figure 4.22, the prime implicants of the function represented are the 0-cube $ABC = 111$ and the 1-cubes $\overline{A}\,\overline{C} = 0x0$ and $\overline{B}\,\overline{C} = x00$. The prime implicant of the function in Figure 4.23 is the 2-cube $B = x1x$, and the prime implicants of the function in Figure 4.24 are the 1-cubes $A\overline{C}\,\overline{D} = 1x00$ and $\overline{B}C\overline{D} = x010$ and the 2-cube $\overline{A}B = 01xx$.

It can be seen from the above discussion that the problem of identifying subcubes of a function is the same as the problem of identifying prime implicants on a Karnaugh map. The minterms that combine on the K-map are the minterms that form subcubes of a function in n dimensions.

Define a filled cube as one for which every minterm covered by that cube has the value 1 or is a don't care that can be treated as a 1. When using the complementary approach, a filled cube would be one for which every minterm that it contains is either a 0 or a don't care term. If the cube under discussion is a subcube of a larger cube, the same definition applies to subcubes.

The implicants of a function correspond to the filled subcubes of that function, and prime implicants correspond to subcubes which are not wholly contained in larger filled subcubes. Prime implicants can also be referred to as *prime cubes,* and there is a one-to-one correspondence between EPIs, NPIs, OPIs, and RPIs, and essential, necessary, optional, and redundant prime cubes.

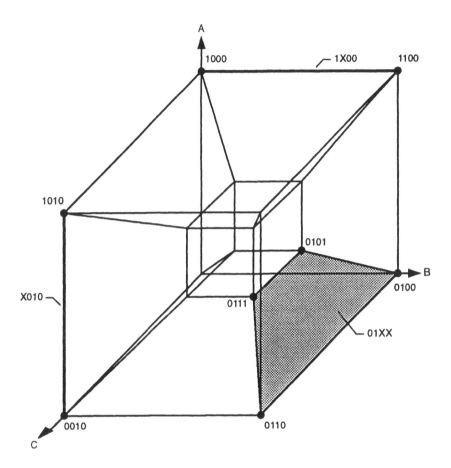

FIGURE 4.24.   A cubic representation of the function $J(A,B,C,D) = A\overline{C}\,\overline{D} + \overline{B}\,C\overline{D} + \overline{A}\,B$.

4.9 EXCLUSIVE-OR AND EXCLUSIVE-NOR MAPPING

The EOR gate finds frequent application in design and is available as a discrete gate package. For these reasons, it is worthy of consideration. EOR and ENOR functions have distinctive Karnaugh-map patterns, characterized by kitty-corner adjacencies and offset adjacencies, as shown in Figure 4.25. The function is

$$F_1(A, B, C, D) = \overline{AB}\left(\overline{CD} + C\overline{D}\right) + \overline{A}B\left(\overline{CD} + CD\right)$$

$$= \overline{A}\left[\overline{B}\cdot C \oplus D + B \cdot \overline{C \oplus D}\right] = \overline{A}(B \oplus C \oplus D) \qquad (4.7)$$

The function F_1 represented by the above map is realized with two EOR gates and an AND gate, as shown in Figure 4.26. Another example of an EOR pattern

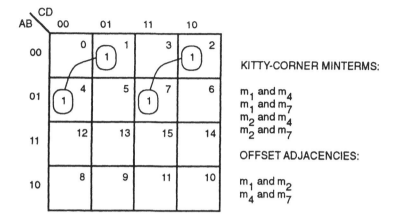

FIGURE 4.25. $F_1 = \overline{A}[B \oplus C \oplus D] = \overline{A}[(B \oplus C) \oplus D]$.

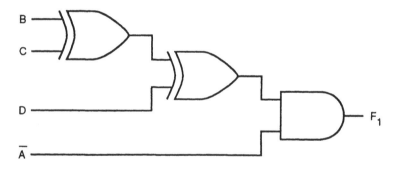

FIGURE 4.26.   Realization of $F_1 = \overline{A}[(B \oplus C) \oplus D]$.

is shown in Figure 4.27. In this case, m_1 and m_5 are kitty-corner adjacencies to m_8 and m_{12}, while m_6 and m_7 are offset adjacencies to m_{10} and m_{11}. The function realized by the mapping in Figure 4.27 is

$$F_2(A,B,C,D) = A\overline{C}\,\overline{D} + \overline{A}\,\overline{C}D + \overline{A}\,BC + A\overline{B}C = \overline{C}(A \oplus D) + C(A \oplus B) \quad (4.8)$$

The function F_2 is implemented with EOR gates in Figure 4.28 and requires five gates and ten inputs. It should be compared to the minimum SOP implementation shown in Figure 4.29, which requires 5 gates and 16 inputs.

EOR trees are commonly used as odd-parity checking circuits. An EOR tree is shown in Figure 4.30. To obtain a Karnaugh mapping of the function $F_3(A,B,C,D)$, place a 1 in all the cells for which A is asserted and B is not and in all the cells for which A is not asserted and B is. This maps the function $A \oplus B = A\overline{B} + \overline{A}B$.

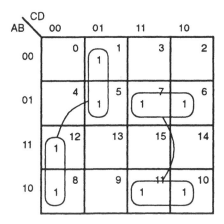

FIGURE 4.27. Map of $F_2 = \overline{C}(A \oplus D) + C(A \oplus B)$.

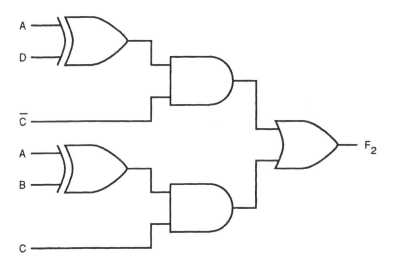

FIGURE 4.28. $F_2 = \overline{C}(A \oplus D) + C(A \oplus B)$.

Next, place a 1 in all the cells for which C is asserted and D is not and in all the cells for which C is not asserted and D is. This maps the function $C \oplus D = C\overline{D} + \overline{C}D$.

The output of the third gate is asserted if one of its inputs is asserted and the other is not. Thus, to obtain the output, F_3, read the set of all minterm cells which contain a single 1 but not two 1s. The minterms are circled on the map in Figure 4.31. Working in reverse order, it is seen that if a checkerboard pattern of cells contains an odd number of 1s, the function mapped is given by $F_3(A,B,C,D)$,

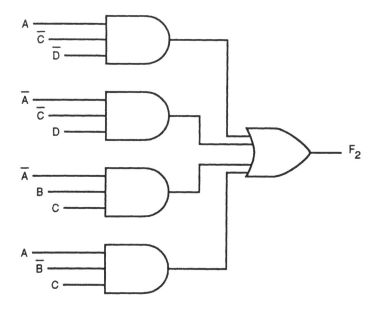

FIGURE 4.29. $F_2 = A\overline{C}\,\overline{D} + \overline{A}\,\overline{C}D + \overline{A}BC + A\overline{B}C.$

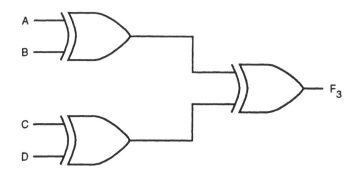

FIGURE 4.30.   An EOR tree realization of $F_3 = (A \oplus B) \oplus (C \oplus D)$.

which consists of minterms 1, 2, 4, 7, 8, 11, 13, and 14. This circuit is called an odd-parity checker because the minterms asserted all have an odd number of 1s in them.

By the same token, the inclusive-OR of $A \oplus B$ with $C \oplus D$ is the set of all cells containing at least one 1 (i.e., the inclusive-OR contains all the EOR minterms plus minterms 5, 6, 9, and 10).

Changing the output gate to an ENOR gate changes the circuit from an odd-parity circuit to an even-parity circuit. The tree circuit is shown in Figure 4.32, and the map in Figure 4.33 is seen to consist of the minterms 0, 3, 5, 6, 9, 10, 12, and 15.

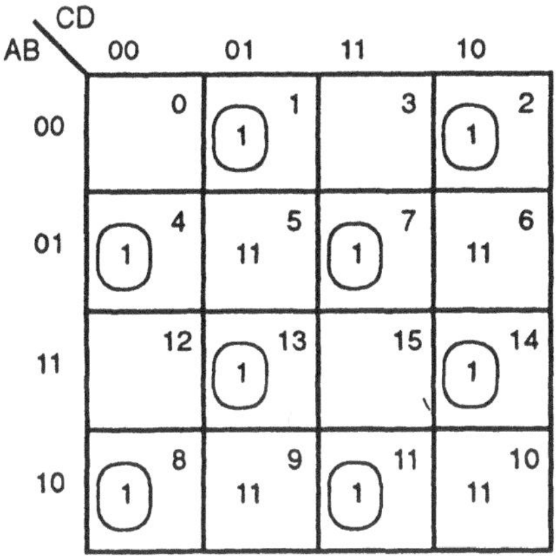

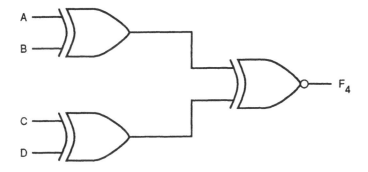

FIGURE 4.31.   Map of the function $F_3 = A \oplus B \oplus C \oplus D$.

A

B

C

D

F_4

FIGURE 4.32.   Even-parity EOR-ENOR realization of $F_4 = (A \oplus B) \odot (C \oplus D)$.

The map is obtained by proceeding to map $A \oplus B$ and $C \oplus D$, as before. This time the output gate is a coincidence gate, and the minterms asserted all have an even number of 1s in them. One now looks for minterm cells that contain either zero 1s or two 1s. These minterms are circled in Figure 4.33. Comparison with Figure 4.31 shows F_3 and F_4 to be complements of each other, as they must be by definition. If a checkerboard pattern of cells contains an even number of 1s, the function mapped is given by F_4.

EXAMPLE 4.6:   Map $F_5(A,B,C,D)$, the output of the circuit shown in Figure 4.34.

SOLUTION: The function is mapped in Figure 4.35. The output of the ENOR gate is only asserted when both inputs are the same. Since the output gate is an EOR gate, only the minterms

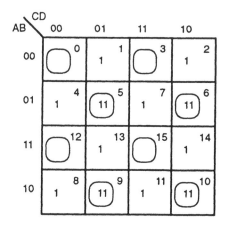

FIGURE 4.33.   Map of the function $F_4 = (A \oplus B) \odot (C \oplus D)$.

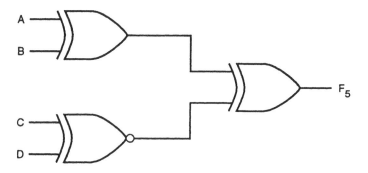

FIGURE 4.34. $F_5 = (A \oplus B) \oplus (C \odot D)$.

counted an odd number of times are used to obtain the answer. F is the sum of minterms 0, 3, 5, 6, 9, 10, 12, and 15. This output is the same as that of Figure 4.33.

4.10 BOOLEAN DIFFERENCES

Another application of EOR functions occurs in testing digital circuits. First, note that the sum or difference function and the EOR function perform modulo-2 arithmetic. That is, if one neglects the carry in binary addition or the borrow in binary subtraction, the answer is the same as that for the EOR. This is shown in Table 4.1.

TABLE 4.1   MOD 2 Arithmetic

A	B	A minus B	A plus B	A ⊕ B
0	0	0	0	0
0	1	1	1	1
1	0	1	1	1
1	1	0	0	0

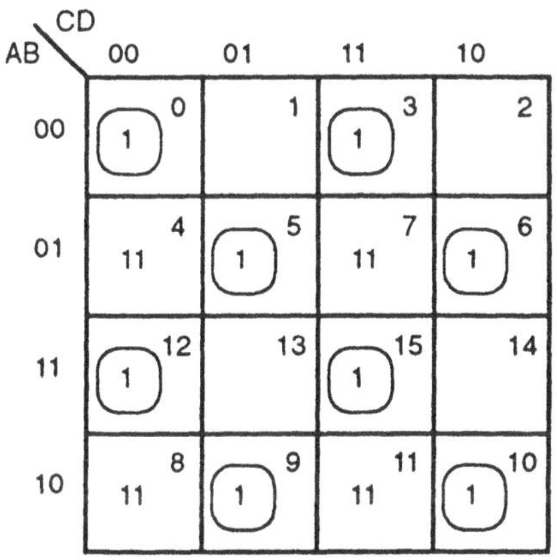

FIGURE 4.35.   Map of $F_5 = \sum m\,(0,3,5,6,9,10,12,15)$.

In calculus, the partial derivative of a function Y(A,B,C,D) with respect to variable A can be written as

$$Y_A = \frac{\partial Y}{\partial A} = \frac{\lim}{\Delta A \to 0} = \frac{Y(A + \Delta A, B, C, D) - Y(A, B, C, D)}{\Delta A} \qquad (4.9)$$

From Table 4.1 it can be seen that $A + \Delta A = A \oplus \Delta A$. In binary boolean algebra, A can only change by 1. With $\Delta A = 1$, $A \oplus \Delta A = A \oplus 1 = \overline{A}$, and the partial derivative becomes

$$Y_A = \frac{\partial Y}{\partial A} = Y(\overline{A}, B, C, D) \oplus Y(A, B, C, D) \qquad (4.10)$$

Physically, Equation 4.10 states that the partial of Y with respect to A is 1 if $Y(\overline{A},B,C,D)$ is different from Y(A,B,C,D). In other words, if the partial of Y with respect to A is 1, then toggling A causes Y to toggle. This means that a path from input A to output Y has been sensitized. On the other hand, if the partial of Y with respect to A is 0, then $Y(\overline{A},B,C,D)$ is the same as Y(A,B,C,D) and the effect of changing A cannot be observed at the output.

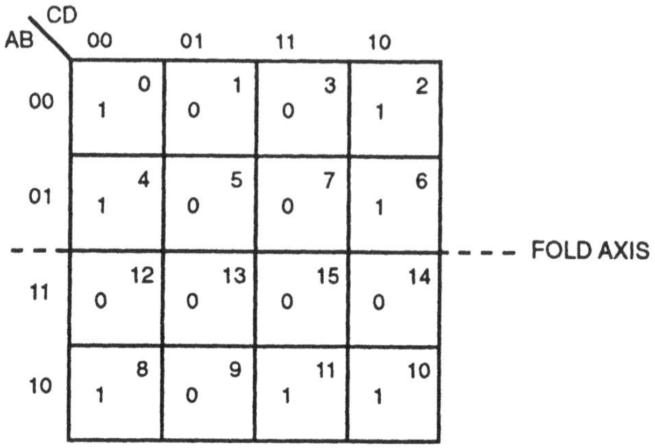

AB \ CD	00	01	11	10
00	0 **1**	1 **0**	3 **0**	2 **1**
01	4 **1**	5 **0**	7 **0**	6 **1**
11	12 **0**	13 **0**	15 **0**	14 **0**
10	8 **1**	9 **0**	11 **1**	10 **1**

--- --- FOLD AXIS (between rows 01 and 11)

FIGURE 4.36. Map for $Y = A\overline{B}C + \overline{A}\,\overline{D} + \overline{B}\,\overline{D}$.

EXAMPLE 4.7: Sensitize a path from input A to output Y where
$$Y = A\overline{B}C + \overline{A}\,\overline{D} + \overline{B}\,\overline{D}.$$

SOLUTION: First map the function Y as shown in Figure 4.36. If the map is now folded such that $A = 0$ and $A = 1$ are superimposed, the map of Figure 4.37 results. On the folded map one can easily see where A and $\overline{A}$ do and do not differ. Figure 4.38 is a map of variables B, C, and D, with a 1 entered where A and $\overline{A}$ are different and a 0 where they are the same. The function mapped on Figure 4.38 is:

$$Y_A(A,B,C,D) = Y(\overline{A},B,C,D) \oplus Y(A,B,C,D) = B\overline{D} + \overline{B}CD = 1$$

The original function is realized in Figure 4.39. A path can be sensitized from input A to output Y either by letting $B = 1$ and $D = 0$, so that $B\overline{D} = 1$, or by letting $B = 0$, $C = 1$, and $D = 1$, so that $\overline{B}CD = 1$. Letting $BD = 10$ enables gate 1 and disables gate 3 of Figure 4.39, allowing gate 4 to respond to input A at gate 1. Letting $BCD = 011$ enables gates 2 and 3 while disabling gate 1 and allows gate 4 to respond to the A input at gate 2.

4.11   THE BINARY ADDER

A third example of EOR design is the binary adder. A circuit that adds two bits and outputs a *sum* and a *carry* is referred to as a *half adder*. The truth table and

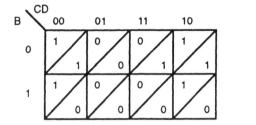

FIGURE 4.37. The map in Figure 4.36 folded on variable A.

FIGURE 4.38. The cubes for which $Y(A)$ and $Y(\overline{A})$ are different.

the conventional symbol for a half adder, as well as a simple circuit to implement it, are shown in Figure 4.40.

A *full adder* adds two bits and a *carry-in* from the next lower position, and outputs a *sum* and a *carry-out*. The truth table and the symbol for a full adder are shown in Figure 4.41, where C_i is the carry-in and C_o is the carry-out. A comparison of Figure 4.40(a) and 4.41(a) shows the half adder to have half as many states as the full adder.

The sum, S, and carry out, C_o, of the full adder are mapped from the truth table in Figure 4.42. The S map is seen to have a pattern indicative of an EOR map. The C_o map can be interpreted as an inclusive-OR or as an EOR pattern, as shown in Figure 4.42(b) and (c).

$$S = A \oplus B \oplus C_i$$

and

$$C_o = AB + C_i(A + B) = AB + C_i(A \oplus B) \tag{4.11}$$

The full adder can be implemented directly from the above maps, as shown in Figure 4.43. A full adder can also be constructed from two half adders and an OR gate as shown in Figure 4.44. In this case,

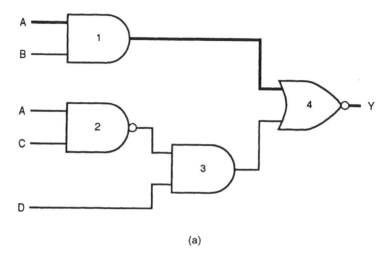

(a)

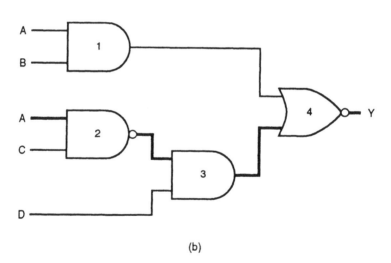

(b)

FIGURE 4.39.   The circuit of Example 4.7, sensitized paths darkened for (a) $BD = 10$ and (b) $BCD = 011$.

$$C_o = C_{o1} + S_1 C_i$$

and

$$S = S_1 \oplus C_i \qquad (4.12)$$

Finally, the full adder can be implemented in AND-OR-INVERT, or AOI, form as a function of A, B, C_i, and C_o. To do this, the truth table for the full adder can be written as a function of A, B, C_i, and C_o, as shown in Figure 4.45. Figure

A	B	S	C
0	0	0	0
0	1	1	0
1	0	1	0
1	1	0	1

(a)

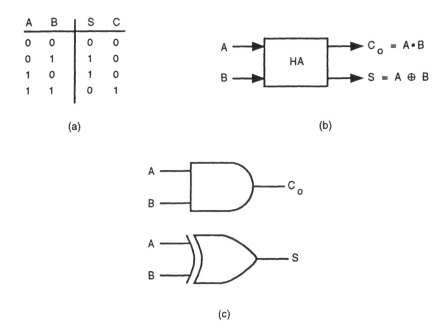

(b)

(c)

FIGURE 4.40. The half adder. (a) The truth table, (b) the symbol, and (c) an implementation.

A	B	C_i	S	C_o
0	0	0	0	0
0	0	1	1	0
0	1	0	1	0
0	1	1	0	1
1	0	0	1	0
1	0	1	0	1
1	1	0	0	1
1	1	1	1	1

(a)

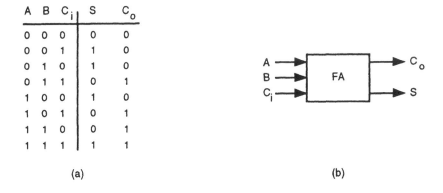

(b)

FIGURE 4.41. The full adder. (a) The truth table and (b) the symbol.

4.45(a) is the same truth table as shown in Figure 4.41, except that C_o is listed as one of four inputs. S as written is seen to be a function of inputs A, B, C_i, and C_o, but such a table has 16 rows. Eight entries are missing, and are therefore don't cares. The full truth table is shown in Figure 4.45(b).

The sum map is then formed for inputs A, B, C_i and C_o, as shown in Figure 4.46. The carry-out, C_o, as obtained from Figure 4.42(b) in the inclusive-OR form is implementable in an AND-OR form and can be complemented to give $\overline{C}_o$ in

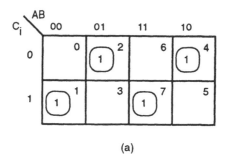

(a)

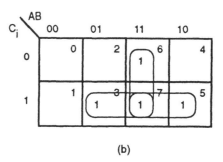

(b)

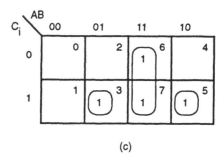

(c)

FIGURE 4.42. The full adder sum and carry. (a) $S = A \oplus B \oplus C_i$, (b) $C_o = AB + C_i(A + B)$, and (c) $C_o = AB + C_i(A \oplus B)$.

the AOI form, and the complemented sum, $\bar{S}$, can also be obtained in AOI form. One then has the AOI implementation of a full-adder circuit, as shown in Figure 4.47.

4.12 SUMMARY

In general, the cost of implementing a switching function in hardware is related directly to the number of terms and the total number of literals in the function.

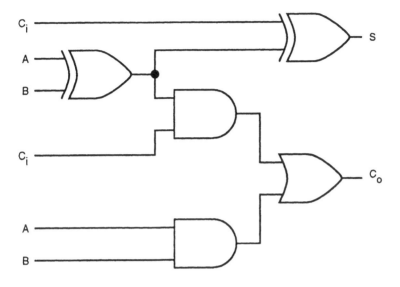

FIGURE 4.43. Implementation of a full adder.

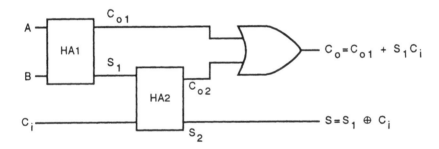

FIGURE 4.44. The full adder constructed from half adders.

Methods of reducing the complexity of switching functions were studied in this chapter. A function is considered simplified if it contains a minimal number of terms and literals, either in an SOP or in a POS form. By implication, any other SOP or POS expression having less terms and literals will not represent the original function.

Any function can be simplified by a judicious application of the Huntington postulates and theorems. This is a purely mathematical approach which, to be successful, requires a high degree of proficiency with the theorems. One can easily obtain functions which appear to be minimal without any guarantee that they are, and one can have a minimal solution without being able to verify that it is.

An approach more suited to engineers is the graphic technique of Karnaugh, which allows a visual simplification and guarantees a minimum solution if certain basic rules of grouping minterms are followed. Due to the difficulty of drawing

A	B	C_i	C_o	S
0	0	0	0	0
0	0	0	1	X
0	0	1	0	1
0	0	1	1	X
0	1	0	0	1
0	1	0	1	X
0	1	1	0	X
0	1	1	1	0
1	0	0	0	1
1	0	0	1	X
1	0	1	0	X
1	0	1	1	0
1	1	0	0	X
1	1	0	1	0
1	1	1	0	X
1	1	1	1	1

A	B	C_i	C_o	S
0	0	0	0	0
0	0	1	0	1
0	1	0	0	1
0	1	1	1	0
1	0	0	0	1
1	0	1	1	0
1	1	0	1	0
1	1	1	1	1

(a)   (b)

FIGURE 4.45. The full-adder truth table for S as a function of C_o.

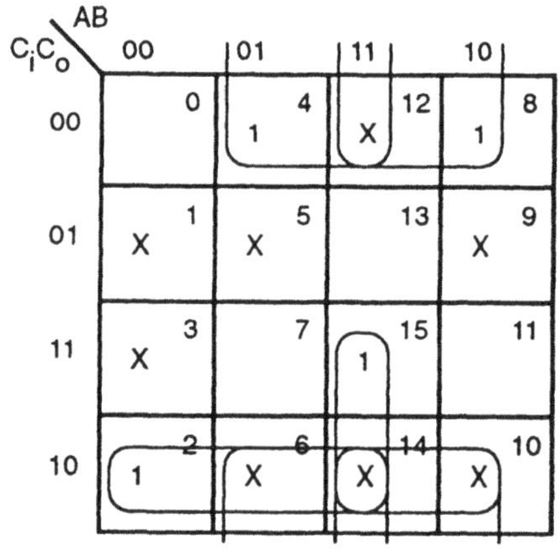

FIGURE 4.46. The map of $S = \overline{C}_o \cdot (A + B + C_i) + A \cdot B \cdot C_i$.

maps in more than three dimensions, this technique is best suited to functions of four variables or less. Many applications of switching theory involve functions of four or less variables, however, and the Karnaugh mapping technique is a very important tool of the digital design engineer. VEM allows one to map a function of more than four variables onto a four-variable map.

The graphic approach requires the identification of a minimum number of terms which can represent a given function. In an SOP expression, each of the product terms is called an implicant of the function because it implies the function; i.e., if the product term is true then the function is true. Simplification of a function involves finding the set of prime implicants (implicants which do not imply any

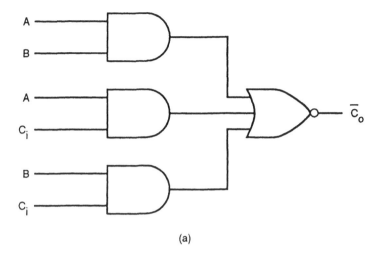

(a)

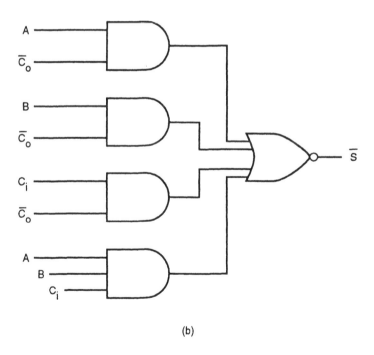

(b)

FIGURE 4.47.   (a) $\overline{C}_o$ and (b) $\overline{S}$ implemented in AOI form.

other implicants) of the function. In terms of Karnaugh maps, a prime implicant is the largest correct grouping of minterms or maxterms. Any given switching function has one unique set of prime implicants since the set is derived from a unique set of minterms.

The problem of determining a minimum SOP representation of a boolean or switching function reduces to the problem of determining a minimum set of prime implicants of the function. This subset of prime implicants is called a minimal cover for the function.

An EPI must be in any minimum representation of the function because it covers at least one minterm that is not covered by any other prime implicant of the function. When two choices of prime implicant exist, the two prime implicants are optional. If one is chosen to represent the function, it becomes an NPI and the other becomes redundant.

A don't care minterm can be treated as either a 0 or a 1, whichever simplifies the answer. Since a don't care minterm can have no effect on the function, choosing it to be a 0 or a 1 can have no effect upon the answer.

Variables can be entered on Karnaugh maps also, and can reduce the size of the map. An n-variable function can be mapped in n–1 space by entering the remaining variable on the map. Thus, one can enter 0s, 1s, don't cares, and variables on the map. More than one variable can be entered on a map, but reading the map correctly can become very difficult.

EOR functions can be recognized on a Karnaugh map by their "kitty-corner" adjacency or offset adjacency pattern. EOR gates are useful in constructing binary adders and parity checking circuits.

The half adder adds two bits and outputs a sum and a carry. It has a truth table with four rows for the four possible input combinations. The full adder adds two bits and a carry-in from the next lower position and outputs a sum and a carry-out. It has a truth table with eight rows for the possible combinations of two bits plus a carry-in bit. The full adder can be constructed of two half adders and an OR gate.

The sum of a half adder or of a full adder is 1 only if there are an odd number of input 1s and it is generated with an odd-parity checking circuit which consists of the EOR of the input bits.

PROBLEMS

Use the appropriate Karnaugh map to simplify each of the following functions. In problems 4.1 through 4.22, list the prime implicants of the function in SOP solutions, or of the complementary function in POS solutions, that are essential, and for which minterm.

4.1.   Give a minimum SOP solution to the following:

(a) $U(A,B,C) = \overline{A}\,\overline{B}\,\overline{C} + \overline{A}\,B\overline{C} + \overline{A}\,BC + ABC$

(b) $V(A,B,C) = \overline{A}\,\overline{B}C + \overline{A}\,B\overline{C} + \overline{A}\,BC + AB\overline{C}$

(c) $W(A,B,C) = A\overline{B}C + AB\overline{C} + ABC$

4.2. Repeat Problem 4.1 for POS solutions.

4.3. Give a minimum SOP solution to the following:

(a) $U(A,B,C) = \overline{A}\,\overline{B}\,\overline{C} + AB\overline{C} + ABC$

(b) $V(A,B,C) = \overline{A}\,\overline{B}\,\overline{C} + A\overline{B}C + AB\overline{C} + ABC$

(c) $W(A,B,C) = \overline{A}\,\overline{B}C + \overline{A}\,B\overline{C} + AB\overline{C} + ABC$

4.4. Repeat Problem 4.3 for POS solutions.

4.5. Give two minimum SOP solutions to the function $X(A,B,C) =$ $\sum m\,(0,2,3,4,5,7)$.

4.6. Give two minimum POS solutions to the function $X(A,B,C) =$ $\prod M\,(0,2,3,4,5,7)$.

4.7. Give two minimum SOP solutions to the function $X(A,B,C) =$ $\sum m\,(0,1,2,5,6,7)$

4.8. Give two minimum POS solutions to the function $X(A,B,C) =$ $\prod M\,(0,1,2,5,6,7)$

4.9. Give two minimum SOP and one minimum POS solution to the function $Y(A,B,C) = \sum m\,(1,2,3,4,5)$.

4.10. Give two minimum SOP and one minimum POS solution to the function $Y(A,B,C) = \sum m\,(0,1,3,6,7)$.

4.11. Give a minimum SOP solution to the following:

(a) $Z(A,B,C,D) = \overline{A}\,B\overline{C}D + \overline{A}\,BCD + A\overline{B}\,\overline{C}\,\overline{D} + AB\overline{C}\,\overline{D} + \overline{A}\,C\overline{D}$

(b) $Z(A,B,C,D) = \overline{A}\,\overline{B}CD + \overline{A}\,B\overline{C}D + AB\overline{C}D + ABC\overline{D} + BCD$

(c) $Z(A,B,C,D) = BC\overline{D} + \overline{A}\,B\overline{C}\,\overline{D} + A\overline{B}CD + AB\overline{C}\,\overline{D}$

4.12. Repeat Problem 4.11 for POS solutions.

4.13. Give a minimum SOP solution to the following:

(a) $Z(A,B,C,D) = A\overline{B}C\overline{D} + A\overline{B}\,\overline{C}\,\overline{D} + \overline{A}\,BCD + \overline{A}\,\overline{B}CD + A\overline{C}D$

(b) $Z(A,B,C,D) = AB\overline{C}\,\overline{D} + A\overline{B}C\overline{D} + \overline{A}\,\overline{B}C\overline{D} + \overline{A}\,\overline{B}\overline{C}D + \overline{B}\,C\overline{D}$

(c) $Z(A,B,C,D) = \overline{B}\,\overline{C}D + A\overline{B}CD + \overline{A}\,BC\overline{D} + \overline{A}\,BCD$

4.14. Repeat Problem 4.13 for POS solutions.

4.15. Give a minimum SOP solution to the following:

(a) $F(A,B,C,D) = \sum m\,(0,2,3,5,7,8,12,13)$

(b) $F(A,B,C,D) = \sum m\,(1,3,4,5,8,9,13,15)$

(c) $F(A,B,C,D) = \sum m\,(1,2,4,5,6,7,12,15)$

4.16. Repeat Problem 4.15 for POS solutions.

4.17. Give a minimum SOP solution to the following:

(a) $F(A,B,C,D) = \sum m\,(1,3,5,8,9,11,15) + \sum d\,(2,13)$

(b) $F(A,B,C,D) = \sum m\,(4,7,12,14,15) + \sum d\,(8,9,10)$

(c) $F(A,B,C,D) = \sum m\,(3,8,12,13,15) + \sum d\,(7,9,14)$

4.18. Repeat Problem 4.17 for POS solutions.

4.19. Give a minimum SOP solution to the following:

(a) $F(A,B,C,D) = \prod M\,(4,5,6,14,15)$

(b) $F(A,B,C,D) = \prod M\,(0,11,14,15)\cdot \prod D\,(1,3,5,7,8,9,12,13)$

(c) $F(A,B,C,D) = \prod M\,(0,2,7,11,12)\cdot \prod D\,(3,8,9,10)$

4.20. Repeat Problem 4.19 for POS solutions.

4.21. Give a minimum SOP solution to the following:

(a) $F(A,B,C,D) = \prod M\,(0,1,8,9,11)$

(b) $F(A,B,C,D) = \prod M\,(0,4,5,15)\cdot \prod D\,(2,3,6,7,10,12,14)$

(c) $F(A,B,C,D) = \prod M\,(3,4,8,13,15)\cdot \prod D\,(5,6,7,12)$

4.22. Repeat Problem 4.21 for POS solutions.

4.23. $Z(A,B,C,D) = \sum m\,(0,1,3,6,7,8,13,14) + \sum d\,(2,9)$. Realize (implement) a minimum two-level AND-OR network to output Z.

4.24. $Z(A,B,C,D) = \sum m\,(1,2,7,8,9,12) + \sum d\,(6,13)$. Realize (implement) a minimum two-level OR-AND network to output Z.

4.25. (a) Map, simplify, and realize $Y(A,B,C) = \sum m\,(1,4,5,7)$ in NAND-NAND logic,

(b) Map, simplify, and realize $Y(A,B,C) = \sum m\,(1,4,5,7)$ in NOR-NOR logic.

4.26. (a) Map, simplify, and realize $Y(A,B,C) = \sum m\,(0,2,3,6)$ in NAND-NAND logic,

(b) Map, simplify, and realize $Y(A,B,C) = \sum m\,(0,2,3,6)$ in NOR-NOR logic.

4.27. $U(A,B,C,D) = \sum m\,(0,1,3,4,5,6,7,8,10,12,14)$. Give three SOP solutions.

4.28. $V(A,B,C,D) = \sum m\,(0,2,4,6,8,9,11,12,13,14,15)$. Give three SOP solutions.

4.29. $W(A,B,C,D) = \sum m\,(0,1,3,5,6,7,8,10,14,15)$. Give three SOP solutions.

4.30. $X(A,B,C,D) = \sum m\,(0,1,2,4,6,7,8,9,13,15)$. Give three SOP solutions.

4.31. Plot the following functions on a three-variable Karnaugh map, and to VEMs as a function of A and B.

(a) $F(A,B,C) = \sum m\,(2,3,6,7)$

(b) $G(A,B,C) = \sum m\,(0,1,2,3,7)$

(c) $H(A,B,C) = \sum m\,(1,2,5,6)$

4.32. Plot the following functions on a three-variable Karnaugh map, and to VEMs as a function of A and B.

(a) $F(A,B,C) = \sum m\,(1,2,3,4)$

(b) $G(A,B,C) = \sum m\,(3,4,6,7)$

(c) $H(A,B,C) = \sum m\,(3,4,5,6)$

4.33. Plot the following functions on a four-variable Karnaugh map, and to a three-variable map as a function of A, B, and C.

(a) $K(A,B,C,D) = \sum m\,(0,1,3,11,12,14,15)$

(b) $L(A,B,C,D) = \sum m\,(0,2,3,7,8,9,10,13)$

4.34. Plot the following functions on a four-variable Karnaugh map, and to a three-variable map as a function of B, C, and D.

(a) $M(A,B,C,D) = \sum m\,(0,1,6,7,9,11,13,14)$

(b) $N(A,B,C,D) = \sum m\,(1,2,4,7,8,9,12,14)$

4.35. $Y = ABC + \overline{A}\,BD + CD$. Find the boolean difference $Y_D \oplus Y_{\overline{D}}$ (a) algebraically and (b) by plotting the function on a four-variable map and folding it on D.

4.36. $Y = A\overline{B}C + \overline{A}\,\overline{D} + \overline{B}\,\overline{D}$. Find the boolean difference $Y_D \oplus Y_{\overline{D}}$ (a) algebraically and (b) by plotting the function on a four-variable map and folding it on D.

4.37. $Y = AB + \overline{A}\,C + CD + BD$. Find the boolean difference $Y_A \oplus Y_{\overline{A}}$ (a) algebraically and (b) by plotting the function on a four-variable map and folding it on A.

4.38. $Y = AC + BC + \overline{A}\,D + A\overline{D}$. Find the boolean difference $Y_A \oplus Y_{\overline{A}}$ (a) algebraically and (b) by plotting the function on a four-variable map and folding it on A.

4.39. Represent the three functions on 3-cubes and simplify them.

(a) $Z(A,B,C) = \sum m\,(0,3,7)$

(b) $Z(A,B,C) = \sum m\,(0,5,6,7)$

(c) $Z(A,B,C) = \sum m\,(1,2,6,7)$

4.40. Represent the three functions on 3-cubes and simplify them.

(a) $Z(A,B,C) = \sum m\,(1,2,6)$

(b) $Z(A,B,C) = \sum m\,(0,1,3,6)$

(c) $Z(A,B,C) = \sum m\,(0,3,4,5)$

4.41. Represent the three functions on 4-cubes and simplify them.

(a) $X(A,B,C,D) = \sum m\,(0,2,3,5,7,8,12,13)$

(b) $X(A,B,C,D) = \sum m\,(1,3,4,5,8,9,13,15)$

(c) $X(A,B,C,D) = \sum m\,(1,2,4,5,6,7,12,15)$

4.42. Represent the three functions on 4-cubes and simplify them.

(a) $X(A,B,C,D) = \sum m\,(2,3,7,8,10,12,13,15)$

(b) $X(A,B,C,D) = \sum m\,(0,2,6,7,10,11,12,14)$

(c) $X(A,B,C,D) = \sum m\,(0,3,8,9,10,11,13,14)$

4.43. Use EOR mapping techniques to find a minterm expansion of W in Figure 4.48.

4.44. Use EOR mapping techniques to find a minterm expansion of X in Figure 4.49.

4.45. Use EOR mapping techniques to find a minterm expansion of Y in Figure 4.50.

4.46. Use EOR mapping techniques to find a minterm expansion of Z in Figure 4.51.

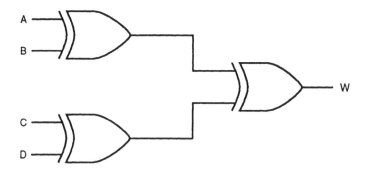

FIGURE 4.48.

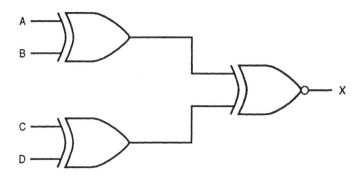

FIGURE 4.49.

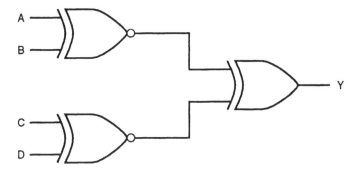

FIGURE 4.50.

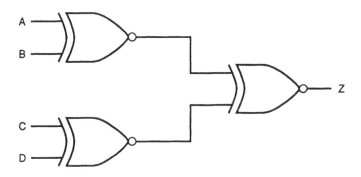

FIGURE 4.51.

SPECIAL PROBLEMS

4.47. It is desired to convert BCD inputs to seven-segment visible displays as outputs. Since inputs numerically greater than 9_{10} are not available, the corresponding outputs are don't cares. Let an excited or luminous state of a segment be identified as state 0, and a dark state of a segment as state 1. The truth table is given below, and the seven segments are labeled in Figure 4.52. Obtain the Karnaugh map, loop the 1s, and write the simplest SOP expression for each Y_i output.

BCD inputs				Minterm	7-segment outputs						
A	**B**	**C**	**D**		Y_6	Y_5	Y_4	Y_3	Y_2	Y_1	Y_0
0	0	0	0	m_0	1	0	0	0	0	0	0
0	0	0	1	m_1	1	1	1	1	0	0	1
0	0	1	0	m_2	0	1	0	0	1	0	0
0	0	1	1	m_3	0	1	1	0	0	0	0
0	1	0	0	m_4	0	0	1	1	0	0	1
0	1	0	1	m_5	0	0	1	0	0	1	0
0	1	1	0	m_6	0	0	0	0	0	1	1
0	1	1	1	m_7	1	1	1	1	0	0	0
1	0	0	0	m_8	0	0	0	0	0	0	0
1	0	0	1	m_9	0	0	1	1	0	0	0
1	0	1	0	m_{10}	x	x	x	x	x	x	x
1	0	1	1	m_{11}	x	x	x	x	x	x	x
1	1	0	0	m_{12}	x	x	x	x	x	x	x
1	1	0	1	m_{13}	x	x	x	x	x	x	x
1	1	1	0	m_{14}	x	x	x	x	x	x	x
1	1	1	1	m_{15}	x	x	x	x	x	x	x

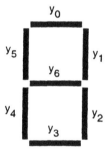

FIGURE 4.52.

4.48. Repeat Problem 4.47 for a seven-segment display to output hexadecimal numbers from 0 to F. Minterms 10 through 15 must be changed to display the letters "A" through "F", and a minor change is required in m_6 of the truth table above in order to distinguish the numeral "6" from the lower-case letter "b". To be consistent, m_9 should be changed also, so that the numbers "6" and "9" are more compatible.

The upper-case letter "B" is indistinguishable from the numeral "8", and the capital letter "D" is indistinguishable from the numeral "0"; therefore, m_{11} must output a lower-case "b" and m_{13} must output a lower case "d". Let the other letters be upper case. One then obtains the truth table below. Obtain the Karnaugh map, loop the 1s, and obtain the simplest SOP expression for each Y_i output.

Hex inputs				Minterm	7-segment outputs						
A	B	C	D		Y_6	Y_5	Y_4	Y_3	Y_2	Y_1	Y_0
0	0	0	0	m_0	1	0	0	0	0	0	0
0	0	0	1	m_1	1	1	1	1	0	0	1
0	0	1	0	m_2	0	1	0	0	1	0	0
0	0	1	1	m_3	0	1	1	0	0	0	0
0	1	0	0	m_4	0	0	1	1	0	0	1
0	1	0	1	m_5	0	0	1	0	0	1	0
0	1	1	0	m_6	0	0	0	0	0	1	0
0	1	1	1	m_7	1	1	1	1	0	0	0
1	0	0	0	m_8	0	0	0	0	0	0	0
1	0	0	1	m_9	0	0	1	0	0	0	0
1	0	1	0	m_{10}	0	0	0	1	0	0	0
1	0	1	1	m_{11}	0	0	0	0	0	1	1
1	1	0	0	m_{12}	1	0	0	0	1	1	0
1	1	0	1	m_{13}	0	1	0	0	0	0	1
1	1	1	0	m_{14}	0	0	0	0	1	1	0
1	1	1	1	m_{15}	0	0	0	1	1	1	0

4.49. The truth table for a two-bit binary adder with carry is shown below.
 (a) Map the sum S_0, the sum S_1, and the carry-out C_2 as a function of A_1, A_0, B_1, and B_0. Let the low-order carry-in, C_0, be a map-entered variable. Show that the sums and carry can be expressed as

$$S_0 = (A_0\overline{B}_0 + \overline{A}_0 B_0)\overline{C}_0 + (A_0 B_0 + \overline{A}_0\overline{B}_0)C_0$$
$$S_1 = (\overline{A}_1\overline{B}_1 + A_1 B_1)[A_0 B_0\overline{C}_0 + (A_0 + B_0)C_0]$$
$$+ (\overline{A}_1 B_1 + A_1\overline{B}_1)[(\overline{A}_0 + \overline{B}_0)\overline{C}_0 + \overline{A}_0\overline{B}_0 C_0]$$
$$C_2 = A_1 B_1 + (A_1 + B_1)A_0 B_0 + (A_1 + B_1)(A_0 + B_0)C_0$$

 (b) Use EOR mapping techniques to show that the answers to part a can also be written in the form

$$S_0 = A_0 \oplus B_0 \oplus C_0$$
$$S_1 = (A_1 \oplus B_1) \oplus [A_0 B_0 + (A_0 \oplus B_0)C_0]$$
$$C_2 = A_1 B_1 + (A_1 \oplus B_1)[A_0 B_0 + (A_0 \oplus B_0)C_0]$$

A_1	A_0	B_1	B_0	C_0	C_2	S_1	S_0
0	0	0	0	0	0	0	0
0	0	0	0	1	0	0	1
0	0	0	1	0	0	0	1
0	0	0	1	1	0	1	0
0	0	1	0	0	0	1	0
0	0	1	0	1	0	1	1
0	0	1	1	0	0	1	1
0	0	1	1	1	1	0	0
0	1	0	0	0	0	0	1
0	1	0	0	1	0	1	0
0	1	0	1	0	0	1	0
0	1	0	1	1	0	1	1
0	1	1	0	0	0	1	1
0	1	1	0	1	1	0	0
0	1	1	1	0	1	0	0
0	1	1	1	1	1	0	1
1	0	0	0	0	0	1	0
1	0	0	0	1	0	1	1
1	0	0	1	0	0	1	1
1	0	0	1	1	1	0	0
1	0	1	0	0	1	0	0
1	0	1	0	1	1	0	1
1	0	1	1	0	1	0	1
1	0	1	1	1	1	1	0
1	1	0	0	0	0	1	1
1	1	0	0	1	1	0	0
1	1	0	1	0	1	0	0
1	1	0	1	1	1	0	1
1	1	1	0	0	1	0	1
1	1	1	0	1	1	1	0
1	1	1	1	0	1	1	0
1	1	1	1	1	1	1	1

REFERENCES

1. **Veitch, E. W.,** A chart method for simplifying truth functions, in Proc. of the ACM, 1952, 127–133.
2. **Karnaugh, M.,** The map method for synthesis of combinational logic circuits, *Trans. AIEE Commun. Electron.,* 72, 593–599, 1953.
3. **Sandige, R. S.,** Modern Digital Design, McGraw-Hill, New York, 1990, 167–182.
4. **Hill, F. J. and Peterson, G. R.,** *Introduction to Switching Theory and Analog Design,* 2nd ed., John Wiley & Sons, New York, 1974, chap. 7.

CHAPTER 5

Advanced Simplification Techniques

5.1 INTRODUCTION

The Karnaugh-map method is convenient for humans because they can readily visualize the minimum solution. It is an excellent tool for simplifying functions of four or fewer variables, and it is quite useful for simplifying five- or six-variable functions. Maps of five or six variables can be created by combining two- or four 4-variable maps, respectively. For functions of more than six variables, a Karnaugh-map approach requires more than three dimensions and is difficult to visualize. For functions of seven variables, a six-variable mapping with one variable entered on the maps is a possibility, but for more than seven variables, it then becomes necessary to formalize the reduction techniques and use a different approach.

Tabular methods are easily implemented on computers, and the number-crunching ability of the computer can be used to solve many-variable problems. One of the simpler tabular methods is the *Quine-McCluskey algorithm*. In this approach the problem is separated into two steps: finding all the prime cubes and then finding the minimum number of prime cubes required.

As in mapping, the minterms that are 1 and don't care can be used to give a minimum SOP solution, or the minterms that are 0 and don't care can be used to give a minimum POS solution to the function. Both approaches will be discussed.

Often the digital designer is required to simplify more than one function. The simultaneous simplification of more than one function will be introduced using both a mapping procedure and a tabular procedure based upon the Quine-McCluskey algorithm. This situation is referred to as multi-output logic and is an extension of the single-output minimization problems discussed up to now.

The *directed-search algorithm (DSA)* is an approach which requires a minterm expansion, but only computes the required prime cubes. Starting from a selected minterm, and a knowledge of which remaining minterms can combine with it, the algorithm develops a *required adjacency direction (RAD)* tree, which terminates

Axis of
Symmetry

ABC DE	000	001	011	010	110	111	101	100
00	0	4	12	8	24	28	20	16
01	1	5	13	9	25	29	21	17
11	3	7	15	11	27	31	23	19
10	2	6	14	10	26	30	22	18

FIGURE 5.1.   A five-variable K-map obtained by folding one four-variable map over another four-variable map (minterms labeled).

in an essential or necessary prime cube of the function. No redundant prime cubes are generated, and the computation time is kept to a minimum.

Modern digital logic functions can have many variables and an extremely large number of minterms. The functions are typically expressed with a relatively small number of cubes, and techniques such as Quine-McCluskey are too time consuming. *Iterative consensus* is an approach which avoids the determination of the minterms or 0-cubes first. Iterative consensus starts with a list of cubes (an SOP expression) and generates all of the prime cubes by repeatedly applying the consensus theorem.

The *generalized iterative consensus algorithm* is a streamlined form of iterative consensus which requires less iteration and is therefore faster.   .

5.2 KARNAUGH MAPS FOR FIVE AND SIX VARIABLES

Maps of five variables can be displayed in three basic ways. In Figure 5.1, the two four-variable maps are symmetric about a central axis, as indicated. In this display, one has to visualize the maps as folded along the axis of symmetry, much the same way that an open book can be folded closed. When the map is folded, adjacent terms from each half of the mapping can be visualized as falling one above the other.

A second approach is to consider the five-variable map to be made up of two separate standard four-variable maps. In this case, one map is placed on top of the other map to achieve a three-dimensional mapping of the function. An example is shown in Figure 5.2. A third approach consists of displaying the above information on one four-variable map by the simple expedient of splitting each cell

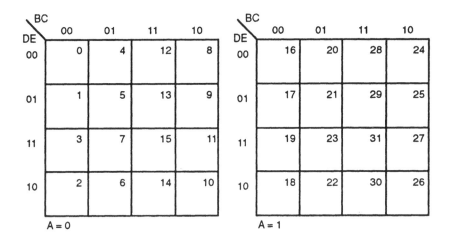

FIGURE 5.2.   A five-variable K-map displayed as two adjacent four-variable maps (minterms labeled).

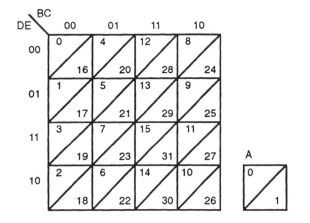

FIGURE 5.3.   A five-variable K-map obtained by splitting cells of a four-variable map to obtain the overlapping maps (minterms labeled).

diagonally and visualizing one half of the cell as behind (or below) the other half of the same cell. This is shown in Figure 5.3.

A function of six variables can be mapped by combining four four-variable maps or two five-variable maps. As was true of five-variable maps, the six-variable maps can be represented in three ways. The first method requires folding the mapping along both a horizontal and a vertical axis of symmetry, and is almost never used.

The two standard methods of displaying a function of six variables are shown in Figures 5.4 and 5.5. In Figure 5.4, four maps are laid out in a two-dimensional

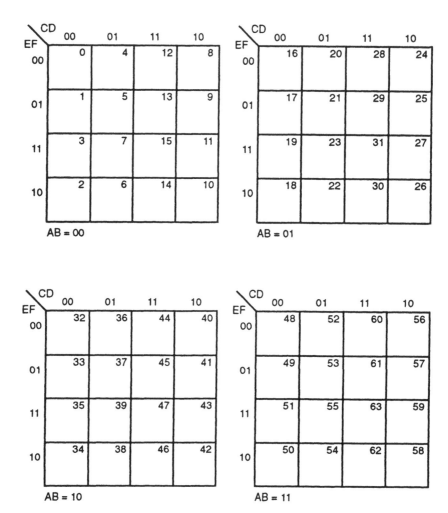

FIGURE 5.4. A six-variable map obtained from four four-variable maps, stacked to form a three-dimensional array (minterms labeled).

array, but one can visualize them stacked one upon the other, forming a three-dimensional array. Likewise, in Figure 5.5, one can visualize a three-dimensional array of four four-variable maps stacked vertically to form the six-variable map.

All the techniques that apply to four-variable maps also apply to mappings of completely or incompletely specified functions of more than four variables. One can loop 1s and don't cares to obtain an SOP solution, or 0s and don't cares to obtain a POS solution, or variables entered on the map, etc.

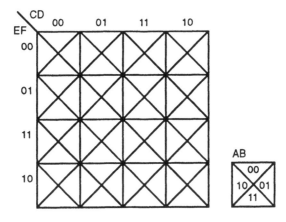

FIGURE 5.5.   A six-variable map obtained by splitting each cell of a four-variable map into four parts.

5.3   THE QUINE-McCLUSKEY TABULAR TECHNIQUE

The tabular method of Quine and McCluskey[1,2] provides a systematic approach to simplifying functions of any number of variables and is ideally suited to computer simplification. The technique utilizes the standard reduction theorem, $XY + X\overline{Y} = X$, to provide a minimum SOP solution to a function that is in the canonical SOP form.

The process consists of two separate operations. First, starting from the minterms (0-cubes) all the prime cubes of the function are determined. Second, a prime-cube table is created to determine the minimum cover of prime cubes for the function. The Quine-McCluskey method proceeds as follows:

1.   Obtain the minterm expansion of the function to be simplified.
2.   Systematically and repeatedly apply the reduction theorem to obtain the prime cubes. A prime cube is hereby defined as an irreducible term.
3.   Systematically select a minimum set of prime cubes which will represent the given function with a minimum number of gates and literals.

Starting with the canonical SOP expression for the function, the reduction theorem is applied to all possible pairs of minterms to obtain all the terms corresponding to 1-cubes of the function. The reduction theorem is next applied to all pairs of minterms (1-cubes) to obtain all the 2-cubes of the function, and so on. The process continues until the largest cubes have been obtained. At each pass, the terms that combined to generate larger cubes are deleted. The result is the set of all possible prime cubes (prime implicants) of the function.

The minterms must be arranged in a table of 1s and 0s in such a manner that each variable can be associated with one column of the table. To assure that all possible cubes are generated, the minterms are arranged by the number of 1s in the minterm, starting with minterm 0 if it is present, and ending with the minterm with the maximum number of 1s in it. The number of 1s in a binary vector is referred to as the *vector magnitude* or just *magnitude*. For two rows to combine, they must differ in only one column (i.e., one variable). Terms with the same number of 1s cannot combine, and terms that differ in two or more columns cannot combine, so each cube must be compared to cubes whose magnitude is greater or lesser by one.

Starting from the top of the table, minterm 0 (if present) is combined with any minterms that have only a single 1 in them, minterms 1, 2, 4, 8, 16, etc. Since the minterms with any given magnitude have already been compared to those above them in the table which have a smaller magnitude, they need only be compared with the minterms below them whose magnitude is larger by one. When all the minterms have been compared, any that did not combine are 0-cube prime cubes and are recorded as such. Those that combined into 1-cubes are next compared to see which combine to form 2-cubes, etc. This process is continued until no further application of the reduction theorem is possible.

Once the prime cubes have been determined, the minimal cover of the function must be determined. The prime cubes are arranged in rows of a table and the minterms of the function to be minimized are listed in columns of the table. If a cube covers a minterm, a check mark is placed in the column of the minterm and the row of the cube. The rows of check marks indicate the minterms covered by a prime cube, and the columns of check marks indicate the cubes that cover a given minterm.

The minimum number of rows required to cover every minterm is the minimum number of prime cubes required to specify the function. A minterm that has only one check in its column is covered by only one prime cube, and that prime cube is essential to the answer. If more than one cube covers a minterm, there are optional prime cubes, and the problem reduces to that of determining the least number of optional cubes that are necessary to specify the function.

5.3.1   Completely Specified Functions

The tabular method will be explained step by step with an example, using the Quine-McCluskey method to simplify the function

$$F(A,B,C,D) = AB + A\overline{B}\,C\overline{D} + A\overline{B}\,\overline{C}D + \overline{A}BCD + \overline{A}B\overline{C}\,\overline{D} \qquad (5.1)$$

AB is not a minterm, and must be expanded first.

$$AB = AB(\overline{C} + C)(\overline{D} + D) = AB\overline{C}\,\overline{D} + AB\overline{C}D + ABC\overline{D} + ABCD$$

The function can be written in a minterm expansion as

$$F(A,B,C,D) = \sum m\,(4,7,9,10,12,13,14,15) \tag{5.2}$$

The minterms can now be tabulated in descending order of increasing magnitude. This is done in Table 5.1.

TABLE 5.1

A	B	C	D	0-cubes	
0	1	0	0	m_4	*
1	0	0	1	m_9	*
1	0	1	0	m_{10}	*
1	1	0	0	m_{12}	*
0	1	1	1	m_7	*
1	1	0	1	m_{13}	*
1	1	1	0	m_{14}	*
1	1	1	1	m_{15}	*

Each minterm that combines with another minterm is checked off with an asterisk. Any minterm not checked off in Table 5.1 cannot combine with any other minterm and is a prime cube. Combine appropriate minterms into 1-cubes, while checking off the 0-cubes in Table 5.1 which are not prime cubes. Table 5.2 is developed in this way, using "x" to represent a missing variable, i.e., a variable which was eliminated when two cubes were combined.

This procedure is repeated until no more cubes combine. One then has the prime cubes of the function. Examination of Table 5.2 reveals that minterms 12, 13, 14, and 15 can combine to give prime cube 11xx. The remaining 1-cubes cannot combine and are prime. This is indicated by their not being checked off with an asterisk. One thus obtains five prime cubes:

$$F(A,B,C,D) = x100 + 1x01 + 1x10 + x111 + 11xx$$
$$= B\overline{C}\,\overline{D} + A\overline{C}D + AC\overline{D} + BCD + AB \tag{5.3}$$

TABLE 5.2

	A	B	C	D	1-cubes	
	x	1	0	0	$m_4 + m_{12}$	←V
	1	x	0	1	$m_9 + m_{13}$	←W
	1	x	1	0	$m_{10} + m_{14}$	←X
	1	1	0	x	$m_{12} + m_{13}$	*
	1	1	x	0	$m_{12} + m_{14}$	*
	x	1	1	1	$m_7 + m_{15}$	←Y
	1	1	x	1	$m_{13} + m_{15}$	*
	1	1	1	x	$m_{14} + m_{15}$	*

The next task is to determine the necessary and essential prime cubes of the function. To select in an optimum manner those prime cubes that account for all of the original minterms, form a new table with a column for each of the minterms and a row for each prime cube. A prime cube with no missing variables accounts for one minterm only, a prime cube with one missing variable accounts for two minterms, and so forth.

The essential prime cubes can be determined by inspection of the table. Any column that has only one check mark in it represents a minterm that can be covered in only one way. From Table 5.3, it is seen that x100 is essential to m_4, x111 is essential to m_7, 1x01 is essential to m_9, and 1x10 is essential to m_{10}. These prime cubes cover the remaining minterms also, and it can be seen that 11xx is a redundant prime cube. The function reduces to $F(A,B,C,D) = B\overline{C}\,\overline{D} + A\overline{C}D + AC\overline{D} + BCD$.

TABLE 5.3

m_4	m_7	m_9	m_{10}	m_{12}	m_{13}	m_{14}	m_{15}	PC	
√				√				x100	V ← E
		√			√			1x01	W ← E
			√			√		1x10	X ← E
	√						√	x111	Y ← E
				√	√	√	√	11xx	Z

If the necessary prime cubes are not obvious, one needs a systematic method of determining them. One approach is to label the prime cubes, as shown to the right in Table 5.3, and form a new table which specifies which prime cubes cover which minterms. This is shown in Table 5.4.

TABLE 5.4

Minterm	PCs that account for the minterm
m_4	$V = x100$
m_7	$Y = x111$
m_9	$W = 1x01$
m_{10}	$X = 1x10$
m_{12}	$V + Z \quad (Z = 11xx)$
m_{13}	$W + Z$
m_{14}	$X + Z$
m_{15}	$Y + Z$

To account for all the minterms simultaneously, form the fictitious function H and simplify, recalling that $V(V + Z) = V$. H expresses in boolean algebra those prime cubes needed to cover all the minterms of the function F, and when simplified, gives the minimum SOP solution to the function F. In the above example,

$$H = VYWX(V + Z)(W + Z)(X + Z)(Y + Z) = VYWX \tag{5.4}$$

and the minimum sum of products solution must include prime cubes V, Y, W, and X. Thus,

$$F(A,B,C,D) = B\overline{C}\,\overline{D} + A\overline{C}D + AC\overline{D} + BCD \tag{5.5}$$

Before computing the function H, one should reduce the prime cube table by deleting all essential prime cubes and the minterms they cover. In the above example, the first four prime cubes are essential and cover all the minterms. This guarantees that prime cube Z is redundant.

Since F is a function of only four variables, it could have been mapped also. The Karnaugh map allows one to envision the various steps required to simplify the function, as shown in Figure 5.6. From the map it is obvious that AB is a redundant prime cube.

5.3.2   Incompletely Specified Functions

The tabular method can be applied to incompletely specified functions also. In applying the Quine-McCluskey method, start by treating all the don't cares as 1s. Unnecessary prime cubes will be removed later. Consider the function

$$F(A,B,C,D) = \sum m \,(1,3,5,10,11,12,13,14,15) + \sum d \,(0,4) \tag{5.6}$$

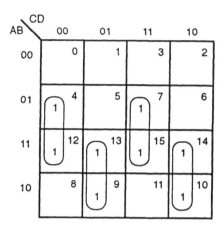

FIGURE 5.6. Karnaugh-map simplification of F(A,B,C,D) of Equation 5.1.

At this point one doesn't know whether the don't care terms will be treated as 1s or as 0s. They must be treated as 1s in the first part of the Quine-McCluskey method in order to see if they will simplify the function.

Form a table, Table 5.5, of all the cover minterms (0-cubes) and don't care terms and, as before, systematically search for .ninterms that combine to form 1-cubes of the function. This is done in Table 5.6. None of the 0-cubes of Table 5.5 are prime cubes. Table 5.6 shows all the 1-cubes of the function.

TABLE 5.5

	A	B	C	D	0-cubes	
	0	0	0	0	m_0	*
	0	0	0	1	m_1	*
	0	1	0	0	m_4	*
	0	0	1	1	m_3	*
	0	1	0	1	m_5	*
	1	0	1	0	m_{10}	*
	1	1	0	0	m_{12}	*
	1	0	1	1	m_{11}	*
	1	1	0	1	m_{13}	*
	1	1	1	0	m_{14}	*
	1	1	1	1	m_{15}	*

TABLE 5.6

A	B	C	D	1-cubes	
0	0	0	x	$m_0 + m_1$	*
0	x	0	0	$m_0 + m_4$	*
0	0	x	1	$m_1 + m_3$	←U
0	x	0	1	$m_1 + m_5$	*
0	1	0	x	$m_4 + m_5$	*
x	1	0	0	$m_4 + m_{12}$	*
x	0	1	1	$m_3 + m_{11}$	←V
x	1	0	1	$m_5 + m_{13}$	*
1	0	1	x	$m_{10} + m_{11}$	*
1	x	1	0	$m_{10} + m_{14}$	*
1	1	0	x	$m_{12} + m_{13}$	*
1	1	x	0	$m_{12} + m_{14}$	*
1	x	1	1	$m_{11} + m_{15}$	*
1	1	x	1	$m_{13} + m_{15}$	*
1	1	1	x	$m_{14} + m_{15}$	*

The process is repeated once more to find the 1-cubes that combine to produce 2-cubes. This is done in Table 5.7. Two of the 1-cubes do not combine to form 2-cubes. They are prime cubes and are labeled U and V on Table 5.6. Since there is only one entry per subgroup in Table 5.7, no more cubes can be combined, and the prime cubes are all the cubes in Table 5.7 as well as the two cubes in Table 5.6.

To determine the essential and necessary prime cubes, proceed as before to determine which minterms are covered by which prime cubes. Don't care terms do not have to be accounted for and must not be included in this step. Thus, in Table 5.8, only minterms asserted true are listed. This automatically removes loops consisting solely of don't care terms.

TABLE 5.7

A	B	C	D	2-cubes	
0	x	0	x	0,1,4,5	←W
x	1	0	x	4,5,12,13	←X
1	x	1	x	10,11,14,15	←Y
1	1	x	x	12,13,14,15	←Z

It can be seen from Table 5.8 that Y is essential to minterm 10, and Y also covers minterms 11, 14, and 15. A reduced prime cube table can be formed at this point by removing the essential prime cube and the columns containing minterms it covers. One can then form the fictitious function H again, which accounts for all the possible remaining covers of the function.

TABLE 5.8

	1	3	5	10	11	12	13	14	15	PC	
	√	√								00x1	U
		√			√					x011	V
	√		√							0x0x	W
			√			√	√			x10x	X
				√	√			√	√	1x1x	Y ←E
						√	√	√	√	11xx	Z

If prime cube Y is not removed, the function H is

$$H = (U + W)(U + V)(W + X)Y(V + Y)(X + Z)(Y + Z)$$
$$= (U + VW)(X + WZ)Y$$
$$H = UXY + UWYZ + VWXY + VWYZ \qquad (5.7)$$

If Y and the columns for minterms 10, 11, 14, and 15 had been removed, the function H would have been

$$H = (U + W)(U + V)(W + X)(X + Z)Y = (U + VW)(X + WZ)Y$$

which reduces to the same value as before.

The function H gives four choices of prime cubes, each of which requires one OR gate. Three of the choices have four prime cubes each and require four AND gates, while one choice has three prime cubes and is the minimum solution, requiring only three AND gates. Hence, the minimum SOP solution to this function is

$$F(A,B,C,D) = U + X + Y = \overline{A}\,\overline{B}D + B\overline{C} + AC \qquad (5.8)$$

If two solutions have the same number of prime cubes, one can easily determine which solution has the least number of literals. In the event two solutions have the same number of literals, one should use the solution requiring the least number of inverters.

Again, since the example uses four variables, it can easily be mapped (Figure 5.7) to visualize the reduction process, and $\overline{A}\,\overline{C}$, $\overline{B}CD$, and AB are seen to be redundant prime cubes.

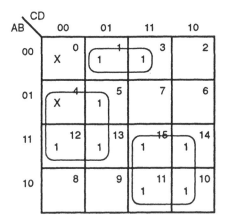

FIGURE 5.7.   Karnaugh-map simplification of F(A,B,C,D) of Equation 5.6.

5.3.3   The Complementary Approach

As with Karnaugh mapping, the 0s can be used to obtain a minimum POS solution. Since one doesn't know which don't care minterms are needed to simplify the function, all the don't cares must again be included in the first step. They will be omitted from the prime cube table in the second step.

The complement of the previous function is

$$\overline{F}(A,B,C,D) = \sum m\,(2,6,7,8,9) + \sum d\,(0,4) \tag{5.9}$$

This function is reduced in Tables 5.9 through 5.12, below. From Tables 5.9 through 5.11, the prime cubes or prime implicants are seen to be x000, 100x, 011x, and 0xx0. Next, determine which prime cubes are essential and which are necessary to account for all the 0s.

TABLE 5.9

A	B	C	D	0-cubes	
0	0	0	0	m_0	*
0	0	1	0	m_2	*
0	1	0	0	m_4	*
1	0	0	0	m_8	*
0	1	1	0	m_6	*
1	0	0	1	m_9	*
0	1	1	1	m_7	*

TABLE 5.10

A	B	C	D	1-cubes	
0	0	x	0	0,2	*
0	x	0	0	0,4	*
x	0	0	0	0,8	←W
0	x	1	0	2,6	*
0	1	x	0	4,6	*
1	0	0	x	8,9	←X
0	1	1	x	6,7	←Y

TABLE 5.11

A	B	C	D	2-cubes	
0	x	x	0	0,2,4,6	←Z

TABLE 5.12

2	6	7	8	9	PC	
			√		x000	W
			√	√	100x	X ←E
	√	√			011x	Y ←E
√	√				0xx0	Z ←E

Inspection of Table 5.12 shows X to be essential to m_9, Y to be essential to m_7, and Z to be essential to m_2. Since these prime cubes cover all the minterms that are asserted 0, the complement of the function F is

$$\overline{F}(A,B,C,D) = X + Y + Z = A\overline{B}\,\overline{C} + \overline{A}\,BC + \overline{A}\,\overline{D} \tag{5.10}$$

and prime cube $W = \overline{B}\,\overline{C}\,\overline{D}$ is seen to be redundant. The function F can now be obtained in a minimum POS form, viz

$$F(A,B,C,D) = (\overline{A} + B + C)(A + \overline{B} + \overline{C})(A + D) \tag{5.11}$$

Again, since F is a function of four variables, a Karnaugh map of the function F is instructive. This is done in Figure 5.8.

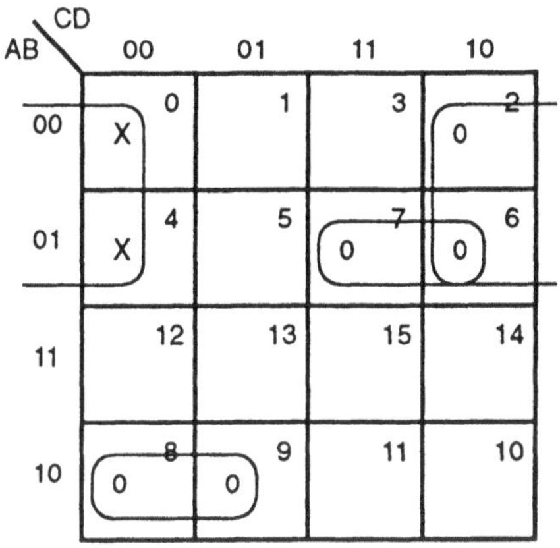

FIGURE 5.8. Karnaugh-map simplification of F(A,B,C,D) of Equation 5.9 in a minimum POS form.

A modified approach to the above Quine-McCluskey tabular method uses the minterm indices. In this procedure, the table of minterms is organized as above, but the minterm comparisons are done differently. Starting with minterms, a cube in a group is compared to those under it by subtracting it from them one at a time. If the difference in the two cubes is a power of two, they combine and the bit position missing in the combined term is the power of two they share.[3]

For example, given minterms 0, 2, 4, and 12, minterms 0 and 2, 0 and 4, and 4 and 12 combine. The combination of minterms 0 and 2 is written 0,2(2), the 2 in parantheses being the weight of the bit position that is missing in the combined 1-cube; 0 and 4 combine to give 0,4(4), meaning the bit having position weight 4 is missing; and 4 and 12 combine giving 4,12(8), since 4 and 12 differ by 8. If the function being minimized consists of four variables, these reductions would have been written in the previous notation as $0000 + 0010 = 00x0$, $0000 + 0100 = 0x00$, and $0100 + 1100 = x100$.

5.4   MULTIPLE OUTPUTS

Sometimes it is economically desirable to minimize more than one function simultaneously. If so, one should first try to minimize the number of gates needed. If two or more solutions require the same number of gates, one should try to minimize the number of inputs. When realizing multiple-output networks, the minimum overall solution may not involve a minimum SOP form for any or all of the functions separately. In fact, some terms in the simultaneous solution may not even be prime cubes of any single function. Methods of mapping and tabulating to simultaneously minimize more than one output will both be considered.

5.4.1 Multiple Outputs by Karnaugh Mapping

Prime cubes essential to an individual function may not be essential to the multiple-output realization. The minterm for which the prime cube is essential to realize one function may also appear on the map of another function and might be covered by a term which is shared by both functions. Sophisticated techniques do exist for finding essential multiple-output terms, but they will not be covered here.

To obtain the prime cubes which are essential to one of the functions under discussion and to the multiple-output realization both, proceed by checking a minterm for prime cubes *only* if that minterm does not appear on other maps. Next, look for cubes that appear on more than one map. These cubes represent gates whose outputs can be used to realize more than one output function.

Consider the functions represented by the Karnaugh maps shown in Figure 5.9. It is desired to minimize the two functions Y_1 and Y_2 simultaneously. Independent minimum SOP solutions are

$$Y_1(A,B,C,D) = \overline{C}D + ABD \quad \text{(3 gates and 7 inputs)}$$
$$Y_2(A,B,C,D) = B\overline{D} + ABC \quad \text{(3 gates and 7 inputs)}$$

The independent minimization of each function leads to a total of 6 gates and 14 inputs.

$\overline{C}D$ is essential to Y_1, $B\overline{D}$ is essential to Y_2, and $ABCD$ is common to both functions. Treated in this manner, the minterms are looped as shown in Figure 5.9 and the simultaneous solution for Y_1 and Y_2 can be achieved with 5 gates and 12 inputs. The realization of the circuit is shown in Figure 5.10, and the SOP solutions are

$$Y_1(A,B,C,D) = \overline{C}D + ABCD$$

and

$$Y_2(A,B,C,D) = B\overline{D} + ABCD \quad (5.12)$$

It is not always advantageous to split prime cubes in order to use one gate in the realization of two or more functions. A case in point is shown in Example 5.1, where the cube $\overline{A}CD$ is common to both functions, but if prime cube CD on the map of Y_3 is split, the solution has the same number of gates but requires replacing AND gate CD with a three-input AND gate and adding one more input to the OR gate of Y_3. This solution would add 1 literal and 2 gate inputs, giving a realization with 20 inputs to 6 gates. Leaving CD integral yields a solution with 18 inputs to 6 gates, which is clearly a superior solution.

EXAMPLE 5.1: Simultaneously simplify the two functions $Y_3(A,B,C,D)$ and $Y_4(A,B,C,D)$, shown mapped in Figure 5.11.

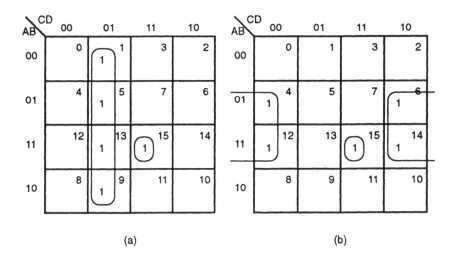

FIGURE 5.9. Simultaneous minimization of $Y_1(A,B,C,D)$ and $Y_2(A,B,C,D)$.

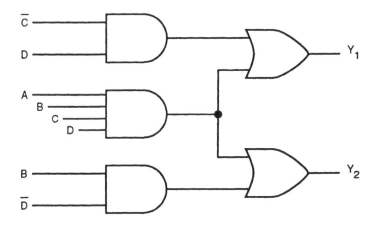

FIGURE 5.10. Realization of $Y_1(A,B,C,D)$ and $Y_2(A,B,C,D)$ with 5 gates and 12 inputs.

SOLUTION: In this example, both functions are best realized in a minimum SOP form. $\overline{A}\,\overline{B}\,\overline{C}\,\overline{D}$ and $AB\overline{C}$ are common to both functions, $\overline{A}\,CD$ is required for Y_4, and CD is required for Y_3. Since 2 gates are common to each function, the 2 functions can be realized with 6 gates and 18 inputs, as shown in Figure 5.12.

In Example 5.2, a simultaneous minimization of three functions having don't care minterms is examined.

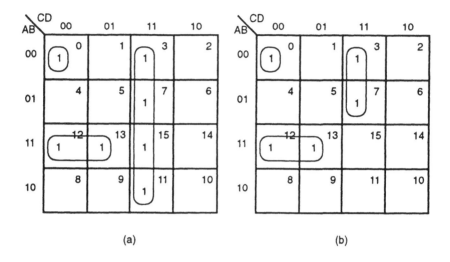

(a)   (b)

FIGURE 5.11. Maps of (a) $Y_3(A,B,C,D)$ and (b) $Y_4(A,B,C,D)$.

EXAMPLE 5.2: Simultaneously simplify the functions $X(A,B,C,D)$, $Y(A,B,C,D)$, and $Z(A,B,C,D)$, shown mapped in Figure 5.13.

SOLUTION: The functions can be separately minimized to obtain

$$X(A,B,C,D) = \overline{C}D + BD$$
$$Y(A,B,C,D) = \overline{A}\,\overline{B}\,\overline{C} + BCD$$
$$Z(A,B,C,D) = \overline{A}\,\overline{B}\,\overline{C} + B\overline{C}D$$

These functions can be realized separately with 9 gates, 16 literals, and 22 inputs. $\overline{A}\,\overline{B}\,\overline{C}$ is common to all three functions, BCD is common to X and Y, and $B\overline{C}D$ is common to X and Z. Implementing the functions in this manner one obtains

$$X(A,B,C,D) = \overline{A}\,\overline{B}\,\overline{C} + B\overline{C}D + BCD$$
$$Y(A,B,C,D) = \overline{A}\,\overline{B}\,\overline{C} + BCD$$
$$Z(A,B,C,D) = \overline{A}\,\overline{B}\,\overline{C} + B\overline{C}D.$$

This corresponds to looping minterms as shown in Figure 5.13. Realizing the functions in this manner, one obtains a solution requiring a total of 6 gates, 9 literals, and 16 inputs, as shown in Figure 5.14.

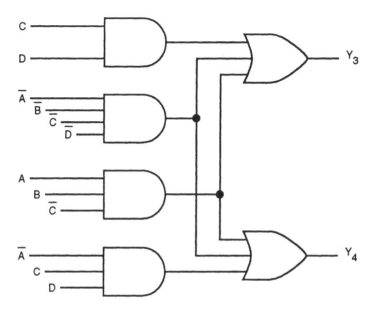

FIGURE 5.12. Realization of $Y_3(A,B,C,D)$ and $Y_4(A,B,C,D)$ with 6 gates and 18 inputs.

5.4.2 Multiple Outputs by the Quine-McCluskey Method

There are several approaches to expanding the Quine-McCluskey tabular method to multiple outputs. They all require a scheme for keeping track of which prime cubes are required for which functions.

Two simple schemes will be discussed here. Both procedures are done in two steps, as was the single-output minimization: the determination of all possible prime cubes, followed by a determination of which prime cubes are essential, necessary, and redundant.

The first method requires a label to identify the function or functions to which each minterm belongs, after which the basic Quine-McCluskey tabular method for single functions must be augmented by three new rules.[4]

1. Two cubes can be combined only if they have at least one label (output function) in common.
2. The label of combined products only contains letters common to the labels of both original cubes.
3. The products whose entire label is contained in the label of the larger cube are checked off.

The procedure is best explained by an example as follows.

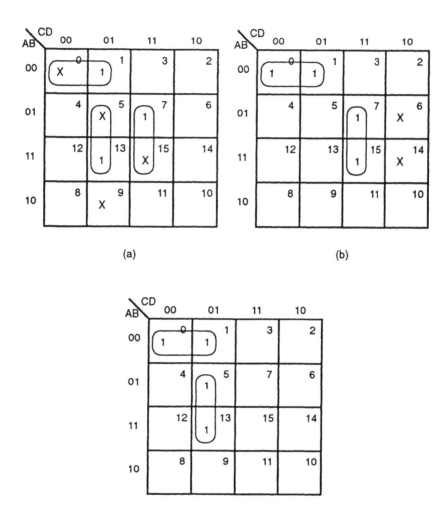

FIGURE 5.13. Maps of (a) X(A,B,C,D), (b) Y(A,B,C,D), and (c) Z(A,B,C,D).

EXAMPLE 5.3: Three functions X, Y, and Z are to be implemented with the least number of logic gates. The functions are

$$X(A,B,C,D) = \sum m\,(2,4,10,11,12,13)$$
$$Y(A,B,C,D) = \sum m\,(4,5,10,11,13)$$
$$Z(A,B,C,D) = \sum m\,(2,10,11,12) + d(3)$$

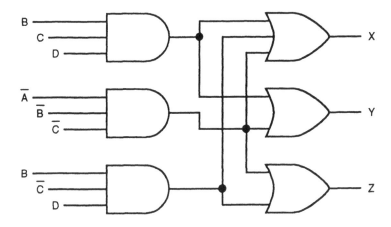

FIGURE 5.14. The AND-OR realization of X(A,B,C,D), Y(A,B,C,D), and Z(A,B,C,D) of Example 5.2.

SOLUTION: The minterms are again ordered by their vector magnitude (the number of 1s they contain), with appended labels that identify the function which contains that minterm. This gives Table 5.13.

TABLE 5.13

	0-cubes	Label		
	0010 (2)	XZ		
	0100 (4)	XY	*	(a)
	0011 (3)	Z		
	0101 (5)	Y		
	1010 (10)	XYZ		
	1100 (12)	XZ	*	(b)
	1011 (11)	XYZ		
	1101 (13)	XY	*	(c)

Minterms 2 and 5 cannot be combined because they occur in different functions (rule 1). Minterms 2 and 3 can combine because they both belong to function Z (rule 1), and the label of the combined pair is Z since both minterms occur in function Z (rule 2). When minterms 2 and 3 are combined to form a new cube, only minterm 3 can be checked off since minterm 2 also belongs to function X and must still be accounted for in the cover of X (rule 3). Proceeding in this manner, all the minterms except 4, 12, and 13 combine into larger cubes. Minterms 4, 12, and 13 are prime cubes. They are flagged with an asterisk and are labeled (a), (b), and (c) in Table 5.13. The remaining minterms are combined into 1-cubes and arranged in Table 5.14. Minterms 2, 3, 10, and 11 combine into a 2-cube for function Z. This is shown in Table 5.15.

TABLE 5.14

	1-cubes	New label		
001x	(2, 3)	Z		
x010	(2,10)	XZ	*	(d)
010x	(4,5)	Y	*	(e)
x100	(4,12)	X	*	(f)
x101	(5,13)	Y	*	(g)
101x	(10,11)	XYZ	*	(h)
110x	(12,13)	X	*	(i)

TABLE 5.15

	2-cubes	Label		
x01x	(2,3,10,11)	Z	*	(j)

All ten prime cubes are now determined. The prime table must next be set up to determine essential and necessary prime cubes. This is done in Table 5.16, where tables for X, Y, and Z are simultaneously compared.

From Table 5.16, it can be seen that the following prime cubes are essential: (b) is essential for minterm 12 in Z, (d) is essential for minterm 2 in X, and (h) is essential for minterm 11 in X and minterms 10 and 11 in Y. The remaining prime cubes are all optional and must be examined for necessary prime cubes. Prime cube (b) also covers minterm 12 in X, (d) covers minterm 10 in X, as well as minterms 2 and 10 in Z, and minterm (h) covers 10 in X, as well as minterms 10 and 11 in functions Y and Z. Minterm (j) is seen to be redundant because function Z is already covered. Minterms 4 and 13 must be covered in X, and minterms 4, 5, and 13 must be covered in Y.

The prime cubes that must be examined further are shown in Table 5.17, along with the minterms they cover. This is a reduced prime table for the three functions. There are no clear-

TABLE 5.16

		X						Y					Z			
		2	4	10	11	12	13	4	5	10	11	13	2	10	11	12
a	4		√					√								
b	12					√										√
c	13						√					√				
d	2,10	√		√									√	√		
e	4,5		√					√	√							
f	4,12		√			√										
g	5,13								√			√				
h	10,11			√	√					√	√			√	√	
i	12,13					√	√									
j	2,3,10,11												√	√	√	

cut choices in Table 5.17. To examine all the possibilities, form the function H, which contains all possible permutations of cover sets for the remaining prime cubes for X and Y both. From Table 5.17,

$$H = (a + f)(c + i)(a + e)(e + g)(c + g) = (a + ef)(c + ig)\ (e + g)$$
$$= ace + acg + agi + cef + efgi$$

The last term consists of four prime cubes and is not a minimal solution. This still leaves four possible sets of solutions, namely

$$a + c + e = \overline{A}\,B\overline{C}\,\overline{D} + AB\overline{C}D + \overline{A}\,B\overline{C}$$
$$a + c + g = \overline{A}\,B\overline{C}\,\overline{D} + AB\overline{C}D + B\overline{C}D$$
$$a + g + i = \overline{A}\,B\overline{C}\,\overline{D} + B\overline{C}D + AB\overline{C}$$
$$c + e + f = AB\overline{C}D + \overline{A}\,B\overline{C} + B\overline{C}\,\overline{D}$$

The first two choices have 11 literals each and are not minimal. The last two possibilities have ten literals each and require the same number of inverters. Prime cubes c, e, and f are arbitrarily chosen as necessary, making the other prime cubes redundant. Table 5.18 lists all the essential and necessary prime cubes and which functions are covered by each.

TABLE 5.17

		X		Y		
		4	13	4	5	13
a	4	√		√		
c	13		√			√
e	4,5			√	√	
f	4,12	√				
g	5,13				√	√
i	12,13		√			

TABLE 5.18

	Prime cube	Output
EPC	$b = AB\overline{C}\,\overline{D}$	X, Z
NPC	$c = AB\overline{C}D$	X, Y
EPC	$d = \overline{B}C\overline{D}$	X, Z
NPC	$e = \overline{A}\,B\overline{C}$	Y
NPC	$f = B\overline{C}\,\overline{D}$	X
EPC	$h = A\overline{B}C$	X, Y, Z

The individual functions must now be examined to see if prime cubes that are essential or necessary to one function are also essential or necessary to other functions. Examination of Table 5.18 shows that prime cube $b = AB\overline{C}\,\overline{D}$ is contained in $f = B\overline{C}\,\overline{D}$ and is redundant in output X while necessary to output Z. The three functions are found to be

$$X(A,B,C,D) = AB\overline{C}D + \overline{B}C\overline{D} + B\overline{C}\,\overline{D} + A\overline{B}C$$

$$Y(A,B,C,D) = AB\overline{C}D + \overline{A}\,B\overline{C} + A\overline{B}C$$

$$Z(A,B,C,D) = AB\overline{C}\,\overline{D} + \overline{B}C\overline{D} + A\overline{B}C$$

The minimum simultaneous solution to the three functions requires 6 AND gates and 3 OR gates, or a total of 9 gates and 20 literals, plus 10 secondary inputs, or 30 total inputs. The solution is shown in Figure 5.15.

The alternative solution in Example 5.3 consists of necessary prime cubes a, g, and i, plus essential prime cubes b, d, and h. Prime cubes a, g, and i are $\overline{A}\,B\overline{C}\,\overline{D}$,

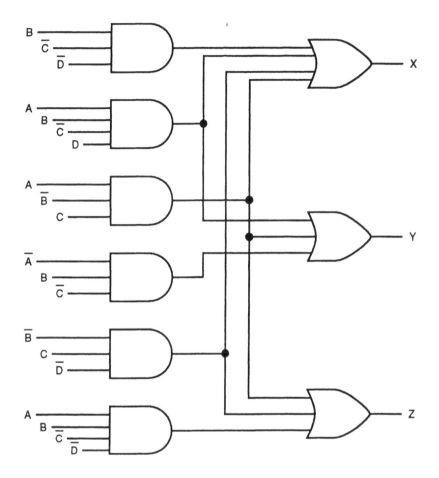

FIGURE 5.15. The AND-OR simultaneous realization of X(A,B,C,D), Y(A,B,C,D), and Z(A,B,C,D) of Example 5.3.

$B\overline{C}D$, and $AB\overline{C}$, respectively, giving the simultaneous solutions

$$X = \overline{A}B\overline{C}\,\overline{D} + AB\overline{C} + AB\overline{C}\,\overline{D} + \overline{B}C\overline{D} + A\overline{B}C$$
$$= \overline{A}B\overline{C}\,\overline{D} + AB\overline{C} + \overline{B}C\overline{D} + A\overline{B}C$$
$$Y = \overline{A}B\overline{C}\,\overline{D} + B\overline{C}D + A\overline{B}C$$
$$Z = AB\overline{C}\,\overline{D} + \overline{B}C\overline{D} + A\overline{B}C \qquad (5.13)$$

Cube b is not a prime cube of X since it is contained in cube i. This reduces X to a sum of 4 prime cubes again and gives a solution with the same structure as the solution in Figure 5.15, consisting of 6 AND gates, 3 OR gates, 20 literals, and 30 total inputs.

An alternative approach to the multi-output Quine-McCluskey solution consists of first finding all the prime cubes of each function, followed by the prime cubes common to pairs of functions, etc. until one has the prime cubes of the products of all the given functions. A table is then formed for the individual variables and all combinations of the variables, after which a minimum cover is found as before.[5] This technique is shown in Example 5.4 for the same three functions as were simplified in Example 5.3. The results of the two approaches can then be compared. This method is perhaps easier to visualize, whereas the previous method is easier to program.

EXAMPLE 5.4:   Three functions X, Y, and Z are to be implemented with the least number of logic gates using the second Quine-McCluskey approach. The functions are

$$X(A,B,C,D) = \sum m\,(2,4,10,11,12,13)$$
$$Y(A,B,C,D) = \sum m\,(4,5,10,11,13)$$
$$Z(A,B,C,D) = \sum m\,(2,10,11,12) + d(3)$$

SOLUTION: The minterm expansions of XY, XZ, YZ, and XYZ are first found.

$$XY = \sum m\,(4,10,11,13)$$
$$XZ = \sum m\,(2,10,11,12)$$
$$YZ = XYZ = \sum m\,(10,11)$$

Simplify each of these 7 functions separately by the Quine-McCluskey method for single-output functions, and one obtains

$$X = \overline{B}C\overline{D} + B\overline{C}\,\overline{D} + A\overline{B}C + AB\overline{C}$$
$$Y = \overline{A}B\overline{C} + B\overline{C}D + A\overline{B}C$$
$$Z = AB\overline{C}\,\overline{D} + \overline{B}C$$
$$XY = \overline{A}B\overline{C}D + AB\overline{C}D + A\overline{B}C$$
$$XZ = AB\overline{C}\,\overline{D} + \overline{B}C\overline{D} + A\overline{B}C$$
$$YZ = XYZ = A\overline{B}C$$

To determine a minimum cover for the original three functions, make a table with columns for the minterms that must be covered in X, Y, and Z, with rows for all the prime cubes in X, Y, Z, XY, XZ, YZ, and XYZ. This is shown in Table 5.19, which must be examined for a minimum cover as before.

TABLE 5.19

		X						Y					Z				
		2	4	10	11	12	13	4	5	10	11	13	2	10	11	12	
a	2,10	√		√													
b	4,12		√			√	‚										X
c	10,11			√	√												
d	12,13					√	√										
e	4,5							√	√								
f	5,13								√			√					Y
g	10,11									√	√						
h	2,3,10,11												√	√	√		Z
i	12															√	
j	4		√					√									
k	10,11			√	√					√	√						XY
l	13						√					√					
m	12					√										√	
n	2,10	√		√									√	√			XZ
o	10,11			√	√									√	√		
p	10,11									√	√			√	√		YZ
q	10,11			√	√					√	√			√	√		XYZ

One row of the Quine-McCluskey table is said to "dominate" another row if it covers at least the same minterms as the dominated row and thus has no more literals than the dominated row. If the dominant row covers other minterms also, it will have less literals than the dominated row. If the dominated row is removed, the resulting solution will be no more complicated than any other expression for the function. Removing dominated rows will yield a minimum solution, if not all minimum solutions.[5] Examination of Table 5.19 shows that row q dominates rows c, g, k, o, and p; hence, these five rows are redundant and can be deleted. Once minterms 10 and 11 are covered in Z, it is seen that row n dominates rows a and h, and both of them can be deleted. Row m dominates row i, which can be deleted. Rows m, n, and q are essential and must be in all solutions. Upon keeping rows m, n, and q, and deleting rows a, c, g, h, i, k, o, and p, one is left with a reduced table consisting of the six rows: b, d, e, f, j, and l. Minterms that are covered by rows m, n, and q are of no

TABLE 5.20

		X			Y			
		4	12	13	4	5	13	
b	4,12	√	√					X
d	12,13		√	√				
e	4,5				√	√		Y
f	5,13					√	√	
j	4	√			√			XY
l	13			√			√	

further concern either, and the columns for minterms 2, 10, 11, and 12 in X and Z can be omitted as can columns for 10 and 11 in Y. This leads to Table 5.20, consisting of six rows and six columns, and all solutions found will also contain essential cubes m, n, and q. Proceeding as before, one can form the function H for Table 5.20 from which all minimum solutions, which include cubes m, n, and q, can be found.

$$H = (b + j)(b + d)(d + l)(e + j)(e + f)(f + l)$$
$$= bel + bfjl + dejl + bdef + dfj$$

Of the five solutions obtained, only bel and dfj are minimum solutions. If one takes b, e, and l as necessary prime cubes, the six prime cubes of Table 5.18 are again obtained, and the three solutions are identical to those obtained in Example 5.3. If one takes d, f, and j as necessary prime cubes, the alternate solution following Example 5.3 is obtained.

5.5 THE DIRECTED SEARCH ALGORITHM

The *Directed Search Algorithm*, or *DSA*, is an approach which calculates only those prime cubes that are members of a minimal cover of the function. This reduces the computation time as compared to Quine-McCluskey and other approaches, which first compute all prime cubes and then determine the minimum required set of prime cubes.

As with the Quine-McCluskey approach, the DSA is useful in a situation where the minterms (0-cubes) are known. Given a listing of the true and don't care minterms of a function, a selected set of true 0-cubes are used as starting points with which to compare all adjacent minterms in order to determine the largest cover for the original 0-cube. This generates a tree structure which terminates with the largest prime cube (highest cube) that covers the given minterm.[6]

Rhyne et al.[6] consider which minterms differ in only one bit from other minterms of the function, and hence can combine (i.e., minterms of distance 1). Two 0-cubes form adjacent vertices of an n-cube if their input combinations differ in only one variable and the *adjacency direction (AD)* from a vertex of the n-cube is the set of signed integers describing how to get from one vertex to an adjacent vertex.

The weight of a bit in the kth position from the LSB is 2^k, starting with k = 0. If the 0-cube under consideration has a 1 in its kth bit position, the weight of that position must be subtracted to obtain the 0-cube with a 0 in that position. If the 0-cube has a 0 in the kth bit position, the weight of that position must be added to obtain a 0-cube with a 1 in its kth bit. The AD for 0-cubes of an n-cube will consist of the sequence of numbers $\{a^k\} = \{\pm 2^k | k = 0, 1, 2, ..., n-1\}$.[6]

For example, minterm 5 of a 4-cube is 0101 and it has an AD of 8, –4, 2, –1. By subtracting 1 from m_5, one obtains $0100 = m_4$; by adding 2, one obtains $0111 = m_7$. By subtracting 4, one obtains $0001 = m_1$, and by adding 8 to minterm 5, one obtains $1101 = m_{13}$. If any of vertices (0-cubes) 1, 4, 7, or 13 are present in the function, they can combine to form a 1-cube.

For a function of variables A, B, C, and D, the 16 0-cubes form the vertices of a 4-cube and can combine with (are adjacent to) 4 other 0-cubes of the 4-cube. Table 5.21 is a listing of all 16 0-cubes, followed by 4 columns for the 0-cubes which combine with each given 0-cube and eliminate the variable in that column. The AD of the given vertex is shown in the last four columns. For instance, minterm 1 can combine with minterm 9 by eliminating the most significant variable A; it can combine with minterm 5 by eliminating variable B, etc.

TABLE 5.21

Minterm	Variable eliminated				Adjacency direction			
	A	B	C	D				
0	8	4	2	1	+8	+4	+2	+1
1	9	5	3	0	+8	+4	+2	–1
2	10	6	0	3	+8	+4	–2	+1
3	11	7	1	2	+8	+4	–2	–1
4	12	0	6	5	+8	–4	+2	+1
5	13	1	7	4	+8	–4	+2	–1
6	14	2	4	7	+8	–4	–2	+1
7	15	3	5	6	+8	–4	–2	–1
8	0	12	10	9	–8	+4	+2	+1
9	1	13	11	8	–8	+4	+2	–1
10	2	14	8	11	–8	+4	–2	+1
11	3	15	9	10	–8	+4	–2	–1
12	4	8	14	13	–8	–4	+2	+1
13	5	9	15	12	–8	–4	+2	–1
14	6	10	12	15	–8	–4	–2	+1
15	7	11	13	14	–8	–4	–2	–1

If two vertices combine, they form a 1-cube; if four vertices combine, they form a 2-cube; and if eight vertices combine, they form a 3-cube, etc. The highest cube obtained in this manner is a prime cube cover for the original 0-cube. The signed numbers, which indicate the AD from one cube to any other cube with which it can combine, are called the *required adjacency directions*, or *RADs*. The RADs can be mapped in a tree structure whose root is a given minterm, and whose leaves are the prime cubes that can cover that minterm. The complete RAD tree for m_0 of a 3-cube is shown below.

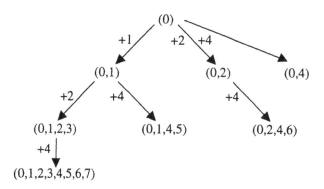

From the tree it can be seen that if m_4 is missing, m_0 is covered by prime cube (0,1,2,3). If m_2 is missing, m_0 is covered by prime cube (0,1,4,5), and if m_1 is missing, m_0 is covered by prime cube (0,2,4,6).

The first step in applying the DSA is to divide the vertices of the function to be simplified into the set of true 0-cubes $C\{1\}$, the set of false 0-cubes $C\{0\}$, and the set of don't care 0-cubes $C\{\times\}$. One element of the set $C\{1\}$ is then chosen, and the direction from this 0-cube to another 0-cube of either the set $C\{1\}$ or the set $C\{\times\}$ gives the desired RAD tree. Starting with an element of the set $C\{1\}$ avoids the possibility of obtaining a prime cube consisting only of don't care terms.

The RADs are applied one at a time to this chosen cube to determine the largest number of vertices with which it can combine. The addition of each RAD doubles the number of vertices that combine until all the true 0-cubes that combine are covered.

If a 0-cube of the set $C\{1\}$ has no RAD, it is an essential prime 0-cube. If a 0-cube of the set $C\{1\}$ has only one RAD, then it combines with only that one other 0-cube, and the pair form an essential prime 1-cube. These prime cubes can be removed before starting the search for higher cubes. If an element of $C\{1\}$ is covered once already, it can be moved to the set $C\{\times\}$ since it need not be covered again.

In working with a Karnaugh map, it is advisable to start with minterms that combine in small loops first, and work up to the larger loops. This locates small prime cubes quickly and avoids the problem of obtaining redundant prime cubes.

By the same token, the computational time involved in applying the DSA is less if one chooses those 0-cubes that have a small number of RADs and expands them first, since they have simpler RAD trees.

The DSA determines one minimum solution to a problem, with a small amount of computing. No attempt is made to identify all possible minimal solutions to a problem. If there are multiple solutions, the solution obtained depends upon the sequence in which the 0-cubes were chosen as roots of the RAD trees generated, since the prime cubes generated depend upon the order in which vertices are chosen as roots of the search trees.

The rules for searching RAD trees can be summarized as follows:

1.   Separate all vertices into three sets, $C\{1\}$ consisting of the set of all true 0-cubes, $C\{0\}$ consisting of the set of all false 0-cubes, and $C\{\times\}$ consisting of the set of all unspecified 0-cubes.
2.   Develop the RADs for all the true 0-cubes.
3.   Initiate a search from a true minterm having the fewest RADs because it has a simple tree. Search down the left side of the tree first. If this does not exhaust the RADs, move to the right and continue the search.
4.   Change all covered true minterms from set $C\{1\}$ to set $C\{\times\}$ and reorder the list in accordance with the increasing number of RADs each element of $C\{1\}$ has.
5.   Select the next 0-cube for expansion and repeat the process until all the minterms are covered.

Two examples follow.

EXAMPLE 5.5:   Use the DSA to find a minimum solution to the function

$$F(A,B,C) = \sum m (0,1,2,3,4,6).$$

SOLUTION: The on-set is $C\{1\} = \{0,1,2,3,4,6\}$.
The RAD table is

Minterm	RADs	Combines with
0	+1, +2, +4	1, 2, 4
1	−1, +2	0, 3
2	−2, + 1, + 4	0, 3, 6
3	−2, −1	1, 2
4	−4, +2	0, 6
6	−4, −2	2, 4

Minterms 0 and 2 each have three RADs, and the rest of the minterms have only two. Omit minterms 0 and 2, and take the remaining four in numerical order. Starting with minterm 1,

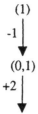

(1)

-1

(0,1)

+2

(0,1,2,3)   Prime cube $(0,1,2,3) = \overline{A}$ is essential.

Move these four minterms to the don't care set, and the function now consists of sets $C\{1\} = \{4,6\}$ and $C\{\times\} = \{0,1,2,3\}$. The next 0-cube in numerical order is m_4, and it generates the RAD tree

(4)

-4

(0,4)

+2

(0,2,4,6)   Prime cube $(0,2,4,6) = \overline{C}$ is essential.

All the minterms are covered, and the function has been simplified to $F(A,B,C) = \overline{A} + \overline{C}$

EXAMPLE 5.6:   Simplify the function $G(A,B,C,D) = \sum m\,(3,7,9,14,15) +$

$\sum d\,(0,1,8)$

SOLUTION: The on-set is $C\{1\} = \{3,7,9,14,15\}$, and the don't care set is $C\{\times\} = \{0,1,8\}$. The RAD table is given below for the on-set.

Minterm	RADs	Combines with
3	−2, +4	1, 7
7	−4, +8	3, 15
9	−8, −1	1, 8
14	+1	15
15	−8, −1	7, 14

Minterm 14 is seen to combine with minterm 15 only, and cube (14,15) = ABC is essential to G. Move these two minterms to the don't care set C{×}, and the function G now consists of sets $C\{1\} = \{3,7,9\}$, and $C\{\times\} = \{0,1,8,14,15\}$. The remaining minterms in C{1} have two RADs each. Minterm 3 is first in numerical order, and the following RAD tree is generated.

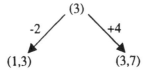

Prime cubes $(1,3) = \overline{A}\,\overline{B}D$ and $(3,7) = \overline{A}CD$. (There are two optional prime cubes.)
Since m_1 is a don't care, take the cube $(3,7) = \overline{A}CD$, which covers two minterms of the set $C\{1\}$. The function now consists of sets $C\{1\} = \{9\}$ and $C\{\times\} = \{0,1,3,7,8,14,15\}$.
Starting with minterm 9, the remaining RAD tree is

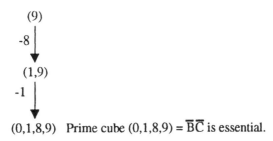

$(0,1,8,9)$ Prime cube $(0,1,8,9) = \overline{B}\,\overline{C}$ is essential.

All the minterms are covered, and the function simplifies to G $= ABC + \overline{A}CD + \overline{B}\,\overline{C}$.

5.6 THE ITERATIVE CONSENSUS ALGORITHM

Many modern logic functions contain a large number of variables. The number of potential minterms grows rapidly with the number of variables, there being 2^n possible minterms of n variables. A function of 10 variables can have up to 1024 minterms, and a function of 20 variables can have upwards of 1 million minterms. Techniques which rely on a minterm list as a starting point are not practical for functions of many variables.

However, functions of many variables often have a small number of product terms. An approach which starts from the given cubes or product terms, and avoids the need to determine the minterms, will be faster than a technique which requires a minterm expansion as a starting point. Iterative consensus is an algorithm whose computational complexity grows with the number of product terms, not with the number of input variables.

Iterative consensus starts with a set of cubes which represent the function to be realized. These cubes need not be minterms or prime cubes, but can be of any dimension between 0 and n, for n variables. The algorithm generates all the prime cubes (prime implicants) from the given set of cubes. The two theorems used in this determination are the consensus theorem, $AB + \overline{A}C = AB + \overline{A}C + BC$, and the absorption theorem, $A + AB = A$.[7]

As with many other techniques, iterative consensus contains a second step which searches for a minimal cover set once the prime cube set is known. This is done by first determining the essential prime cubes and then determining the necessary prime cubes.

It is easy to show that starting with the cubes of a function which have a non-zero consensus term, the cubes that are not prime can be replaced by the consensus terms which are prime cubes. The consensus of two combinable 0-cubes is the 1-cube covering both minterms. For example, the consensus term of ABCD and $\overline{A}BCD$ is BCD, and both 0-cubes can be replaced by the 1-cube BCD. If there is another 1-cube $\overline{B}CD$ which can combine with BCD, the consensus term is seen to be CD, which is the 2-cube obtained by combining the two 1-cubes.

This can be generalized to the consensus of cubes of any dimension. Consider the function

$$X(A,B,C,D) = \overline{B}D + BCD + ABC\overline{D} \qquad (5.14)$$

The consensus term of cubes $\overline{B}D$ and BCD is CD, and the consensus term of BCD and $ABC\overline{D}$ is ABC. After a consensus term is obtained, it is added to the list of cubes under investigation. The function X, above, can be augmented by adding the two prime cubes CD and ABC to give the new function

$$X(A,B,C,D) = \overline{B}D + BCD + ABC\overline{D} + CD + ABC \qquad (5.15)$$

The absorption theorem is next applied to remove the nonprime subcubes, BCD is absorbed by CD, and $ABC\overline{D}$ is absorbed by ABC to give X as a function of three prime cubes.

$$X(A,B,C,D) = \overline{B}D + CD + ABC \qquad (5.16)$$

By removing one prime cube at a time and examining the remaining function for redundancy, it can be determined which cubes are essential. In the function above, the only remaining consensus term is ACD, which is covered by CD. Thus, each prime cube is essential to X, and the function has been reduced to a minimal form. Two more examples of the technique are shown below.

EXAMPLE 5.7: Use iterative consensus to obtain a minimal solution to the function

$$Y(A,B,C,D) = \overline{A}\,\overline{C}\overline{D} + \overline{A}\,\overline{C} + \overline{A}BD + BCD + ABC\overline{D}$$
$$+ \overline{A}BCD$$

SOLUTION: Application of the absorption theorem shows that $\overline{A}\,\overline{C}D$ is covered by $\overline{A}\,\overline{C}$, and $\overline{A}BCD$ is covered by both BCD and $\overline{A}BD$. Remove these two terms, and the function reduces to

$$Y(A,B,C,D) = \overline{A}\,\overline{C} + \overline{A}BD + BCD + ABC\overline{D}$$

The consensus of $\overline{A}\,\overline{C}$ and BCD is $\overline{A}BD$ which is already present. The consensus of BCD and $ABC\overline{D}$ is ABC. Add this to Y, and

$$Y(A,B,C,D) = \overline{A}\,\overline{C} + \overline{A}BD + BCD + ABC\overline{D} + ABC$$

$ABC\overline{D}$ is covered by ABC and is removed when the absorption theorem is applied. At this stage, Y is given as

$$Y(A,B,C,D) = \overline{A}\,\overline{C} + \overline{A}BD + BCD + ABC$$

The consensus theorem must now be applied to ABC and the remaining cubes of Y. The consensus of $\overline{A}BD$ and ABC is BCD, and the consensus of $\overline{A}\,\overline{C}$ and BCD is $\overline{A}BD$. Both are already present, and no new terms are generated. Y must next be tested for essential prime cubes.

Delete $\overline{A}\,\overline{C}$ and apply the consensus theorem. It is not generated again as a consensus of two other prime cubes. Hence, $\overline{A}\,\overline{C}$ is essential.

Delete $\overline{A}BD$ and it is generated again as a consensus of $\overline{A}\,\overline{C}$ and BCD. $\overline{A}BD$ is not an essential prime cube if BCD is used in the cover of Y.

Delete BCD and it is generated again as a consensus of $\overline{A}BD$ and ABC. BCD is not an essential prime cube if $\overline{A}BD$ is used in the cover of Y.

Delete ABC and it is seen to be an essential prime cube since it is not generated again as a consensus term.

$\overline{A}\,\overline{C}$ and ABC are essential prime cubes, and $\overline{A}BD$ and BCD are optional prime cubes. Either $\overline{A}BD$ or BCD is a necessary prime cube, the other being redundant. There are two minimal solutions for Y.

$$Y(A,B,C,D) = \overline{A}\,\overline{C} + ABC + \overline{A}BD$$
and
$$Y(A,B,C,D) = \overline{A}\,\overline{C} + ABC + BCD$$

5.7 THE GENERALIZED ITERATIVE CONSENSUS ALGORITHM

The *generalized consensus algorithm* is an improved form of the iterative consensus algorithm, which is faster because it involves less iterations than the original iterative consensus algorithm. It consists of five steps, as follows.[8]

1. Remove any redundant terms, using the absorption theorem.
2. Partition the cubes of the function into three sets, $C\{1\}$, $C\{0\}$, and $C\{\times\}$, according to whether the cube contains a 1, a 0, or a don't care for variable i.
3. Test every cube in set $C\{1\}$ with every cube in set $C\{0\}$ for consensus terms. If a consensus cube exists, add it to a set $C\{C\}$ of consensus cubes.
4. Compare each consensus cube in set $C\{C\}$ with all the other cubes in set $C\{C\}$ and all the cubes in set $C\{\times\}$. Remove any redundant consensus cubes.
5. Compare the remaining consensus cubes in $C\{C\}$ with the cubes in $C\{0\}$, $C\{1\}$, and $C\{\times\}$, removing any of the latter that are covered by a consensus cube and are redundant. Use the absorption theorem.
6. Combine all four arrays and iterate on the next variable.

EXAMPLE 5.8: Apply the generalized consensus algorithm to the function of Example 5.7.

SOLUTION: After removing redundant cubes, $Y(A,B,C,D) = \overline{A}\,\overline{C} + \overline{A}\,BD + BCD + ABC\overline{D}$.

For the variable, A: $C\{A = 1\} = \{BC\overline{D}\}$, $C\{A = 0\} = \{\overline{C}, BD\}$, and $C\{A = \times\} = \{BCD\}$. There are no consensus terms generated.

For the variable, B: $C\{B = 1\} = \{\overline{A}D, CD, AC\overline{D}\}$, $C\{B = 0\} = \emptyset$, and $C\{B = \times\} = \{\overline{A}\,\overline{C}\}$. There are no consensus terms generated.

For the variable, C: $C\{C = 1\} = \{BD, AB\overline{D}\}$, $C\{C = 0\} = \{\overline{A}\}$, and $C\{C = \times\} = \{\overline{A}BD\}$. The consensus term generated is $\overline{A}BD$, which is already present in $C\{C = \times\}$.

For the variable, D: $C\{D = 1\} = \{\overline{A}B, BC\}$, $C\{D = 0\} = \{ABC\}$, and $C\{D = \times\} = \{\overline{A}\,\overline{C}\}$. The consensus term generated is ABC, which is a new term. Upon comparing with the previous cubes, $ABC\overline{D}$ is covered by ABC and is removed. Y is found to consist of the set of four prime cubes: $\overline{A}\,\overline{C}$, $\overline{A}BD$, BCD, and ABC. The essential prime cubes are obtained as before.

EXAMPLE 5.9: Apply the generalized consensus algorithm to simplify the function

$$Y(A,B,C,D) = \overline{A}\,B\overline{C} + \overline{A}\,CD + \overline{A}\,\overline{C}\,\overline{D} + \overline{B}\,\overline{C}\,\overline{D} + \overline{A}\,BC\overline{D}$$
$$+ ABC\overline{D}$$

SOLUTION: There are no redundant cubes.

For the variable, A: $C\{A = 1\} = \{BC\overline{D}\}$, $C\{A = 0\} = \{B\overline{C}$, CD, $\overline{C}\,\overline{D}$, $BC\overline{D}\}$, and $C\{A = \times\} = \{\overline{B}\,\overline{C}\,\overline{D}\}$. The consensus term generated is $BC\overline{D}$, which is a new term. Upon comparison it is seen that $BC\overline{D}$ covers $\overline{A}\,BC\overline{D}$ and $ABC\overline{D}$. Replace these two terms, and Y simplifies to

$$Y_1(A,B,C,D) = \overline{A}\,B\overline{C} + \overline{A}\,CD + \overline{A}\,\overline{C}\,\overline{D} + \overline{B}\,\overline{C}\,\overline{D} + BC\overline{D}$$

For the variable, B: $C\{B = 1\} = \{\overline{A}\,\overline{C}, C\overline{D}\}$, $C\{B = 0\} = \{\overline{C}\,\overline{D}\}$, and $C\{B = \times\} = \{\overline{A}\,CD, \overline{A}\,\overline{C}\,\overline{D}\}$. The consensus term generated is $\overline{A}\,\overline{C}\,\overline{D}$, which is already present, and there is no change in Y.

For the variable, C: $C\{C = 1\} = \{\overline{A}\,D, B\overline{D}\}$, $C\{C = 0\} = \{\overline{A}\,B, \overline{A}\,\overline{D}, \overline{B}\,\overline{D}\}$, and $C\{C = \times) = \varnothing$. The new consensus terms generated are $\overline{A}\,BD$, and $\overline{A}\,B\overline{D}$.

Upon adding these two terms, the function becomes

$$Y_2(A,B,C,D) = \overline{A}\,B\overline{C} + \overline{A}\,CD + \overline{A}\,\overline{C}\,\overline{D} + \overline{B}\,\overline{C}\,\overline{D} + BC\overline{D}$$
$$+ \overline{A}\,BD + \overline{A}\,B\overline{D}$$

For the variable, D: $C\{D = 1\} = \{\overline{A}\,C, \overline{A}\,B\}$, $C\{D = 0\} = \{\overline{A}\,\overline{C}, \overline{B}\,\overline{C}, BC, \overline{A}\,B\}$, and $C\{D = \times\} = \{\overline{A}\,B\overline{C}\}$. The consensus terms generated are $\overline{A}\,BC$, $\overline{A}\,B\overline{C}$, and $\overline{A}\,B$. $\overline{A}\,B$ covers $\overline{A}\,BC$, $\overline{A}\,B\overline{C}$, $\overline{A}\,BD$, and $\overline{A}\,B\overline{D}$. Remove the non-prime cubes and

$$Y_3(A,B,C,D) = \overline{A}\,B + \overline{A}\,CD + \overline{A}\,\overline{C}\,\overline{D} + \overline{B}\,\overline{C}\,\overline{D} + BC\overline{D}$$

By removing one prime cube at a time and testing for consensus terms, it is seen that $\overline{A}\,B$ and $\overline{B}\,\overline{C}\,\overline{D}$ generate the term $\overline{A}\,\overline{C}\,\overline{D}$, and it is a redundant prime cube. The remaining prime cubes are essential, and $Y_4(A,B,C,D) = \overline{A}\,B + \overline{A}\,CD + \overline{B}\,\overline{C}\,\overline{D}$ $+ BC\overline{D}$.

A Pascal program which uses iterative consensus to find prime cubes is given in Wakerly.[9]

5.8 SUMMARY

Five- and six-variable Karnaugh maps can be constructed without too much difficulty and are relatively easy for humans to use. Functions of seven variables could be represented with six-variable maps, provided one variable is entered on the map. Beyond this, a tabular method is needed.

The Quine-McCluskey tabular method of finding a minimal SOP or POS expression for a function is one of the simplest for didactic purposes and consists of two steps. The first step is the determination of all the prime cubes of the function, and the second step is the determination of the minimum set of prime cubes necessary to cover the function. If an SOP solution is desired, the minterms that are 1s and don't cares are combined. If a POS solution is desired, the minterms that are 0s and don't cares are combined, and the SOP solution of the complementary function is obtained. When complemented, this yields the desired POS solution.

The first step consists of systematically comparing all minterms with n ones in them to all the minterms with n + 1 ones in them to find which combine to give a larger cube. The second step consists of examining which cubes cover each on-set minterm with don't care minterms excluded and determining the essential, necessary, and redundant prime cubes of the function.

Often the digital designer is required to simplify more than one function at a time. This is referred to as multi-output logic. The simultaneous simplification of more than one function can be done on Karnaugh maps and using a generalized form of the Quine-McCluskey method.

The DSA allows one to simplify digital functions starting from the minterms, but only computing the required prime cubes, rather than calculating all prime cubes as in the Quine-McCluskey algorithm. The RAD specifies which bits of a minterm must be changed in order for it to combine with another minterm. The search generates a tree which terminates in the prime cubes which cover the initial minterm. Except for the case when two or more optional prime cubes are generated, the approach only generates required prime cubes and saves the computational time required to compute redundant prime cubes that must be eliminated later.

For digital logic functions that have many variables and an extremely large number of minterms, it is not economical to obtain a minterm expansion. If the functions contain a small number of cubes, iterative consensus offers an approach which starts with an SOP expression and generates all of the prime cubes directly.

More efficient algorithms exist for computer applications. These are beyond the scope of this text. For more information, the reader is directed to the texts by Dietmeyer[10] and Muroga.[11]

PROBLEMS

Use the appropriate Karnaugh map (five or six variable) to simplify the functions listed in Problems 5.1 through 5.10, inclusive. List the prime cubes that are essential, necessary, optional, and redundant for the functions in Problems 5.1 through 5.6, and for the complements of the functions in Problems 5.7 through 5.10. Give an SOP solution for Problems 5.1 through 5.6, and a POS solution for Problems 5.7 through 5.10. For all ten problems, specify how many gates, literals, and inputs are needed to realize the function in SSI.

5.1. $U(A,B,C,D,E) = \sum m\,(1,9,10,11,15,26,27,30,31)$.

5.2. $V(A,B,C,D,E) = \sum m\,(2,8,9,10,13,18,24,25,26,29)$.

5.3. $W(A,B,C,D,E) = \sum m\,(7,8,10,11,12,14,23,31) + d(15)$.

5.4. $X(A,B,C,D,E) = \sum m\,(1,3,4,5,8,10,15,16,17,19,21,23,26,27) +$
$\qquad \sum d\,(2,7,12,22,31)$.

5.5. $Y(A,B,C,D,E,F) = \sum m\,(0,4,20,32,44,45,46,47,52) + d(36)$.

5.6. $Z(A,B,C,D,E,F) = \sum m\,(6,12,14,22,24,30,44,46,54,56) +$
$\qquad \sum d\,(8,38,40,62)$.

5.7. $W(A,B,C,D,E) = \prod M\,(9,10,11,15,25,26,27,30,31)$.

5.8. $X(A,B,C,D,E) = \prod M\,(0,2,7,8,9,10,16,18,23,24,25,26)$.

5.9. $Y(A,B,C,D,E,F) = \prod M\,(1,8,10,21,23,24,26,29,31,37,40,42,53,56,$
$\qquad 58,61)\prod D\,(5,55,63)$.

5.10. $Z(A,B,C,D,E,F) = \prod M\,(0,3,6,7,8,14,16,24,30,32,34,35,38,40,46,48,$
$\qquad 53,56,62)\prod D\,(2,39,61)$.

Use the Quine-McCluskey tabular method of reduction to simplify the functions in Problems 5.11 through 5.20.

5.11. $Q(A,B,C,D) = \prod M\,(0,2,4,6,7,8,9)$.

5.12. $R(A,B,C,D) = \prod M\,(6,7,8,9,11,13,15)$.

5.13. $S(A,B,C,D) = \sum m\,(1,5,7,9,11,12,14,15)$.

5.14. $T(A,B,C,D) = \sum m\,(0,1,3,5,6,7,8,10,14,15)$.

5.15. $U(A,B,C,D) = \sum m\,(2,3,4,7,9,11,12,13,14) + \sum d\,(1,10,15)$.

5.16. $V(A,B,C,D) = \sum m\,(1,3,4,5,6,7,10,12,13) + \sum d\,(2,9,15)$.

5.17. $W(A,B,C,D,E) = \sum m\,(0,4,14,22,30) + \sum d\,(7,8,12,28)$.

5.18. $X(A,B,C,D,E) = \sum m\,(8,16,17,24,27) + \sum d\,(0,7,19,25)$.

5.19. $Y(A,B,C,D,E) = \sum m\,(4,10,12,13,18) + \sum d\,(5,20,21,26)$.

5.20. $Z(A,B,C,D,E) = \sum m\,(8,16,22,23,30) + \sum d\,(0,6,14,31)$.

Use the DSA to simplify the functions in Problems 5.21 through 5.24 as an SOP, and Problems 5.25 through 5.28 as a POS.

5.21. $S(A,B,C,D) = \sum m\,(1,5,7,9,11,12,14,15)$.

5.22. $T(A,B,C,D) = \sum m\,(0,1,3,5,6,7,8,10,14,15)$.

5.23. $U(A,B,C,D) = \sum m\,(2,3,4,7,9,11,12,13,14) + \sum d\,(1,10,15)$.

5.24. $V(A,B,C,D) = \sum m\,(1,3,4,5,6,7,10,12,13) + \sum d\,(2,9,15)$.

5.25. $W(A,B,C,D) = \prod M\,(1,5,7,9,11,12,13)\prod D\,(0,14)$.

5.26. $X(A,B,C,D) = \prod M\,(2,6,7,8,13)\prod D\,(1,3,4)$.

5.27. $Y(A,B,C,D) = \prod M\,(0,2,3,4,7,8,9,11)\prod D\,(1,10,15)$.

5.28. $Z(A,B,C,D) = \prod M\,(1,3,4,5,6,7,10,11)\prod D\,(2,9,15)$.

Use the generalized iterative consensus algorithm to simplify the functions in Problems 5.29 through 5.36.

5.29. $S(A,B,C,D) = \overline{A}\,\overline{B}\,\overline{C} + \overline{A}\,BD + BCD + AB\overline{C}D$.

5.30. $T(A,B,C,D) = \overline{A}\,\overline{C}\,\overline{D} + A\overline{B}\,\overline{C}\,\overline{D} + \overline{A}\,BC + ABCD + ABC\overline{D}$.

5.31. $U(A,B,C,D) = \overline{A}\,\overline{B}\,\overline{C} + \overline{A}\,\overline{B}D + \overline{A}\,BC + ABC + \overline{A}\,B\overline{C}D + A\overline{B}\,\overline{C}\,\overline{D}$.

5.32. $V(A,B,C,D) = \overline{A}\,BC\overline{D} + \overline{A}\,\overline{B}\,\overline{C} + \overline{A}\,BD + \overline{A}\,CD + ABC + A\overline{B}\,CD$.

5.33. $W(A,B,C,D) = (\overline{A} + \overline{B} + \overline{C})(\overline{A} + B + D)(B + C + D)(A + B + \overline{C} + D)$.

5.34. $X(A,B,C,D) = (\overline{A} + \overline{C} + \overline{D})(A + \overline{B} + \overline{C} + \overline{D})(\overline{A} + B + C)(A + B + C + D)(A + B + C + \overline{D})$.

5.35. $Y(A,B,C,D) = (\overline{A} + \overline{B} + \overline{C})(\overline{A} + \overline{B} + D)(\overline{A} + B + C)(A + B + C)(\overline{A} + B + \overline{C} + D)(A + \overline{B} + \overline{C} + \overline{D})$.

5.36. $Z(A,B,C,D) = (\overline{A} + B + \overline{C} + \overline{D})(\overline{A} + \overline{B} + \overline{C})(\overline{A} + C + D)(A + B + C)(A + \overline{B} + C + D)$.

In Problems 5.37 through 5.40, the functions are to be simplified simultaneously. Find a solution that minimizes the total number of gates for the system of three functions and sketch the circuit implementation in NAND-NAND logic.

5.37. $X = \sum m\,(0,2,5,8) + \sum d\,(7,10,14,15).$

   $Y = \sum m\,(0,1,7,10) + \sum d\,(2,4,5,8,9).$

   $Z = \sum m\,(0,1,2,8,12) + \sum d\,(3,10).$

5.38. $X = \sum m\,(11,12,13,14) + \sum d\,(0,2,15).$

   $Y = \sum m\,(3,7,11,12,13,15) + \sum d\,(0,10).$

   $Z = \sum m\,(3,7,12,13) + \sum d\,(0,8,15).$

5.39. $X = \sum m\,(3,4,6,12,14) + \sum d\,(1,9,11).$

   $Y = \sum m\,(3,6,9,12,13,14) + \sum d\,(1,7,11,15).$

   $Z = \sum m\,(3,4,5,6,9) + \sum d\,(1,11,14).$

5.40. $X = \sum m\,(1,3,4,9,11,14) + \sum d\,(6).$

   $Y = \sum m\,(1,3,4,6,14) + \sum d\,(5,7).$

   $Z = \sum m\,(4,9,11) + \sum d\,(1,6,14).$

SPECIAL PROBLEMS

5.41. A comparator circuit is to be designed as follows: the inputs are two words labeled A and B, each consisting of two bits labeled A_1, A_0 and B_1, B_0. The three outputs are to be asserted for A > B, A = B, and A < B. Make a truth table with inputs A_1, A_0, B_1, and B_0 and outputs for A > B, A = B, and A < B. Map the outputs, simplify the maps, and implement the circuit. HINT: The map for A = B consists of four diagonal terms that can be represented as $(A_1 \odot B_1) \cdot (A_0 \odot B_0)$. Also

$$A < B = \overline{(A > B) + (A = B)} = \overline{(A > B)} \cdot \overline{(A = B)}$$

The three functions can be realized with 8 gates, 4 AND, 1 OR, 1 NOR, 2 ENOR, and 19 inputs.

5.42. The four inputs to a BCD to seven-segment display converter network are A, B, C, D, and their complements which are available from toggle switches. Using the truth table from Problem 4.47, and letting $X_i = Y_i$, one obtains (from the appropriate Karnaugh maps) the following outputs to represent the luminous states of the LED segments.

$$X_0 = A + \overline{B}C + \overline{B}\,\overline{D} + BD$$
$$X_1 = \overline{B} + \overline{C}\,\overline{D} + CD$$
$$X_2 = B + \overline{C} + D$$
$$X_3 = \overline{B}\,\overline{D} + C\overline{D} + \overline{B}C + B\overline{C}D$$
$$X_4 = \overline{B}\,\overline{D} + C\overline{D}$$
$$X_5 = A + \overline{C}\,\overline{D} + B\overline{C} + B\overline{D}$$
$$X_6 = A + \overline{B}C + B\overline{C} + B\overline{D}$$

Design a network to do the above conversion, using only three-input NAND gates. For the sake of simplicity, omit the unused inputs. It will be assumed that they are properly connected and will not interfere with the solution. Show how the function was factored or how the maps were looped to obtain the answer. Any solution which uses 21 or less gates is acceptable.

REFERENCES

1. **Quine, W. V.,** The problem of simplifying truth functions, *Am. Math. Mon.,* 59(8), 521–531, 1952; see also Quine, W. V., A way to simplify truth functions, *Am. Math. Mon.,* 62(9), 627–631, 1955.
2. **McCluskey, E. J.,** Minimization of boolean functions, *Bell Syst. Tech. J.,* 35(5), 1417–1444, 1956.
3. **Breeding, K. J.,** *Digital Design Fundamentals,* Prentice-Hall, Englewood Cliffs, NJ, 60–65, 1989.
4. **Hill, F. J. and Peterson, G. R.,** *Introduction to Switching Theory and Logical Design,* 2nd ed., John Wiley & Sons, New York, 1974, 159–167.
5. **Breeding, K. J.,** *Digital Design Fundamentals,* Prentice-Hall, Englewood Cliffs, NJ, 1989, 66–70.
6. **Rhyne, V. T., Noe, P. S., McKinney, M. H., and Pooch, U. W.,** A new technique for the fast minimization of switching functions, *IEEE Trans. Electron. Comput.,* C-26(8), 759–763, 1977.
7. **Mott, T. H., Jr.,** Determination of the irredundant normal forms of a truth function by iterated consensus of the prime cubes, *IRE Trans. Electron. Comput.,* EC-9(2), 245–252, 1960.
8. **Tison, P.,** Generalization of consensus theory and application to the minimization of boolean functions, *IEEE Trans. Electron. Comput.,* EC-16(4), 446–456, 1967.
9. **Wakerly, J. F.,** *Digital Design Principles and Practices,* Prentice-Hall, Englewood Cliffs, NJ, 1990, 198–206.
10. **Dietmeyer, D. L.,** *Logic Design of Digital Systems,* Allen and Bacon, Boston, 1978, chap. 10.
11. **Muroga, S.,** *Logic Design and Switching Theory,* Wiley-Interscience, New York, 1979, chap. 4.

CHAPTER 6

Multiplexers, Demultiplexers, ROMs, and PLDs

6.1 INTRODUCTION

Multiplexing is the technique of putting multiple signals on the same data path, and demultiplexing is the technique of sorting or decoding the signals back to multiple data paths. This is convenient when interfacing between different parts of a circuit. For instance, a microprocessor can easily be programmed to handle multiplexed data with less input/output or I/O wires. If the data is to be routed to another part of the system, one bus wire can carry many signals, which can be decoded at the other end of the bus line.

Multiplexers are versatile and easy to design and are usually easier to trouble-shoot than a discrete gate layout. They also take up less real estate and require fewer interconnections due to the reduced pin count. Because of their versatility they are sometimes referred to as *universal logic modules* or *ULMs*. Small multiplexers can be connected in tree structures to realize more complicated boolean expressions.

Demultiplexers can be used either to reverse the multiplexing operation or to decode the input signals, depending upon how the inputs are treated. If the input signal is taken on the enable line, the other inputs become select controls which transfer the input to a specific output line. If the inputs are used as data inputs the device decodes the input signal of n bits to one of 2^n outputs.

Small demultiplexers can be connected together to emulate larger configurations. Again, smaller demultiplexers have fewer inputs, but the penalty for cascading them is that each demultiplexer adds another propagation delay to the circuit.

The enable input to either a multiplexer or a demultiplexer can often be ignored until the design is to be implemented. Since the enable input only determines whether or not a given chip is able to respond to its data inputs, no generality is lost in doing this. In writing a truth table for an enabled multiplexer or demultiplexer, one row would indicate no response when the enable signal is not present, and the truth tables discussed below would apply when the multiplexer or demultiplexer is enabled.

A *decoder* is a minterm generator, which converts a binary coded input of n bits to one of 2^n different minterms. When used to drive a memory matrix the decoder plus memory array is referred to as a *ROM* or *read-only memory*. In designing a ROM, 10 address lines can be used with a decoder to select 1 of $2^{10} = 1024$ word lines, and 20 address lines can select 1 of over a million word lines in the ROM.

An *encoder* is a combinational logic circuit that is designed to generate a binary output code for each different data input, with only one of the data inputs enabled at a time. A *priority encoder* is a special encoder which can handle cases where more than one data input at a time is active. The priority encoder is programmed to respond to the highest-priority data input line that is active, all other data inputs being don't cares.

A decoder driving an array of OR gates can realize functions in two-level AND-OR SOP form, the decoder forming the AND plane. The combination is called a ROM. If the OR plane is programmable, the device is called a *programmable ROM (PROM)*. PROMs can be programmed once only with fuses or repeatedly if they are erasable either electrically or with ultraviolet (UV) light.

The PROM concept can be generalized by replacing the decoder with a programmable AND plane, resulting in a *programmable logic array (PLA)*. One can also realize SOP solutions with a fixed OR plane. A device that does this is referred to as a *programmable array logic (PAL)*. (PAL is a registered trademark of Monolithic Memories, Inc.) The PLA, PROM, and PAL are three forms of *programmable array logic devices (PLDs)*.

The PLA is the most versitile of the three PLDs, since both its AND plane and its OR plane can be custom programmed. It is usually more expensive, somewhat harder to program, and a little slower. The PAL device generally requires less real estate than an equivalent PLA chip. The PROM has a fixed AND array, and is the least versatile programmable device.

6.2 MULTIPLEXERS

Multiplexers, also called *data selectors,* are circuits that select one out of n data input lines and transmit it to the output. The multiplexer, abbreviated as *MUX*, is the electrical analog of a mechanical rotary switch, with n positions. A 4-to-1 mechanical switch is shown in Figure 6.1. In an audio system consisting of a tape deck, turntable, radio, and TV, such a switch might select which of these data inputs, D_i, is connected to the preamplifier.

As another example, a 16-character hexidecimal keypad can easily be connected to one I/O line of the microprocessor to which it is attached. The response time of a human is orders of magnitude slower than that of the electronic circuitry, and the multiplexing operation is transparent to the human operator. The keypad need only be scanned once every 20th or 50th of a second to accomplish this.

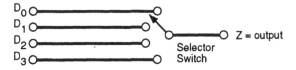

FIGURE 6.1. A 4-to-1-line mechanical selector switch.

The MUX generally has an *enable input,* or *inhibit input,* which allows small MUXs to be connected together to obtain larger configurations. Smaller MUXs have fewer data inputs, but each MUX in a cascade adds an additional propagation delay to the overall circuit.

There are three basic approaches to implementing a boolean function with MUXs. One can map the function and locate the appropriate segments of the Karnaugh map which correspond to the select input variables chosen. If one maps the select input variables, the data inputs are the minterms of the select function. One can apply Shannon's expansion theorem to partition the function appropriately, or one can partition the truth table for the chosen select input variables. All three methods will be examined.

The choice of select input variables is important in determining the size and number of MUXs required to implement a given function. Examination of a function to determine an optimum selection of select inputs will be discussed.

6.2.1   The Basic Multiplexer or Data Selector

An electrical data selector to perform the 4-to-1 multiplexing of Figure 6.1 can be constructed from four AND gates feeding into an OR gate, as shown in Figure 6.2(a). The symbol for this MUX is shown in Figure 6.2(b). The desired data input, D_i, is selected by choosing the appropriate signal on the *select* or *control* lines, labeled S_1 and S_0 in Figure 6.2. The four select input lines enable one AND gate and disable the remaining AND gates, thus passing only the selected data input signal to the output. A description of the MUX can be given by a truth table or by an equation for the output of the MUX. There are 6 inputs and 64 rows in the full truth table. However, this can be condensed into only 4 rows with each row corresponding to 16 rows of the expanded truth table, as shown in Table 6.1.

$$Z(D_i, S_0, S_1) = D_0 \cdot \overline{S}_1 \overline{S}_0 + D_1 \cdot \overline{S}_1 S_0 + D_2 \cdot S_1 \overline{S}_0 + D_3 \cdot S_1 S_0 = \sum D_i m_i \qquad (6.1)$$

where m_i is minterm i in the select variables S_0 and S_1.

If A and B are the select input variables, Equation 6.1 can be written as

$$Z(D_i, A, B) = D_0 \cdot \overline{A}\,\overline{B} + D_1 \cdot \overline{A} B + D_2 \cdot A \overline{B} + D_3 \cdot AB$$

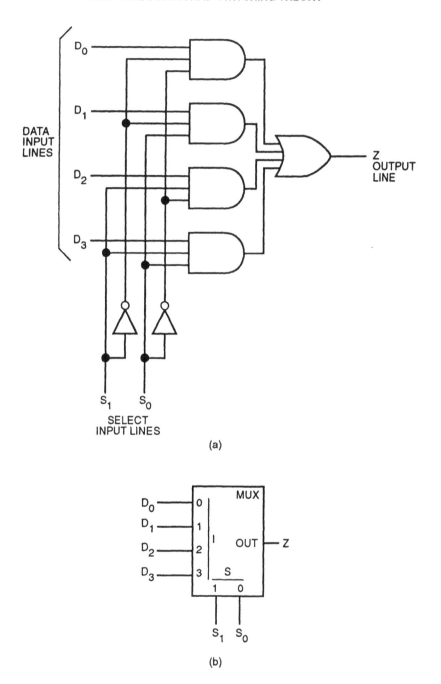

FIGURE 6.2.   (a) A 4-to-1-line MUX and (b) the 4-to-1-line MUX symbol.

This is shown in Table 6.2, which conveys all the essential information in a very condensed form. By adding more AND gates, larger MUXs can be constructed. An 8-to-1-line MUX requires three select inputs, and a 16-to-1-line MUX requires four select inputs.

TABLE 6.1

I_0	I_1	I_2	I_3	S_1	S_0	Z
D_0	x	x	x	0	0	D_0
x	D_1	x	x	0	1	D_1
x	x	D_2	x	1	0	D_2
x	x	x	D_3	1	1	D_3

TABLE 6.2

A	B	Z
0	0	D_0
0	1	D_1
1	0	D_2
1	1	D_3

EXAMPLE 6.1: Use a 4-to-1 line MUX to realize the function $Z(A,B,C) = \overline{A}\,\overline{B} + AC$, with A and B as select input variables.

SOLUTION: The function must be expanded to include both select input variables in each term. Thus,

$$Z(A,B,C) = \overline{A}\,\overline{B} + A(\overline{B} + B)C = \overline{A}\,\overline{B} + A\overline{B}C + ABC.$$

Compare Z to the output Equation 6.1, and the data inputs are $D_0 = 1$, $D_1 = 0$, $D_2 = C$, and $D_3 = C$, as shown in Figure 6.3(a).

EXAMPLE 6.2: Use an 8-to-1-line MUX to realize the function $Z(A,B,C) = \overline{A}\,\overline{B} + AC$, with A, B, and C as select input variables.

SOLUTION: The function must be expanded to include all three select input variables in each term.

$$Z(A,B,C) = \overline{A}\,\overline{B}(\overline{C} + C) + A(\overline{B} + B)C = \overline{A}\,\overline{B}\,\overline{C} + \overline{A}\,\overline{B}C$$
$$+ A\overline{B}C + ABC = \sum m\,(0,1,5,7)$$

The output equation of an 8-to-1 line MUX is

$$Z = D_0 \cdot \overline{S}_2\overline{S}_1\overline{S}_0 + D_1 \cdot \overline{S}_2\overline{S}_1 S_0 + D_2 \cdot \overline{S}_2 S_1 \overline{S}_0 + D_3 \cdot \overline{S}_2 S_1 S_0$$
$$+ D_4 \cdot S_2 \overline{S}_1 \overline{S}_0 + D_5 \cdot S_2 \overline{S}_1 S_0 + D_6 \cdot S_2 S_1 \overline{S}_0 + D_7 \cdot S_2 S_1 S_0$$
$$= \sum D_i m_i$$

Upon comparing the function to the MUX output equation, it is seen that $D_0 = 1$, $D_1 = 1$, $D_2 = 0$, $D_3 = 0$, $D_4 = 0$, $D_5 = 1$, $D_6 = 0$, and $D_7 = 1$.

To program this MUX, simply connect inputs 0,1,5,7 to a logic 1 and inputs 2,3,4,6 to a logic 0.

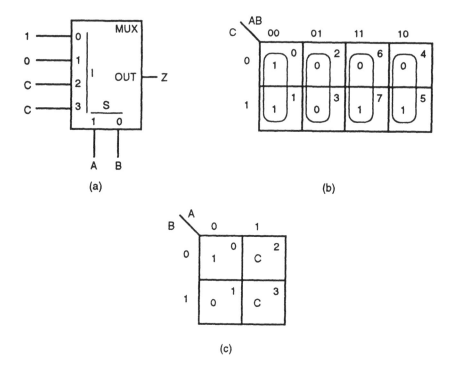

FIGURE 6.3. The 4-to-1-line MUX realization of $Z(A,B,C) = \overline{A}\,\overline{B} + AC$: (a) the MUX, (b) the three-variable mapping of Z, and (c) the reduced mapping of Z.

6.2.2 Mapping and Multiplexer Design

A Karnaugh map can be used to design a MUX circuit. One can map Z and identify the segments of the map which correspond to the data inputs D_i. The MUX of Example 6.1 is shown in Figure 6.3(a), the three-variable map of the function is shown in Figure 6.3(b), with the segments or patches of the map corresponding to the four MUX data inputs identified.

The map can be reduced to a two-variable map by simply mapping the four data input functions on a map of the select input variables. One now has a variable-entered map, with the nonselect input variable, C, an entry on the map. This is shown in Figure 6.3(c). The four cells of the reduced map contain the four input values. Recall that the reduced variable-entered map can have entries 0, 1, and x (don't care) and variables (primed or unprimed).

EXAMPLE 6.3: Realize the function $Z(A,B,C,D) = \sum m\,(0,1,3,6,7,8,11,12,$ $14)$, with select inputs A, B, and C.

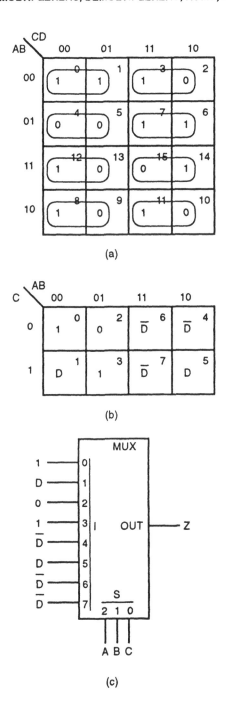

(a)

(b)

(c)

FIGURE 6.4.   (a) Four-variable map, (b) reduced map, and (c) MUX.

SOLUTION: A map of the function Z is shown in Figure 6.4(a), with the
sections or patches of the map corresponding to the eight data
inputs looped on the map. Standard Karnaugh-map reduction
techniques are next applied to each segment of the map to
determine the simplest expression for each of the eight
subfunctions of the map corresponding to the data inputs to the
MUX. A reduced three-variable map can also be used and is
shown in Figure 6.4(b). The 8-to-1-line MUX, with data inputs
labeled is shown in Figure 6.4(c).

A given MUX can be replaced by a MUX with less data inputs and less select
input variables. The penalty for this replacement is an increase in the number of
gates required to augment the MUX. Sandige[1] classifies MUXs as type n, where
n is the number of variables that are required as data inputs to the MUX.

The function Z of Example 6.3 is designed using a 4-to-1-line MUX in
Example 6.4. An inclusive-OR gate and an ENOR gate are required to accomplish
this, as shown in Figure 6.5(c).

EXAMPLE 6.4:   Realize the function of Example 6.3 with a 4-to-1-line
MUX and select inputs A and B.

SOLUTION: The function is mapped and the segments looped in Figure
6.5(a), the reduced map is shown in Figure 6.5(b), and the
MUX with data input circuitry is shown in Figure 6.5(c).

6.2.3   Multiplexers with Enable Inputs

The *enable, inhibit,* or *strobe* input is usually an active-low input. The truth
table for a 4-to-1-line MUX with an enable input that is active low is shown in
Table 6.3, and the block diagram of the MUX is shown in Figure 6.6. The external
label $\overline{EN}$ is used to indicate that EN is an active-low or asserted-low input. The
internal label is EN regardless of the external polarity.

There are seven inputs, corresponding to the enable signal EN, taken to be
active low, the four data inputs, which are active high, the two control or select
inputs, also active high, and there is one output, Y, which is active high. Don't care
inputs are represented by xs. In general, a seven-input function has $2^7 = 128$ rows
in its truth table. However, when the enable signal is high, the output is low for
all 64 rows of the truth table. When the enable signal is low, all but one of the four
data inputs are don't cares, and the truth table can once again be condensed to four
rows, with EN low and one row with EN high.

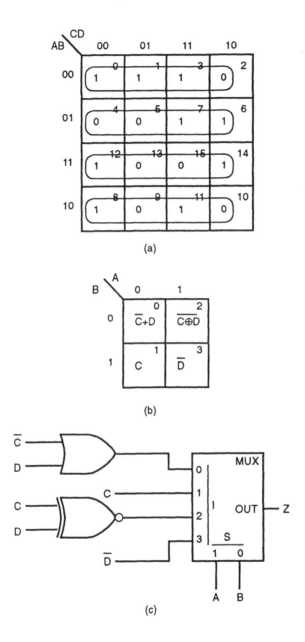

FIGURE 6.5.   (a) Four-variable map, (b) reduced map of two select variables, and (c) MUX with input circuitry.

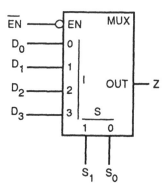

FIGURE 6.6.   Block diagram of a 4-to-1 line MUX with enable input.

TABLE 6.3

EN	I_0	I_1	I_2	I_3	S_1	S_0	Z
1	x	x	x	x	x	x	0
0	D_0	x	x	x	0	0	D_0
0	x	D_1	x	x	0	1	D_1
0	x	x	D_2	x	1	0	D_2
0	x	x	x	D_3	1	1	D_3

6.3 AND-OR-INVERT LOGIC AND MULTIPLEXER TREES

AND-OR-INVERT (AOI) gates reduce the actual number of transistors needed for two levels of NAND or NOR logic by physically merging the logic into one level. This results in faster circuits. Also, with outputs Z and $\overline{Z}$ available, both SOP and POS solutions are available at the output. The basic AOI circuit, with true and complemented outputs, is shown in Figure 6.7.

A 4-to-1 AOI MUX circuit with select inputs A and B, and outputs Z and $\overline{Z}$, is shown in Figure 6.8, with the enable or strobe input omitted. The function realized is $Z(A,B,D_i) = D_0 \overline{A}\,\overline{B} + D_1 \overline{A} B + D_2 A\overline{B} + D_3 AB$ and is in the SOP form. This is analagous to the four-position, single-pole rotary switch shown in Figure 6.1, with both the true and complemented outputs available.

Two AOI circuits can realize a double-pole, double-throw switch as shown in Figure 6.9, and the circuit of Figure 6.8 can be replaced by a tree consisting of three 2-to-1 MUXs, as shown in Figure 6.10. The noninverting outputs are shown in Figure 6.10 and the most significant select bit is controlling the output or highest-level MUX. MUX trees are needed whenever a function requires more data inputs than are available with a single MUX.

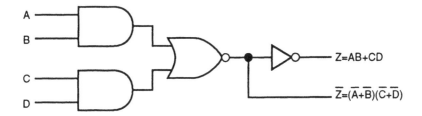

FIGURE 6.7. The AOI structure with complemented (POS) and uncomplemented (SOP) outputs.

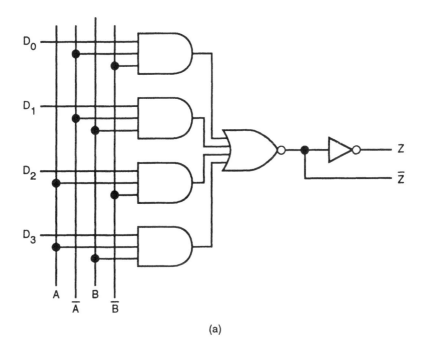

(a)

FIGURE 6.8. (a) A 4-to-1-line MUX structure with outputs Z and $\overline{Z}$ and (b) the MUX symbol.

EXAMPLE 6.5: Design an 8-to-1-line MUX to realize the function

$Z(A,B,C,D) = \sum m\,(2,5,8,9,11,12,14,15)$ with select inputs A, B, and C.

Repeat for two 4-to-1-line MUXs driving a 2-to-1-line MUX and for four 2-to-1-line MUXs driving a 4-to-1-line MUX.

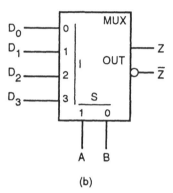

(b)

FIGURE 6.8.   (continued).

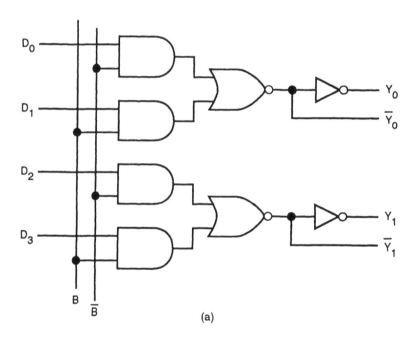

(a)

FIGURE 6.9.   Two two-input MUXs. (a) AOI structure, (b) symbols, and (c) equivalent double-pole, double-throw switch.

SOLUTION: The four-variable Karnaugh map is shown in Figure 6.11(a), and the reduced map is shown in Figure 6.11(b), with D as a map-entered variable.

An 8-to-1 line MUX to realize the function mapped in Figure 6.11 is shown in Figure 6.12. The entire circuit can be replaced

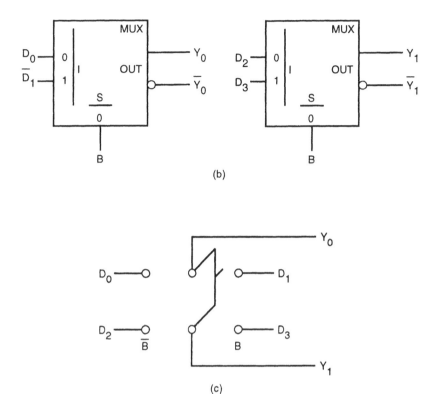

(b)

(c)

FIGURE 6.9. (continued).

by two 4-to-1-line MUXs driving a 2-to-1-line MUX, as shown in Figure 6.12(b), or by four 2-to-1-line MUXs driving a 4-to-1-line MUX, as shown in Figure 6.12(c). The strobe input is omitted in all three realizations.

Figure 6.13 shows the AOI circuits corresponding to the MUX of Figure 6.12. The gate count increases from 9 to 13 to 17 as one progresses from Figure 6.13(a) to Figure 6.13(c), while the MFI drops from 8 to 4. The total number of gate inputs is almost constant, changing from 40 to 38 to 40 inputs, while the number of literals for data and select inputs drops from 32 to 26 to 24, respectively.

If complementary outputs are also available, the MUX of Figure 6.12 can be configured as two 2-to-1-line MUXs driving a 4-to-1-line MUX, as shown in Figure 6.14. The outputs of the 2-to-1-line MUX are

$$Y_0 = C\overline{D} \qquad Y_1 = \overline{C}D \qquad Y_2 = \overline{C} + D \qquad Y_3 = C + \overline{D} \qquad (6.2)$$

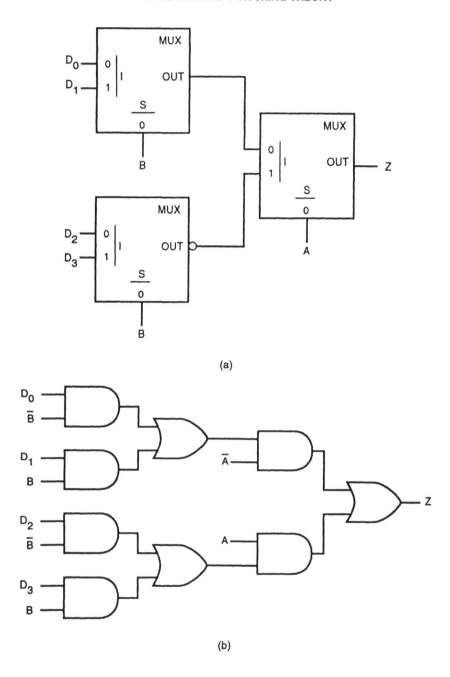

(a)

(b)

FIGURE 6.10.   A 4-to-1-line MUX realized with a MUX tree, consisting of three 2-to-1-line MUXs.
(a) Symbol and (b) AOI structure.

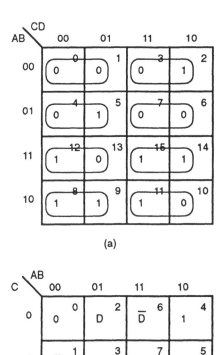

(a)

(b)

FIGURE 6.11. (a) Four-variable mapping and (b) reduced map of Z, above.

Y_2 is the complement of Y_0, and Y_3 is the complement of Y_1; thus, the number of first-level MUXs required is only two if both true and complemented outputs are available.

Lastly, one can replace each two-input MUX by gates. In Figure 6.15(a), the MUX is driven by a quad-NOR chip, and in Figure 6.15(b) the same output, Z, is obtained by driving the MUX with a quad-NAND chip.

6.4 SHANNON'S EXPANSION THEOREM AND MULTIPLEXER DESIGN

Shannon's expansion theorem can be used when designing with MUXs. The theorem was presented in Chapter 3, Section 3.8, and Equations 3.4 and 3.8 from Chapter 3 are repeated below.

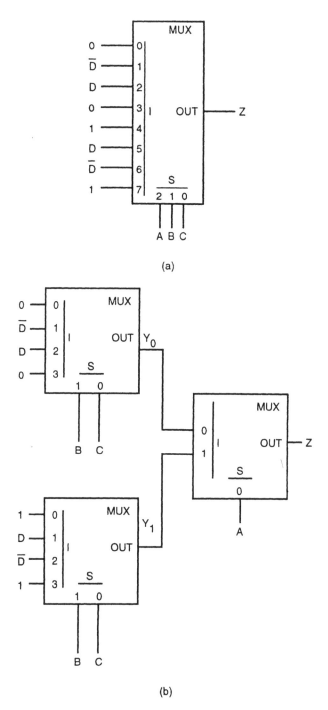

FIGURE 6.12.   (a) An 8-to-1-line MUX, (b) configured as two 4-to-1-line MUXs driving a 2-to-1-line MUX, and (c) configured as four 2-to-1-line MUXs driving one 4-to-1-line MUX. EN is omitted in all three cases.

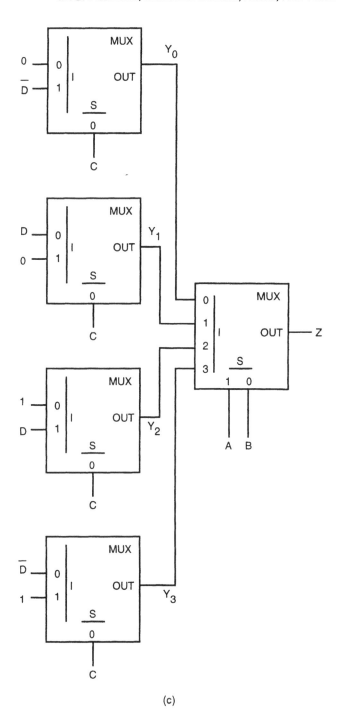

(c)

FIGURE 6.12. (continued).

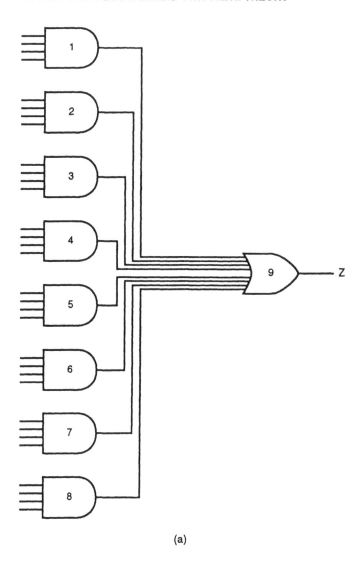

(a)

FIGURE 6.13.   Circuits for (a) an 8-to-1-line MUX, (b) two 4-to-1-line MUXs driving a 2-to-1-line MUX, and (c) four 2-to-1-line MUXs driving a 4-to-1-line MUX. EN is omitted in all three cases.

$$F(X_1, X_2, \ldots, X_n) = \overline{X}_1 \cdot F(0, X_2, \ldots, X_n) + X_1 \cdot F(1, X_2, \ldots, X_n)$$
$$= \overline{X}_1 \cdot F(0) + X_1 \cdot F(1) \qquad (6.3)$$
$$F(X_1, X_2, \ldots, X_n) = \overline{X}_1 \overline{X}_2 \cdot F(0,0) + \overline{X}_1 X_2 \cdot F(0,1) + X_1 \overline{X}_2 \cdot F(1,0)$$
$$+ X_1 X_2 \cdot F(1,1) \qquad (6.4)$$

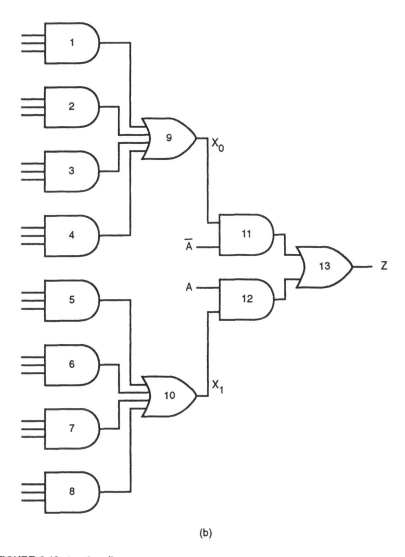

(b)

FIGURE 6.13. (continued).

Each application of the expansion theorem splits any function into two parts, one of which gives the part of the function for which the splitting variable is true, multiplied by the splitting variable, and the other which gives the function for which the splitting variable is complemented, multiplied by the complement of the splitting variable. Shannon's theorem can be applied repeatedly until the function cannot be further split.

Shannon's theorem gives a formal way of designing with MUXs. For example, consider the function

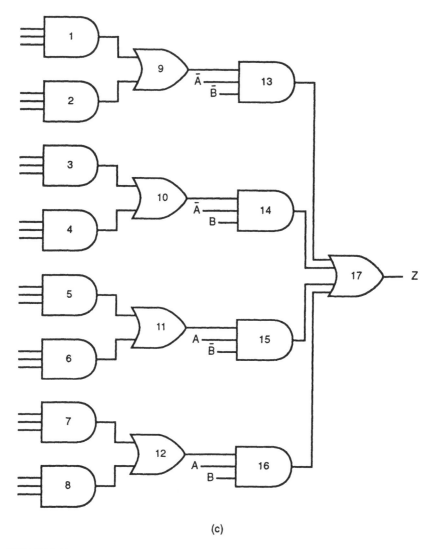

(c)

FIGURE 6.13. (continued).

$$F(A,B,C,D) = \sum m\,(2,5,8,9,11,12,14,15)$$
$$= \overline{A}\,\overline{B}C\overline{D} + \overline{A}\,B\overline{C}D + A\overline{C}\,\overline{D} + A\overline{B}D + ABC \qquad (6.5)$$

The function will first be expanded around variable A. This gives

$$F(A,B,C,D) = \overline{A}\cdot(\overline{B}C\overline{D} + B\overline{C}D) + A\cdot(\overline{C}\,\overline{D} + \overline{B}D + BC) \qquad (6.6)$$

To apply the expansion theorem again to variables B and C, one must first expand the terms in parentheses in the variables B and C.

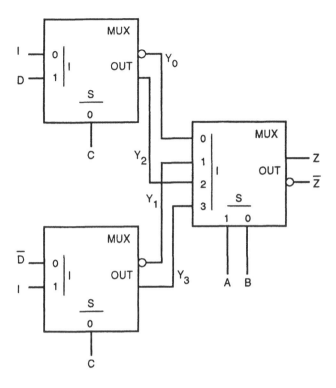

FIGURE 6.14. Realization of the circuit of Figure 6.12, with complementary-output MUX.

$$F(A,B,C,D) = \overline{A} \cdot (\overline{B}\,C\overline{D} + B\overline{C}D) + A \cdot (\overline{B}\,\overline{C} + \overline{B}CD + B\overline{C}\,\overline{D} + BC) \quad (6.7)$$

Now Shannon's theorem can be applied to variables B and C.

$$
\begin{aligned}
F(A,B,C,D) = \overline{A} \cdot [\overline{B}\,\overline{C} \cdot (0) + \overline{B}C \cdot (\overline{D}) + B\overline{C} \cdot (D) + BC \cdot (0)] \\
+ A \cdot [\overline{B}\,\overline{C} \cdot (1) + \overline{B}C \cdot (D) + B\overline{C} \cdot (\overline{D}) + BC \cdot (1)] \quad (6.8)
\end{aligned}
$$

The function in Equation 6.5 is the same as in Example 6.5, and the tree realized by two 4-to-1-line MUXs and one 2-to-1-line MUX is the one shown in Figure 6.12(b).

The above function can be implemented with four two-input and one four-input MUXs by applying Shannon's theorem to variables A and B. From Equation 6.8:

$$
\begin{aligned}
F(A,B,C,D) = \overline{A} \cdot [\overline{B}\,\overline{C} \cdot (0) + \overline{B}C \cdot (\overline{D}) + B\overline{C} \cdot (D) + BC \cdot (0)] \\
+ A \cdot [\overline{B}\,\overline{C} \cdot (1) + \overline{B}C \cdot (D) + B\overline{C} \cdot (\overline{D}) + BC \cdot (1)]
\end{aligned}
$$

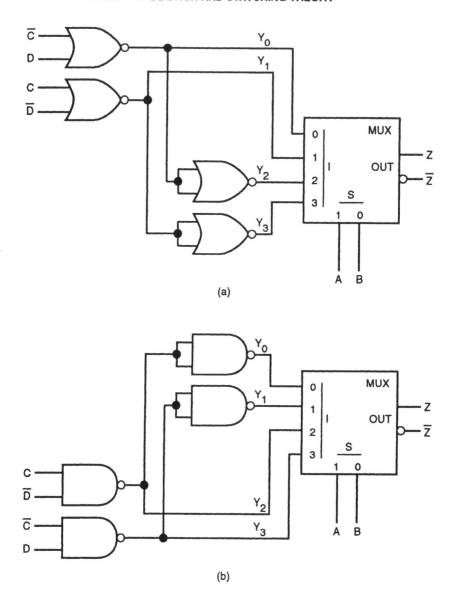

(a)

(b)

FIGURE 6.15. Realization of the MUX of FIGURE 6.12 with the help of (a) four NOR gates and (b) four NAND gates.

TABLE 6.4

A	B	C	D	F
0	0	0	0	0
0	0	0	1	0 $\rangle$ 0
0	0	1	0	1
0	0	1	1	0 $\rangle$ $\overline{D}$
0	1	0	0	0
0	1	0	1	1 $\rangle$ D
0	1	1	0	0
0	1	1	1	0 $\rangle$ 0
1	0	0	0	1
1	0	0	1	1 $\rangle$ 1
1	0	1	0	0
1	0	1	1	1 $\rangle$ D
1	1	0	0	1
1	1	0	1	0 $\rangle$ $\overline{D}$
1	1	1	0	1
1	1	1	1	1 $\rangle$ 1

$$= \overline{A}\,\overline{B}\cdot[\overline{C}\cdot(0) + C\cdot(\overline{D})] + \overline{A}\,B\cdot[(\overline{C}\cdot(D) + C\cdot(0)]$$
$$+ A\overline{B}\cdot[\overline{C}\cdot(1) + C\cdot(D)] + AB\cdot[\overline{C}\cdot(\overline{D}) + C\cdot(1)] \qquad (6.9)$$

The function expanded as given by Equation 6.9 is shown in Figure 6.12(c).

6.5 TRUTH-TABLE PARTITIONING FOR MULTIPLEXER DESIGN

The MUX trees can be realized by partitioning the truth table of the function, as was done in variable-entered mapping (VEM). The truth table for the function represented in Equation 6.5 is shown in Table 6.4, and the function is realized in Figure 6.12(a) as a single 8-to-1-line MUX with D as an input variable. The truth table for the function as represented by Equation 6.8 is shown in Table 6.5, and the function is realized in Figure 6.12(b) as a tree with two 4-to-1-line MUXs driving a 2-to-1-line MUX. The function can also be expanded to give a tree structure with four 2-to-1-line MUXs driving a 4-to-1-line MUX, as shown in Figure 6.12(c). The truth table for this realization is shown in Table 6.6.

As a last example, the truth table can be partitioned to give a 4-to-1-line MUX controlled by select inputs A and B, with C and D the data inputs as given by Equations 6.2 or 6.9 and represented in Figure 6.15. This is shown in Table 6.7.

TABLE 6.5

A	B	C	D	F
0	0	0	0	0 ⎫
0	0	0	1	0 ⎭ 0
0	0	1	0	1 ⎫
0	0	1	1	0 ⎭ $\bar{D}$
0	1	0	0	0 ⎫
0	1	0	1	1 ⎭ D
0	1	1	0	0 ⎫
0	1	1	1	0 ⎭ 0
1	0	0	0	1 ⎫
1	0	0	1	1 ⎭ 1
1	0	1	0	0 ⎫
1	0	1	1	1 ⎭ D
1	1	0	0	1 ⎫
1	1	0	1	0 ⎭ $\bar{D}$
1	1	1	0	1 ⎫
1	1	1	1	1 ⎭ 1

TABLE 6.6

A	B	C	D	F
0	0	0	0	0 ⎫
0	0	0	1	0 ⎭ 0
0	0	1	0	1 ⎫
0	0	1	1	0 ⎭ $\bar{D}$
0	1	0	0	0 ⎫
0	1	0	1	1 ⎭ D
0	1	1	0	0 ⎫
0	1	1	1	0 ⎭ 0
1	0	0	0	1 ⎫
1	0	0	1	1 ⎭ 1
1	0	1	0	0 ⎫
1	0	1	1	1 ⎭ D
1	1	0	0	1 ⎫
1	1	0	1	0 ⎭ $\bar{D}$
1	1	1	0	1 ⎫
1	1	1	1	1 ⎭ 1

TABLE 6.7

A	B	C	D	F	
0	0	0	0	0	
0	0	0	1	0	$C\overline{D}$
0	0	1	0	1	
0	0	1	1	0	
0	1	0	0	0	
0	1	0	1	1	$\overline{C}D$
0	1	1	0	0	
0	1	1	1	0	
1	0	0	0	1	
1	0	0	1	1	$\overline{C} + D$
1	0	1	0	0	
1	0	1	1	1	
1	1	0	0	1	
1	1	0	1	0	$C + \overline{D}$
1	1	1	0	1	
1	1	1	1	1	

MUX trees are particularly useful when implementing a function with inputs that cannot be accommodated by a single MUX. A special case is when a function has many variables and many minterms. Consider a six-variable function which is split on variable F and is given by three sets of minterms as follows:

$$Z(A,B,C,D,E,F) = F \cdot \sum m\,(0,12,27,29) + \overline{F} \cdot \sum m\,(4,15,30)$$
$$+ \sum m\,(9,24,26) \tag{6.10}$$

where the minterms are labeled as functions of A, B, C, D, and E.

The function requires a MUX tree which can accommodate minterms 0 through 30. One MUX can be eliminated since minterms 16 through 23 are all 0. The tree structure can be realized with three 8-to-1-line MUXs controlled by select input variables C, D, and E, driving a 4-to-1-line MUX controlled by select input variables A and B. The circuit is shown in Figure 6.16.

6.6 DETERMINATION OF MINIMAL MULTIPLEXER TREES

A function can be mapped, and the map segments which give the MUX data inputs can be read from the map once the select input variables are determined. The choice of select input variables determines the data inputs to the MUXs. A function of four variables, or a function which can be reduced to four possible select input variables by the use of variable-entered mapping, is easy to realize as a tree of four 4-to-1 MUXs driving a fifth 4-to-1 MUX.

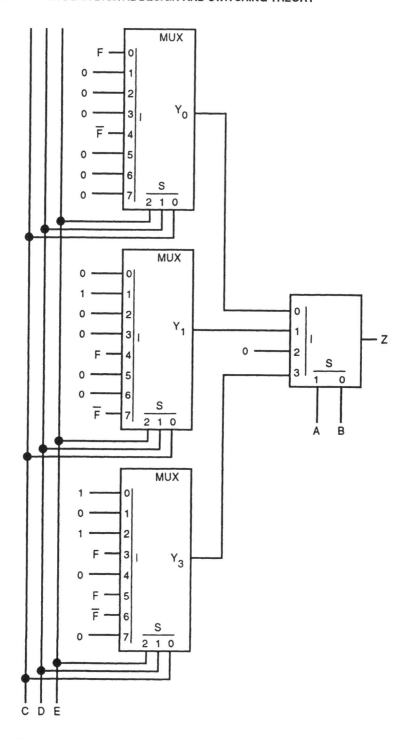

FIGURE 6.16. MUX tree realization of the function given by Equation 6.10.

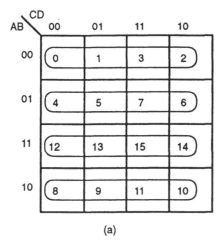

(a)

FIGURE 6.17. (a) Map and (b) MUX for primary select inputs A and B; (c) map and (d) MUX for primary select inputs C and D.

The map location of minterms of any function as data inputs to the first four MUXs depends upon the variables chosen as the two primary select inputs. The six possible combinations of primary select input variables, taken two at a time, are A and B, A and C, A and D, B and C, B and D, and C and D. The other two variables in each case control the first level of MUXs. Maps that chart the minterm locations are referred to as *decomposition charts.*[2]

The maps and data input assignments for these six combinations of primary and secondary select input variables are shown in Figures 6.17 through 6.19, inclusive. Figure 6.17 shows the map and gives the MUX tree for (a) and (b) primary select input variables A and B and (c) and (d) primary select input variables C and D. Figure 6.18 shows the map and gives the MUX tree for (a) and (b) primary select variables B and C and (c) and (d) primary select variables A and C. Figure 6.19 shows the map and gives the MUX tree for (a) and (b) primary select variables B and D and (c) and (d) primary select variables A and D.

The strobe or enable input can be used to select one of two MUXs. This allows the realization of a 2n-to-1-line MUX from two n-to-1-line MUXs and an OR gate. Figure 6.20 shows the realization of a 16-to-1-line MUX from two 8-to-1-line MUXs and an OR gate.

6.7 DEMULTIPLEXERS AND DECODERS

The opposite of a MUX is a circuit which has one data input line, n select lines or control lines, and up to 2^n output lines. This unit is called a *demultiplexer (DMUX),* and it transfers the data on the single-input line to the output line selected by the control lines. A 1-to-4-line DMUX acts as the electrical analog of a 1-to-4-position rotary switch, which is the converse of the switch in Figure 6.1.

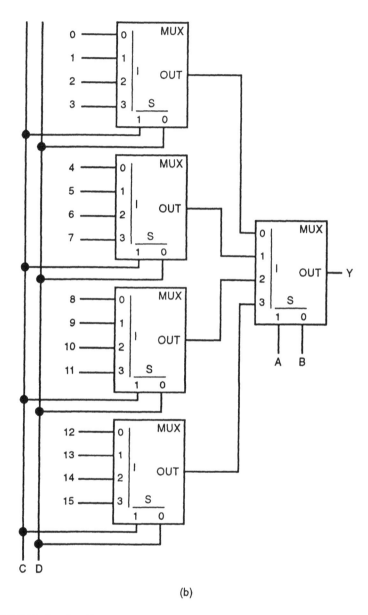

(b)

FIGURE 6.17. (continued).

If the device has n data input lines and up to 2^n output lines, it is called a *decoder*. The same device can be used for both demultiplexing and decoding.

If all 2^n output combinations are present, the device is referred to as a *binary decoder*. Some n-bit binary codes are truncated to represent fewer values. In the BCD and XS3 codes, there are only ten valid output combinations of four bits

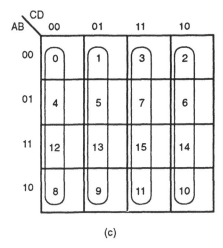

(c)

FIGURE 6.17. (continued).

each. Due to their importance, 4-to-10-line decoders are commercially available as well as 4-to-16-line binary decoders.

The name decoder arises due to the fact that the DMUX can convert various codes to decimal or alphabetic characters. It takes an input code and converts it to a different output code in a one-to-one mapping. If the enable or strobe is not asserted, the decoder outputs a disabled code word. The most common output code is a 1-out-of-n code, which contains n bits, only one of which is asserted at a time.

6.7.1   The Basic Demultiplexer or Decoder

A DMUX performs the inverse operation of a MUX. It acts as the electrical analog of a rotary switch which sends a single input signal to one of 2^n outputs, depending upon the value of the n-bit select inputs which control the DMUX. For example, if it is controlled by the same select inputs as a MUX, it can unscramble the multiplexed signal and send it to the appropriate destination.

In general, an n-to-2^n-line decoder, or DMUX, generates all 2^n minterms, m_i, or maxterms, M_i, of the n input variables. If the outputs, Y_i, are not inverted, $Y_i = m_i$, and if the outputs are inverted, $Y_i = M_i$. A 2-to-4-line decoder with complemented outputs is shown in Figure 6.21(a), and the standard symbol is shown in Figure 6.21(b).

This is the same basic circuitry as the 4-to-1-line MUX, with the output OR gate omitted. The only input common to all the AND gates is the enable, and it can be used as the DMUX input. The other inputs then select the output line to

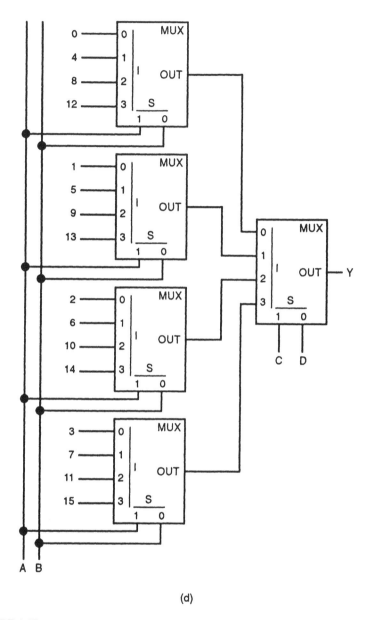

(d)

FIGURE 6.17. (continued).

which the input signal is transferred. If the enable input is connected to logic 0, the other inputs become data inputs and the device is a decoder.

The defining truth table of an active-low output 4-to-10-line decoder with inputs A, B, C, and D, and no enable input, is shown in Table 6.8. Four select lines can accomodate sixteen possible outputs, but since minterms 10 through 15

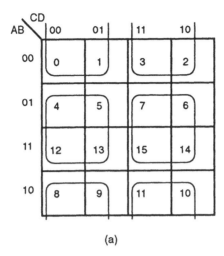

(a)

FIGURE 6.18. (a) Map and (b) MUX for primary select inputs B and C; (c) map and (d) MUX for primary select inputs A and C.

inclusive are never input, these six rows of the truth table would be don't cares. Hardware is saved and fewer total pins are needed when using only 10 output NAND gates, rather than 16. Figure 6.22 shows a 4-to-10-line decoder to convert BCD input to decimal output.

TABLE 6.8

BCD input				Decimal output									
A	B	C	D	0	1	2	3	4	5	6	7	8	9
0	0	0	0	0	1	1	1	1	1	1	1	1	1
0	0	0	1	1	0	1	1	1	1	1	1	1	1
0	0	1	0	1	1	0	1	1	1	1	1	1	1
0	0	1	1	1	1	1	0	1	1	1	1	1	1
0	1	0	0	1	1	1	1	0	1	1	1	1	1
0	1	0	1	1	1	1	1	1	0	1	1	1	1
0	1	1	0	1	1	1	1	1	1	0	1	1	1
0	1	1	1	1	1	1	1	1	1	1	0	1	1
1	0	0	0	1	1	1	1	1	1	1	1	0	1
1	0	0	1	1	1	1	1	1	1	1	1	1	0

Full decoding accounts for all 16 possible inputs and ensures that all outputs remain off for all invalid data input combinations. This is achieved by assuring that all outputs are kept high for the six invalid data input conditions.

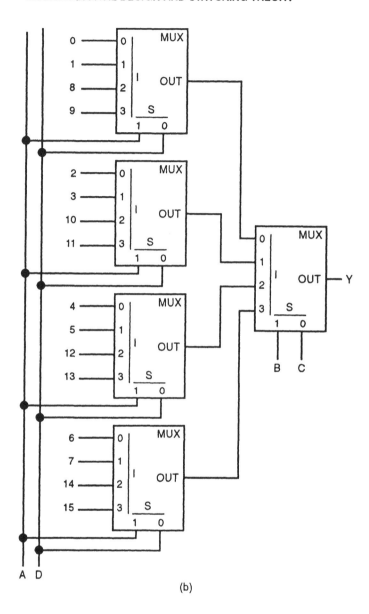

(b)

FIGURE 6.18. (continued).

6.7.2 Demultiplexers with Enable Inputs

While DMUXs require an enable input, decoders can be used without enable inputs. When enable inputs are added to the decoders it is easier to cascade them. The truth table of a 3-to-8-line decoder with an enable input, active low, and outputs active low is shown in Table 6.9.

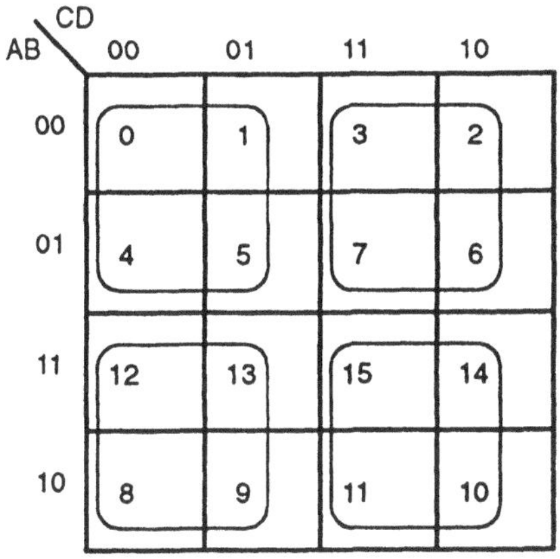

AB\CD	00	01	11	10
00	0	1	3	2
01	4	5	7	6
11	12	13	15	14
10	8	9	11	10

(c)

FIGURE 6.18. (continued).

TABLE 6.9

EN	S_2	S_1	S_0	0	1	2	3	4	5	6	7
1	x	x	x	1	1	1	1	1	1	1	1
0	0	0	0	0	1	1	1	1	1	1	1
0	0	0	1	1	0	1	1	1	1	1	1
0	0	1	0	1	1	0	1	1	1	1	1
0	0	1	1	1	1	1	0	1	1	1	1
0	1	0	0	1	1	1	1	0	1	1	1
0	1	0	1	1	1	1	1	1	0	1	1
0	1	1	0	1	1	1	1	1	1	0	1
0	1	1	1	1	1	1	1	1	1	1	0

A DMUX is a circuit which receives a data signal on a single input line and transfers it to one of 2^n possible output lines. From the truth table, the enable input can be considered as the one data input, with A, B, and C thought of as the three select lines which contain the code for selecting one output line. Treated this way, a suitable symbol for the 1-to-8-line DMUX is shown in Figure 6.23.

If the select inputs are data inputs, the DMUX becomes a 3-to-8-line decoder, as shown in Figure 6.24. Decoders with two enable inputs can be connected to have one common enable and one select enable.

EXAMPLE 6.6: Implement a 4-to-16-line decoder from two 3-to-8-line decoders, with select inputs B, C, and D.

SOLUTION: The enable input, EN, must be the most significant bit; therefore, let it be data input A. The circuit is shown in Figure 6.25.

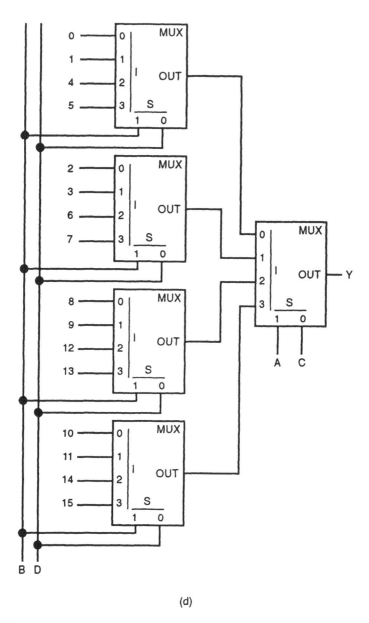

(d)

FIGURE 6.18. (continued).

6.7.3 The Decoder as a Function Generator

Any function represented by an n-variable truth table can be implemented by an n line to 2^n-line decoder, with additional gates to generate the appropriate sum of minterms or product of maxterms. A decoder with uncomplemented outputs

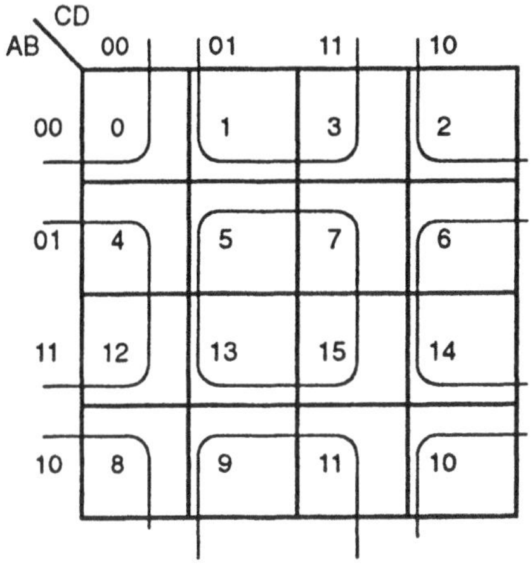

(a)

FIGURE 6.19. (a) Map and (b) MUX for primary select inputs B and D; (c) map and (d) MUX for primary select inputs A and D.

generates a minterm expansion of a given function, while a decoder with complemented outputs generates a maxterm expansion of the same function.

The minterm expansion of a function requires additional OR gates to realize the function in the two-level AND-OR form, while the maxterm expansion requires additional NAND gates to realize the function in the two-level NAND-NAND form.

Multiple functions can be realized also, as shown in the example below. This property of decoders will be used to create ROMs in the next section.

EXAMPLE 6.7: Realize functions $Z_1(A,B,C) = \sum m (1,3,5,7)$ and $Z_2(A,B,C) = \sum m (2,4,5,7)$ with a 3-to-8-line decoder.

SOLUTION: $Z_1 = \overline{\prod M(1,3,5,7)}$ and $Z_2 = \overline{\prod M(2,4,5,7)}$. As expressed, Z_1 and Z_2 can be realized in an SOP form, with a 3-to-8-line decoder with active-low outputs and two NAND gates, as shown in Figure 6.26. This is NAND-NAND logic.

6.8 PRIORITY ENCODERS

A device which has more input lines than output lines is called an *encoder*. To function properly, an encoder must have only one input asserted at a time. A 2^n-to-n encoder is called a *binary encoder* and performs the opposite function of a decoder. Its input code is a 1-out-of-2^n code, its output code is n-bit binary, and it can be constructed from n OR gates, each having 2^{n-1} inputs.[3]

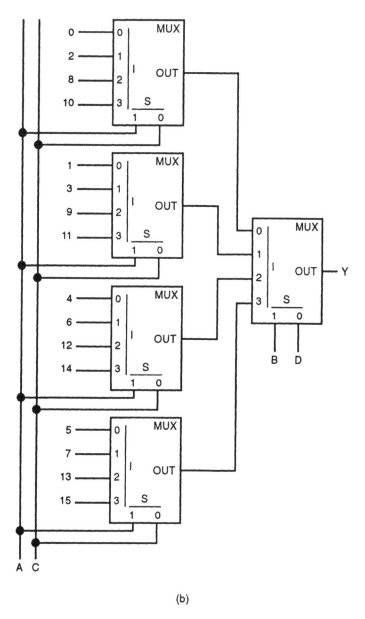

(b)

FIGURE 6.19. (continued).

The *priority encoder* has no restriction on the number of data inputs that can be on at the same time. The inputs are ordered by priority, and the output then corresponds to the highest priority input signal present. This scheme is common in microprocessors and digital computers which must handle multiple interrupts.

The truth table for a simple 3-input priority encoder is shown in Table 6.10, and a circuit realization is shown in Figure 6.27.

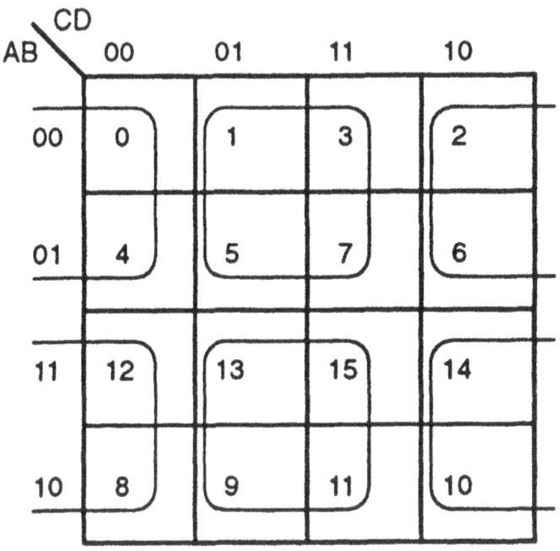

(c)

FIGURE 6.19. (continued).

TABLE 6.10

	A	B	C	Z_1	Z_0
	0	0	0	0	0
	0	0	1	0	1
	0	1	x	1	0
	1	x	x	1	1

6.9 READ-ONLY MEMORY

In the last section, it was shown that a decoder outputs a minterm or maxterm realization of a boolean function and that more than one function can be realized in this manner. This idea can be extended one step further by supplying an array of OR gates at the output of a decoder, and allowing them to be hooked up in an arbitrary manner. This requires a matrix of OR gates whose inputs are all of the minterms of the decoder, and the circuit can be drawn as shown in Figure 6.28. The open circles in Figure 6.28 indicate sites where a suitable connection can be made to program the OR array.

A suitable matrix of rows and columns that can be connected electrically and retain the information stored in it forms a memory array. The simplest memory array consists of rows of electrical lines called *output lines,* which are connected to positive supply through *pull-up* resistors, or connected to negative supply through *pull-down* resistors, and connected to *word lines* through suitably placed diodes or transistors. A decoder followed by such a memory array is called a *Read-Only Memory,* or *ROM.* If the array is user-programmable, the device is called a *Programmable Read-Only Memory* or *PROM.* The two terms will be used interchangeably in the following discussion.

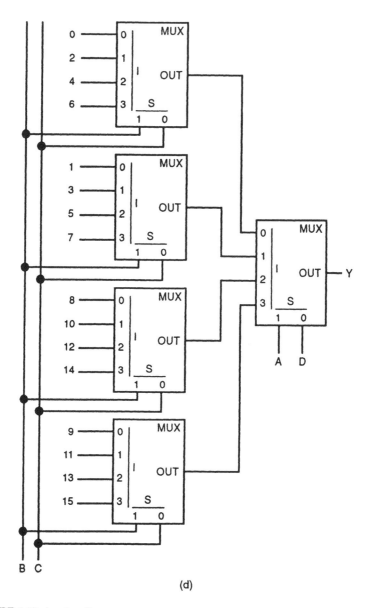

(d)

FIGURE 6.19. (continued).

A ROM is a combinational circuit with n inputs, 2^n word lines, and m outputs. The inputs are called *address inputs,* and are usually labeled $A_0, A_1, \ldots, A_{n-1}$. The outputs are called *data outputs* and will be labeled $Z_0, Z_1, \ldots, Z_{m-1}$. One can think of information as being "stored" in a ROM that has been programmed. A ROM or PROM is a *nonvolatile memory,* meaning that its contents are preserved when power is removed.

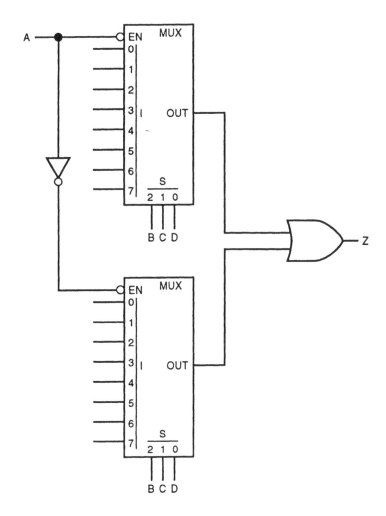

FIGURE 6.20.   Implementation of a 16-to-1-line MUX from two 8-to-1-line MUXs and an OR gate.

To program a ROM it is not important to know just how the word lines and output lines are connected, but it is vital to know where they are connected. Let a large dot at the intersection of a word line and an output line indicate a suitable electrical connection, and the absence of such a dot indicate no connection at that intersection. A three-input to four-output ROM is shown in Figure 6.29. A general ROM or PROM is shown in Figure 6.30, with n inputs, 2^n word lines, and m output lines.

A truth table can be used to specify the contents of a ROM, as can a diagram indicating the electrical connections. The truth table for a three-bit decoder to output one of eight four-bit words is shown in Table 6.11. A data file can be used to specify the truth table to be stored in the ROM when it is programmed. The data

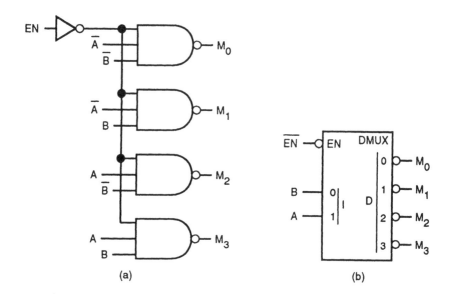

(a)   (b)

FIGURE 6.21. (a) A 2-to-4-line decoder with complemented outputs and (b) the symbol of this 2-to-4-line decoder.

file usually gives the address and data values in hexadecimal. Reading from left to right, from Z_3 to Z_0, a data file could specify Table 6.11 by stating that addresses 0 through 7 store the values C, A, 8, 6, 5, 7, 9, and 3, respectively.

The ROM shown in Table 6.11 has four outputs which are obtained in an SOP form by reading the minterms that are asserted true for each function in turn. Thus,

$$Z_3 = \sum m\,(0,1,2,6),\; Z_2 = \sum m\,(0,3,4,5),$$
$$Z_1 = \sum m\,(1,3,5,7),\; Z_0 = \sum m\,(4,5,6,7) \tag{6.11}$$

TABLE 6.11

A	B	C	Z_3	Z_2	Z_1	Z_0
0	0	0	1	1	0	0
0	0	1	1	0	1	0
0	1	0	1	0	0	0
0	1	1	0	1	1	0
1	0	0	0	1	0	1
1	0	1	0	1	1	1
1	1	0	1	0	0	1
1	1	1	0	0	1	1

The above functions are not the output words. The output words are obtained by observing the values of Z_3, Z_2, Z_1, and Z_0 when a particular input code is presented to the ROM. Thus, the output words, of four bits each, are obtained from

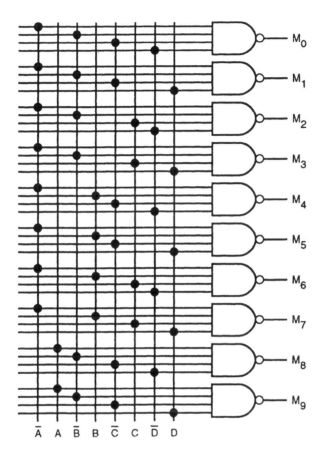

$\overline{A}$ A $\overline{B}$ B $\overline{C}$ C $\overline{D}$ D

FIGURE 6.22. A 4-to-10-line decoder to convert BCD input to decimal output, enable input omitted.

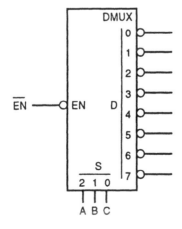

FIGURE 6.23. A 1-to-8-line DMUX with select inputs A, B, and C, and enable input $\overline{EN}$.

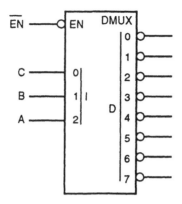

FIGURE 6.24. A 3-to-8-line decoder with an active-low enable input.

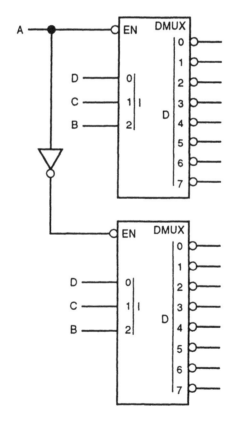

FIGURE 6.25. A 4-to-16-line decoder constructed of two 3-to-8-line decoders.

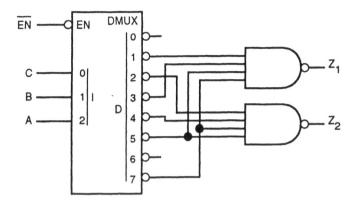

FIGURE 6.26. Realization of the two-output network with a decoder and NAND gates.

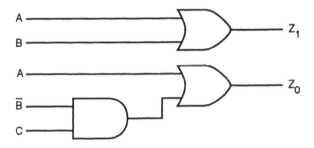

FIGURE 6.27. Circuit to realize the three-input priority encoder of Table 6.10.

Table 6.11 by reading the row of the truth table corresponding to the minterm that was input. A PROM realization of Equation 6.11 is shown in Figure 6.31.

Since the decoder consists of a series of AND gates which output the minterms of the select inputs, and the memory array can be represented as a series of OR gates, the overall ROM can be seen to consist of an AND input plane and an OR output plane. This can be realized with either NAND-NAND logic or NOR-NOR logic, depending upon whether the output is to be asserted high or low.

ROMs are typically used for code converters, character generators, and data storage tables. In each case, tabular data is generally preferred over boolean equations.

6.10 PROGRAMMABLE ARRAY LOGIC DEVICES

The term *programmable (array) logic device, or PLD*, refers to any structured logic device whose function is specified by the user after the device is manufactured. The programmed ROM or PROM is a read-only memory, as discussed

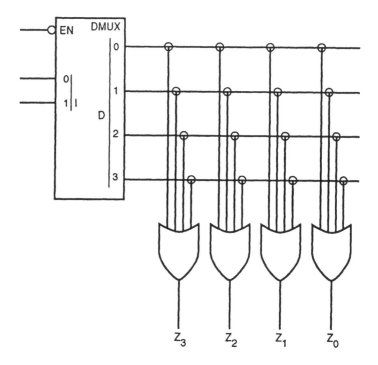

FIGURE 6.28. A 2-to-4-line decoder driving an array of four OR gates.

above, which can be programmed either electrically or with ultraviolet light, and is thus a PLD. The *programmable logic array* or *PLA* and the *programmable array logic* or *PAL* device are also PLDs.

PLDs are two-level combinational logic networks. The first level performs AND logic and the second level performs OR logic. If the outputs are asserted high, the PLD performs AND-OR logic, and the functions should be written in an SOP format. If the outputs are asserted low, the PLD performs AND-OR-IN-VERT or AOI logic, and the functions should be written in a POS format.

In SSI-MSI design, one of the main advantages of PLDs is the reduction in the number of IC chips required in a digital circuit. Chip count is proportional to cost, and fewer chips also lead to simpler printed-circuit designs, which are more reliable.

The three types of PLDs are shown symbolically in Figures 6.32 through 6.34. In each case solid dots represent fixed interconnections and open circles represent programmable sites of the device. Input and output registers are shown also. These registers must be clocked in order for the PLD to work properly. Clocking will be discussed in Chapter 7.

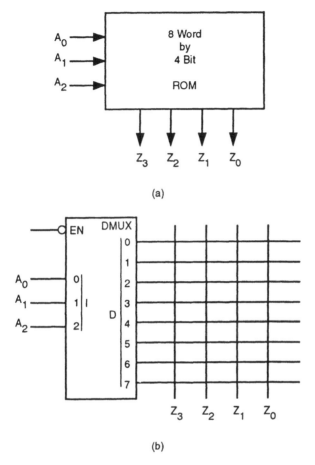

(a)

(b)

FIGURE 6.29. A 3-to-8-line decoder that outputs minterms only, plus an eight-word by four-bit memory array is a ROM. (a) symbol and (b) layout.

6.10.1   The Programmable Read-Only Memory

The field-programmable ROM (FPROM) is identical to the ROM as discussed above, except that it is field programmable by the user. It has a programmed AND plane which outputs all the minterms of the input variables and a programmable OR plane. It can be electrically erasable (EEPROM) or ultraviolet erasable (EPROM) and reusable. The PROM is ideally suited to functions of few variables, since the number of minterms increases rapidly with increasing variable count. It is usually specified by a truth table and is easy to program, but it is not as flexible as a PLA. The EPROM or EEPROM can be quickly reprogrammed when breadboarding circuits and making modifications.

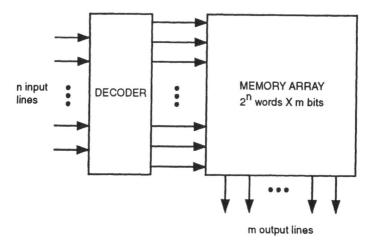

m output lines

FIGURE 6.30. A $2^n \times m$ ROM.

6.10.2 The Programmable Logic Array

The first PLA was introduced in 1975. The PLA performs the same basic function as a PROM, but usually with much less hardware. The PLA consists of an address decoder AND input plane and a data matrix OR output plane, as does the PROM. However, the PLA utilizes a programmable input AND plane, which can realize minimum SOP solutions rather than cannonical minterm SOP solutions. For this reason the PLA is used to implement functions in equation form rather than in tabular form, as was the case for the PROM. The PLA is widely used today in VLSI design, because it realizes digital circuits with a minimum of hardware.

A PROM requires one address line for each input variable, and each added variable doubles the number of minterms generated by the PROM. PLAs are ideally suited to the realization of functions of many variables, but few product terms. Product terms can be shared by PLA output lines since the number of product terms available for each output of a PLA is programmable. The major disadvantage of PLAs as compared to PROMs and PALs is that they are slower due to longer propagation delay.

The size of a PLA is specified by the number of inputs, the number of product lines, and the number of output lines. Each input line can be connected to each product line, and each product line can be connected to each output line. The sites of these potential connections are referred to as *crosspoints,* and the overall size of the PLA is a function of the total number of crosspoints.

Since all PLDs consist of an AND plane driving an OR plane, it is not necessary to show the AND gates and OR gates in a diagram of the PLD. A simplified layout of the PROM of Figure 6.31 is shown in Figure 6.35. The

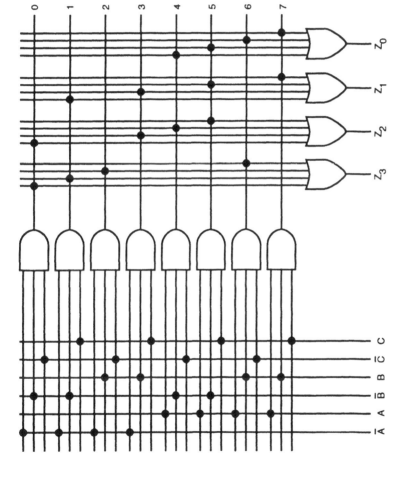

FIGURE 6.31.   The realization of Z_0, Z_1, Z_2, and Z_3 as a PROM.

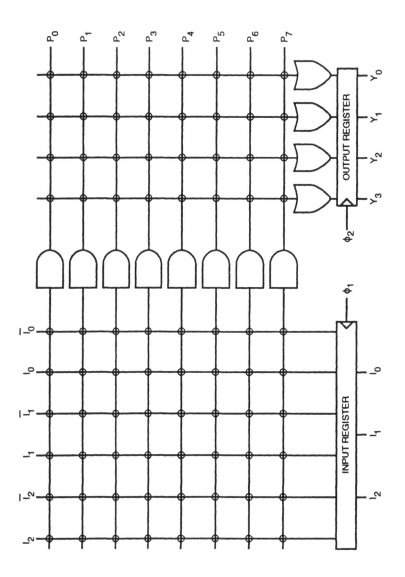

FIGURE 6.32.   Representation of a PLA device with both planes programmable.

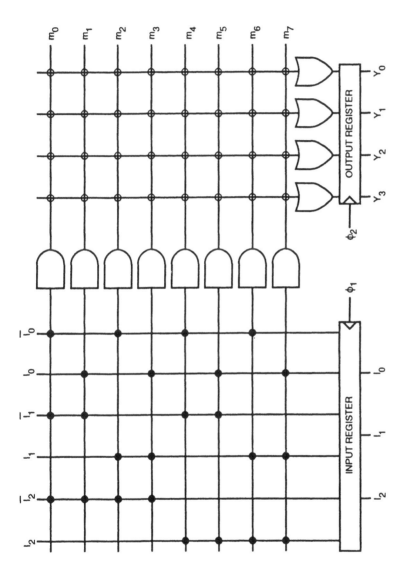

FIGURE 6.33. Representation of a PROM device with programmable OR plane.

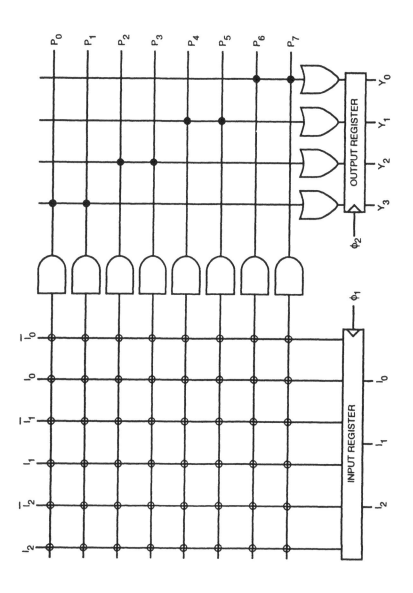

FIGURE 6.34. Representation of a PAL device with programmable AND plane.

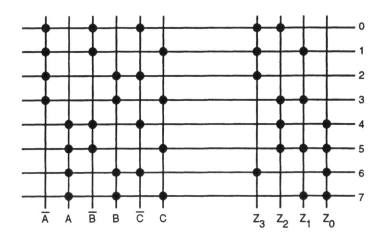

FIGURE 6.35. Reduced or simplified representation of the PROM of Figure 6.31.

pertinent information is contained in the crosspoints that are to be connected, as shown by dots.

A typical small commercial PLA has 16 inputs, 48 product lines, and 8 outputs. It thus has $2 \times 16 \times 48 = 1536$ crosspoints in the AND plane, and $8 \times 48 = 384$ crosspoints in the OR plane, for a total of 1920 crosspoints.

EXAMPLE 6.8:   Implement the following four functions with a PLA.
$$Z_3 = \overline{A}\,\overline{B} + \overline{A}\,C$$
$$Z_2 = BC + AB$$
$$Z_1 = \overline{B}$$
$$Z_0 = A$$

SOLUTION: First the functions are mapped, as shown in Figure 6.36, and rewritten to require the least number of input gates.

$$Z_3 = \overline{A}\,\overline{B} + \overline{A}\,BC \qquad\qquad Z_2 = \overline{A}\,BC + AB$$
$$Z_1 = \overline{A}\,\overline{B} + A\overline{B} \qquad\qquad Z_0 = A\overline{B} + AB$$

The PLA is implemented in Figure 6.37.

In Figure 6.37 the PLA is symbolically represented by an AND plane and an OR plane, with connections indicated by dots, as was done for the output plane of the PROM. The input lines must be clearly labeled.

The PLA only requires four product lines, whereas separately simplified maps for the four functions in Example 6.8 require six product terms, $\overline{A}\,\overline{B}$, $\overline{A}\,C$, BC, AB, $\overline{B}$, and A, and a ROM requires eight word lines (eight minterms) by four output lines.

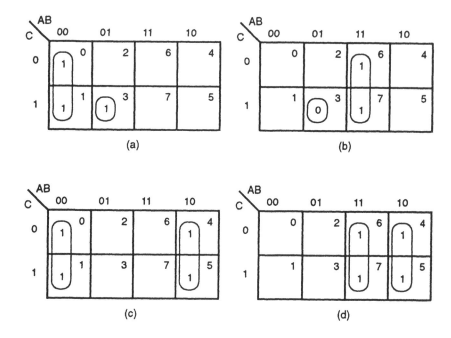

FIGURE 6.36.   Maps of the functions (a) Z_3, (b) Z_2, (c) Z_1, and (d) Z_0.

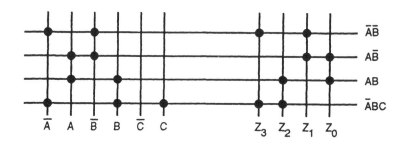

FIGURE 6.37.   PLA realization of the functions Z_3, Z_2, Z_1, and Z_0.

6.10.3   Programmable Array Logic

The PAL device was developed by John Birkner at Monolithic Memories, Inc., in 1976. As with the PLA, PALs are generally used to implement functions in equation form rather than in tabular form. A ROM with 20 inputs has $2^{20} = 1,048,576$ word lines. PLAs and PALs can have up to 20 inputs and as few as 20 to 50 product lines, but the number of product terms available to each PAL OR gate configuration is fixed during manufacture and cannot be changed by the user. This limits the usefulness of PALs compared to PLAs, but it also limits the cost and allows PALs to operate at higher speeds than equivalent PLAs.

EXAMPLE 6.9: Program a PAL to realize the functions $Y_3 = AB + \overline{A}C$, $Y_2 = \overline{A}C + A\overline{B}\,\overline{C}$, and $Y_1 = AB + \overline{A}BC$.

SOLUTION: The PAL requires two product lines for each function, for a total of six product lines, as shown in Figure 6.38. The last two product lines are not used.

In Example 6.9, product $\overline{A}C$ has to be an input to the Y_3 and Y_2 OR gates, and product AB has to be an input to the Y_3 and Y_1 OR gates. It is necessary to duplicate both of these products since no two OR gates are connected to the same product lines. This is a disadvantage of the PAL with respect to the PLA.

Different PALs have different numbers of inputs (product lines) to their OR gates. The PAL16L8 has 64 rows and 32 columns or 2048 crosspoints in its AND array. It can handle a maximum of 16 input variables and a maximum of 8 dedicated OR inputs or product lines for each of 8 OR gates. This device can implement up to eight boolean functions, each with up to ten input variables, and each equation can contain up to seven product terms. Furthermore, up to six of the outputs can be programmed as inputs, giving it much more flexibility than a PROM. This allows the PAL to implement 2 logic functions with up to 16 input variables, with each output containing up to 7 product terms.[4]

6.11 SUMMARY

MUXs, or data selectors, are the logic-circuit equivalent of rotary selector switches that have many input lines and a single output line. They select one of many data input lines and route it to an output line, the routing being done by select or control inputs. They often require fewer interconnections than other designs and they are easy to use. MUXs are among the most versatile design tools available, and they are easy to troubleshoot.

DMUXs can also be viewed as logic-circuit rotary selector switches that have one input and many outputs. They take a single data input and route it to one of many output lines, the routing again being done by select or control inputs.

Binary decoders are circuits that convert binary-coded inputs into a more readily recognized character. The most common commercial decoders are 4-to-10-line (4-line to 1-of-10-line) decoders that convert BCD or XS3 codes to decimal numbers, and 4-to-16-line (4-line to 1-out-of-16-line) decoders that can convert any four-bit character to hexadecimal or Gray code. A BCD-to-seven-segment decoder accepts a four-bit BCD code input and provides the circuitry to drive a seven-segment display.

Binary encoders can convert a single input such as that from a keyboard to a binary output corresponding to the key pressed. When two or more inputs can be activated at the same time, a priority encoder must be used. Most commercially available encoders include a priority encode feature that outputs the input signal of higher priority when two or more inputs are activated simultaneously. Two

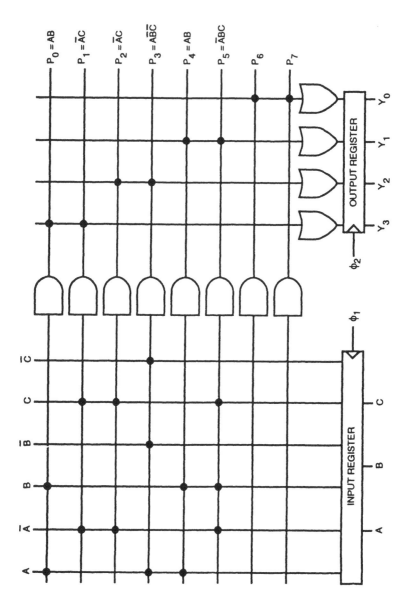

FIGURE 6.38. PAL implementation of Y_3, Y_2, and Y_1.

commercially available priority encoders are an 8-to-3-line and a 10-to-4-line encoder.

Any two-level logic function can be represented in an SOP form which can be realized with an AND array that feeds an OR array. Such an array is called a programmable logic device or PLD. If both the AND plane and the OR plane are programmable, the device is called a programmable logic array or PLA. It is the most versatile PLD, but can be bulky and slow. The programmable array logic device or PAL has a fixed OR plane and a programmable AND plane, while the programmable read-only memory or PROM is the least versatile of the three programmable devices. The PROM realizes a canonical or minterm expansion of the function, which is usually the maximal nonredundant solution, but it is the easiest to program.

An $n \times m$ ROM has n inputs, 2^n word lines, and m outputs, as shown in Figure 6.33. An $n \times m$ PLA usually has far fewer than 2^n product lines and occupies much less space than an $n \times m$ ROM. The PLA realizes a minimal SOP solution of the output functions.

Each input to a ROM is a minterm and selects one output word line. In a PLA or PAL, the inputs are product terms, and zero, one, or more than one word line can be selected by each combination of input values.

When the number of input variables is small, a ROM is easier to program and might be more economical than a PLA, provided enough minterms are required. When the number of input variables is large, the PLA is frequently much smaller and, hence, more ecomonical than the ROM. This is because the ROM has to decode all the minterms and doubles in size for each added input variable. A function of 10 inputs would require a ROM with $2^{10} = 1024$ word lines, and a function of 12 inputs would require a ROM of 4096 word lines. If the function of 12 inputs could be simplified, for example, to 138 product terms, the PLA realization would only require 138 product lines or 3.4% of the word lines of the ROM.

The inputs and outputs of a PLD are normally clocked through an input register and an output register. Clocking and the need for it are discussed in Chapter 7.

PROBLEMS

6.1. Design a 4-to-1-line MUX to implement the functions below. The select variables are A and B. Show the three-variable Karnaugh mapping and a reduced two-variable Karnaugh mapping of each function.

(a) $Z_1 = \sum m (0,1,5,6)$

(b) $Z_2 = \sum m (0,3,5,6)$

6.2. (a) Implement the function $Z_3 = \sum m (0,1,2,4,5,6,7,10,13,15)$ with an 8-to-1-line MUX having select inputs A, B, and C. Show the four-variable Karnaugh mapping and a reduced three-variable Karnaugh mapping.

(b) Repeat part a for select inputs A, B, and D.

6.3. (a) Implement the function $Z_4 = \sum m (0,1,4,5,7,8,9,10,13,15)$ with an 8-to-1-line MUX having select inputs A, B, and C. Show the four-variable Karnaugh mapping and a reduced three-variable Karnaugh mapping.

(b) Repeat part a for select inputs A, B, and D.

6.4. (a) Implement the function Z_3 of Problem 6.2 with a 4-to-1-line MUX with select inputs A and B. Show the MUX and any additional input circuitry needed.

(b) Implement the function Z_4 of Problem 6.3 with a 4-to-1-line MUX with select inputs A and B. Show the MUX and any additional input circuitry needed.

6.5. (a) Implement the function $Z_5 = \sum m (0,1,2,4,5,7,10,15)$ with an 8-to-1-line MUX with select inputs A, B, and C. Show the MUX and any necessary input gates.

(b) Repeat part a for a 4-to-1-line MUX with select inputs A and B.

6.6. (a) Implement the function $Z_6 = \sum m (0,3,4,7,9,10,13,14)$ with a 4-to-1-line MUX and select inputs C and D. Show the MUX and any necessary input gates.

(b) Repeat part a for a 4-to-1-line MUX with select inputs A and B. You can supplement the MUX with one EOR/ENOR gate.

6.7. (a) Implement the function $Z_7 = \sum m (1,3,4,7,9,10,12,14)$ with an 8-to-1-line MUX with select inputs A, B, and D. Show the MUX and any necessary input gates.

(b) Repeat part a for a 4-to-1-line MUX with select inputs A and B.

6.8. (a) Implement the function $Z_8 = \sum m (0,2,5,7,8,10,12,14)$ with an 8-to-1-line MUX and select inputs B, C, and D. Show the MUX and any necessary input gates.

(b) Repeat part a for a 4-to-1-line MUX with select inputs A and B.

6.9. (a) Implement the function $Z_9 = \sum m (0,3,4,6,8,10,12,15)$ with an 8-to-1-line MUX with select inputs B, C, and D. Show the MUX and any necessary input gates.

(b) Repeat part a for a 4-to-1-line MUX with select inputs C and D.

6.10. (a) Implement the function $Y_0 = \sum m (3,4,6,8,10,12,15)$ with an 8-to-1-line MUX with select inputs B, C, and D. Show the MUX and any necessary input gates.

(b) Repeat part a for a 4-to-1-line MUX with select inputs A and B.

6.11. (a) Implement the function $Y_1 = \sum m\,(3,4,10,11,13,14,15)$ with an 8-to-1-line MUX and select inputs A, C, and D. Show the MUX and any necessary input gates.

(b) Repeat part a for an 8-to-1-line MUX and select inputs A, B, and D.

6.12. (a) Implement the function $Y_2 = \sum m\,(0,3,5,7,8,10,13,14)$ with an 8-to-1-line MUX with select inputs A, C, and D. Show the MUX and any necessary input gates.

(b) Repeat part a for an 8-to-1-line MUX with select inputs B, C, and D.

6.13. Realize each 8-to-1-line MUX of Problem 6.2 as a tree of two 4-to-1-line MUXs driving a 2-to-1-line MUX.

6.14. Realize each 8-to-1-line MUX of Problem 6.3 as a tree of two 4-to-1-line MUXs driving a 2-to-1-line MUX.

6.15. Realize the 8-to-1 line MUX of Problem 6.5 as a tree of two 4-to-1-line MUXs driving a 2-to-1-line MUX.

6.16. Realize the 8-to-1-line MUX of Problem 6.7 as a tree of two 4-to-1-line MUXs driving a 2-to-1-line MUX.

6.17. Realize the 8-to-1-line MUX of Problem 6.8 as a tree of two 4-to-1-line MUXs driving a 2-to-1-line MUX.

6.18. Realize the 8-to-1-line MUX of Problem 6.9 as a tree of two 4-to-1-line MUXs driving a 2-to-1-line MUX.

6.19. Implement a full-adder circuit to add A and B, with two 4-to-1-line MUXs. Show the circuits for the two MUXs constructed with NAND-NAND logic.

6.20. Realize (implement) a circuit of AND and OR gates to convert from XS3 code to BCD. The conversion table is shown below.

BCD				XS3			
A	B	C	D	a	b	c	d
0	0	0	0	0	0	1	1
0	0	0	1	0	1	0	0
0	0	1	0	0	1	0	1
0	0	1	1	0	1	1	0
0	1	0	0	0	1	1	1
0	1	0	1	1	0	0	0
0	1	1	0	1	0	0	1
0	1	1	1	1	0	1	0
1	0	0	0	1	0	1	1
1	0	0	1	1	1	0	0

6.21. Realize the functions $F_1 = \prod M\,(0,1,2)$ and $F_2 = \sum m\,(0,2,3)$ with a 2-to-4-line decoder that outputs both minterms and maxterms. One AND gate and one OR gate will be required to augment the MUX.

6.22. (a) Realize (implement) a circuit that uses a 4-to-10 line decoder and NAND gates to convert from BCD to XS3 code. The conversion table is shown above in Problem 6.20.

   (b) Redraw the BCD-to-XS3 code converter of part a as a reduced ROM layout.

6.23. Realize (implement) a circuit that uses EOR gates to convert from binary to Gray code. The conversion table is shown below. To facilitate the design, observe the following:

   1. The MSB of the Gray code is the MSB of the binary number (a = A).
   2. The second MSB in the Gray code is the EOR of the two MSBs in the binary number (b = A $\oplus$ B).
   3. The third MSB in the Gray code is the EOR of the second and third MSBs of the binary number (c = B $\oplus$ C).
   4. Continue until the LSB is reached (Specifically, d = C $\oplus$ D).

Binary				Gray			
A	B	C	D	a	b	c	d
0	0	0	0	0	0	0	0
0	0	0	1	0	0	0	1
0	0	1	0	0	0	1	1
0	0	1	1	0	0	1	0
0	1	0	0	0	1	1	0
0	1	0	1	0	1	1	1
0	1	1	0	0	1	0	1
0	1	1	1	0	1	0	0
1	0	0	0	1	1	0	0
1	0	0	1	1	1	0	1
1	0	1	0	1	1	1	1
1	0	1	1	1	1	1	0
1	1	0	0	1	0	1	0
1	1	0	1	1	0	1	1
1	1	1	0	1	0	0	1
1	1	1	1	1	0	0	0

6.24. Realize (implement) a Gray-code-to-binary converter with EOR gates. The conversion table is shown above. As in Problem 6.23, observe the following:

   1. The MSB in the binary number is the MSB in Gray code (A = a).
   2. The second MSB in binary is the EOR of the two MSBs in the Gray code (B = a $\oplus$ b. Also, B = A $\oplus$ b).

3. The third MSB in binary is the EOR of the second binary bit and the third Gray code bit ($C = B \oplus c$).

4. All succeeding binary bits are equal to the EOR of the corresponding Gray code bit and the next-most significant binary bit (Specifically, $D = C \oplus d$).

6.25. Design a priority encoder to give top priority to I_2, followed by I_1, with lowest priority given to I_0. The truth table for this encoder is given below.

Inputs			Outputs	
I_2	I_1	I_0	X_1	X_0
0	0	0	0	0
0	0	1	1	1
0	1	0	1	0
0	1	1	1	0
1	0	0	0	1
1	0	1	0	1
1	1	0	0	1
1	1	1	0	1

6.26. Construct a ROM look-up table that converts input I to $I^2 + 2I$, for $I = 0$ to 7. Realize the ROM with 8-to-1 multiplexers. The select inputs are A, B, and C.

6.27. Design a PLA to implement the look-up table of Problem 6.26.

6.28. The internal connection diagram of a PLA is shown in Figure 6.39.
   (a) Write the equations realized by the PLA.
   (b) Obtain the truth table for a ROM that would realize the functions of part (a).

6.29. Implement (realize) a BCD-to-XS3 code converter with a PLA. The conversion table is shown below Problem 6.20.

6.30. Design a PLA to implement the following functions:

$$Z_0 = \sum m\,(0,5,10,14,15) + \sum d\,(1,11)$$

$$Z_1 = \sum m\,(0,1,4,10,11,12,15) + \sum d\,(3,14)$$

$$Z_2 = \sum m\,(1,4,7) + \sum d\,(2,5,10,12)$$

$$Z_3 = \sum m\,(1,5,7,11,14,15) + d_{10}$$

SPECIAL PROBLEMS

6.31. Program a PROM that converts BCD inputs to seven-segment display outputs. The truth table and LED segment identification are given in Chapter 4, Special Problem 4.47.

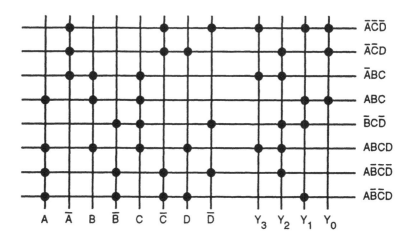

FIGURE 6.39.

6.32. Program a PROM that converts hexadecimal inputs to seven-segment display outputs. The truth table is given in Chapter 4, Special Problem 4.48.

REFERENCES

1. **Sandige, R. S.,** *Modern Digital Design,* McGraw-Hill, New York, 1990, 334.
2. **Breeding, K. J.,** *Digital Design Fundamentals,* Prentice-Hall, Englewood Cliffs, NJ, 1989, 258–263.
3. **Wakerly, J. F.,** *Digital Design Principles and Practices,* Prentice-Hall, Englewood Cliffs, NJ, 1990, 271–278.
4. **Sandige, R. S.,** *Modern Digital Design,* McGraw-Hill, New York, 1990, 384–387.

CHAPTER 7

Latches and Flip Flops

7.1 INTRODUCTION

Thus far only combinational logic has been considered. Combinational circuits can only combine present input signals to produce an output, and the outputs change shortly after the inputs change.

Sequential logic is required if data is to be stored for future use. Sequential circuits have output signals which depend upon the past history of the circuit. A digital clock is a sequential logic system which must remember its present state so that it advances to the correct time every minute. It must also be able to count seconds, keep track of how many seconds have transpired in order to know when to change the minute indicator, and keep track of minutes in order to know when to change the hour indicator.

In ANSI/IEEE* Standard 91-1984, bistable devices are divided into four catagories: transparent latches, edge-triggered flip-flops, pulse-triggered (master/slave) flip-flops, and data lockout flip-flops.

Latches and flip-flops are the basic building blocks of most sequential circuits. An unclocked latch is a sequential device that can change its output at any time, whereas a clocked latch and a flip-flop can only change their outputs after an appropriate clock signal occurs. An asynchronous latch "grabs" or "latches" onto the input information, and the outputs of the latch will change after the signal has had time to propagate through the latch. A *clocked* latch or *gated* latch is a synchronous latch whose outputs change after the clocking or gating signal occurs. The four basic latches are the set-reset (S-R), the data, the toggle, and the J-K latches.

The master/slave flip-flop is similar to a gated S-R latch except that it uses the changing state of the clock signal to change the output. The master/slave flip-flop is preferred to a gated latch because it eliminates the problem of race. A race is

---

* ANSI — American National Standards Institute; IEEE — Institute of Electrical and Electronic Engineers.

said to occur whenever the outputs of a latch switch more than once on a single clock signal.

While the master/slave S-R flip-flop eliminates race problems, there is still the problem of what happens when both input signals are activated at the same time. The JK flip-flop is designed to toggle or switch states when both inputs are activated.

Direct set and clear inputs can be added to a flip-flop. These are asynchronous inputs that override the clock and synchronous data inputs. Direct clear inputs can be used to clear all the flip-flops of a circuit whenever power is applied to the circuit.

The master/slave flip-flop can switch due to noise spikes on the data lines. The data-lockout master/slave and the edge-triggered J-K flip-flop avoid this problem because their outputs change only in response to the input data present at the much shorter time of the positive or negative edge of the clock. They are as close to flawless as flip-flops can be.

7.2 BASIC ASYNCHRONOUS BINARY LATCHES

Any sequential logic system must have memory elements that store past events, as well as combinational logic that allows it to generate new outputs from the stored events and the present data. The most basic digital memory device is the *latch*, a simple circuit, which "latches" or stores one bit of information and remembers it. In its simplest form, the latch is analogous to a toggle switch such as a light switch on the wall. The toggle switch must "remember" the state it is in and leave the lights either on or off until an input signal "toggles" the switch and it switches the lights to the opposite state. The state of the wall switch could be considered to be the state of the light bulb. Thus, the switch could be said to be in the on state when the light is lit, and in the off state when the light is not lit.

A simple asynchronous *set-latch* can be made from a single OR-gate by feeding the output back to one input, as shown in Figure 7.1(a). If the output is initially at logical 0, a logical 1 on the input line will cause the output to transition from 0 to 1. The input must be held high sufficiently long to allow the output of the gate to respond and go high. The time lag before the output changes in response to an input change is referred to as the *propagation delay, t_{pd}*, of the gate. Propagation delay for the output to change from low to high is generally longer than the propagation delay for an output transition from high to low, and the manufacturer specifies a typical and a worst-case propagation delay for each device.

If the input is held high for a time greater than the propagation delay, the timing diagram showing the set input and the Q output as functions of time are shown in Figure 7.1(b). The latch is set at time t_s, and remains in the set or 1 state when the SET signal is removed. A 1 has been latched by the OR gate.

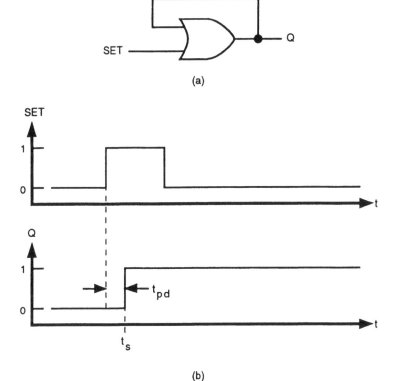

(a)

(b)

FIGURE 7.1. The basic OR latch: (a) the circuit, and (b) the timing diagram.

7.2.1 The Active-High Asynchronous S-R Latch

Two NOR gates can be used to construct a set latch and output Q and $\overline{Q}$. The basic latch is shown in Figure 7.2(a), and the timing diagram in Figure 7.2(b). The pulse-width of the set input must be greater than the time for the signal to propagate through gate 2 and change $\overline{Q}$, plus the time for $\overline{Q}$ to propagate through gate 1 and latch the output. The latch is again set at time t_s, which is now two propagation delays or $2t_{pd}$ after the set signal is applied.

A minimum-width data pulse is required to insure proper response of the latch. This can be thought of as the minimum time for the signal to traverse the feedback loop and latch the output. Since the signal requires two propagation delays to be fed back to the input, the minimum pulse width must be more than $2t_{pd}$.

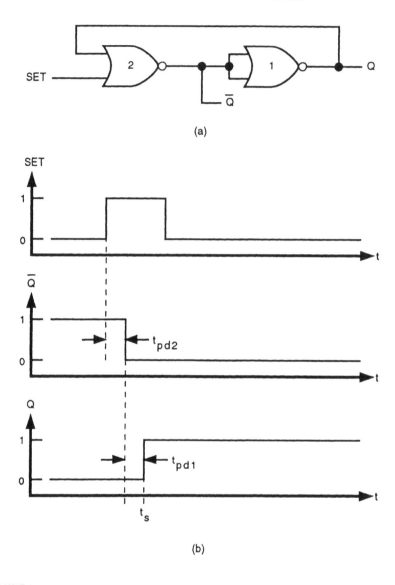

FIGURE 7.2.   The NOR latch: (a) the circuit, and (b) the timing diagram.

The latch still suffers from a serious deficiency though. Once the output goes high, it "latches" there, and the set input has lost control and cannot reset the output back to 0. This can easily be corrected by using one input of gate 1 to reset the latch to the output-0 state. The *set-reset (S-R)* latch is shown in Figure 7.3(a), with the reset input labeled R, and can now be set or reset. A suitable timing diagram is shown in Figure 7.3(b).

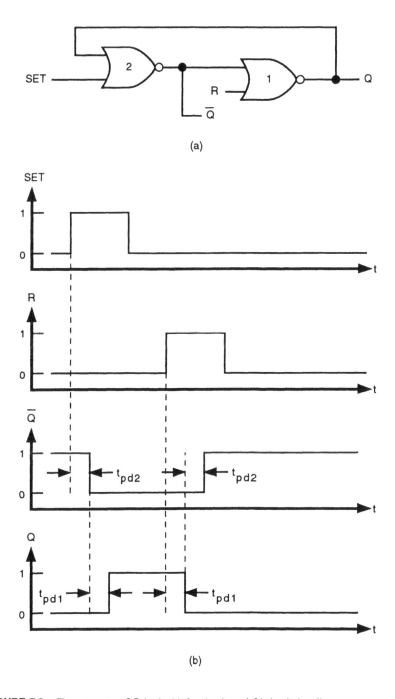

FIGURE 7.3.   The set-reset or S-R latch: (a) the circuit, and (b) the timing diagram.

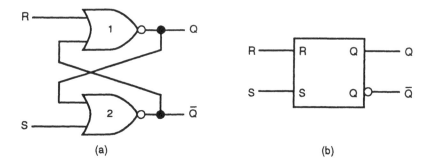

FIGURE 7.4. (a) The cross-coupled NOR latch, and (b) the S-R latch symbol.

The S-R latch is the basic digital static-memory circuit. It is sometimes called a *bistable multivibrator*, since it has two stable output states and can switch between them repeatedly. It is also called a *toggle*, because the output changes by "toggling", and a *binary*, since it has two stable states. The cross-connected NOR gates are usually drawn as in Figure 7.4(a), and the symbol is as shown Figure 7.4(b).

ANSI/IEEE Standard 91-1984 specifies an *internal logic* state which exists inside a symbol outline and an *external logic* state which exists outside a symbol outline. In Figure 7.4(b) there are two latch outputs whose internal logic states are Q, one asserted high and one asserted low. For positive logic, the external logic state which is asserted high is labeled Q, and the external logic state which is asserted low is labeled $\overline{Q}$.

The state of the latch is the value of external output state Q. When Q = 1, the latch is said to be in the set state, and when Q = 0, the latch is said to be in the reset or clear state. The truth table to determine the next state of the latch, labeled Q^+, is shown in Table 7.1, from which it can be seen that the latch remains in whatever state it is in as long as S = R = 0. It is said to be *bistable* in the sense that it is stable in either output state as long as S = R = 0. It is in the *idle* or remember mode and retains the value of 1 or 0 stored in it. A high signal at S sets the latch, and a high signal at R clears the latch.

TABLE 7.1

S	R	Q^+	Comment
0	0	Q	No change
0	1	0	Clear latch
1	0	1	Set latch
1	1	X	Not allowed

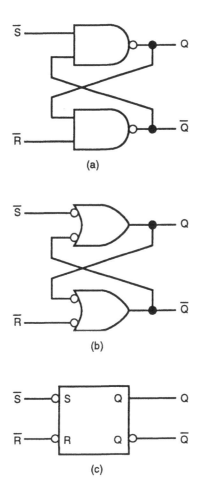

FIGURE 7.5.   The active-low S-R latch: (a) the positive-logic NAND circuit, (b) the negative-logic
NOR circuit, and (c) the symbol.

One cannot simultaneously set and clear the latch; hence, $S = R = 1$ is not
allowed. If $S = R = 1$ is input to the latch, both Q and $\overline{Q}$ will be forced to the 0
state. $Q = \overline{Q}$ is bad boolean logic, but perhaps even worse is the fact that if S and
R both change from 1 to 0 simultaneously, the output state of the latch is
indeterminate. If the gate whose output is Q switches first, the latch will be set and
$\overline{Q}$ will remain 0. If the gate whose output is $\overline{Q}$ switches first, the latch will be
cleared and Q will remain 0. The next state of the latch depends upon which gate
is faster or which input changed from 1 to 0 a nanosecond or two sooner. This is
referred to as a *race* condition.

7.2.2 The Active-Low Asynchronous S-R Latch

If the cross-coupled NOR gates are replaced by cross-coupled positive-logic NAND gates, as shown in Figure 7.5(a), which are negative-logic NOR gates, as shown in Figure 7.5(b), one obtains an active-low latch. The circuit diagram for the active-low latch has active-low inputs labeled $\overline{S}$ and $\overline{R}$ in Figure 7.5, and the symbol requires invert bubbles at both inputs, as shown in Figure 7.5(c). In Figure 7.4(b), the external and internal logic states of the inputs are the same, whereas in Figure 7.5(c), they differ for positive logic, but would be the same for negative logic.

The positive-logic truth table for the latch in Figure 7.5 is shown in Table 7.2. The behavior of an active-low S-R latch for negative logic is the same as that of an active-high S-R latch for positive logic. In particular, minimum pulse-width requirements are the same. The memory state or idle state is now SR = 00 or $\overline{S}\,\overline{R}$ = 11, the not-allowed inputs are SR = 11 or $\overline{S}\,\overline{R}$ = 00, a low signal at S sets the latch, and a low signal at R clears it.

Table 7.2 is the negative-logic equivalent of Table 7.1. If two inverters are used with the cross-coupled NAND gates, the latch is changed into a positive-logic S-R latch, as shown in Figure 7.6. In this case the next-state table becomes the same as that of Table 7.1.

The *characteristic table* or *present state/next state table* of a sequential system lists its inputs, its present state, and its next state, which completely describes its behavior. The characteristic table for an S-R latch is shown in Table 7.3. This is basically the same as Tables 7.1 and 7.2, with an additional input column for the present state, Q, and eight rows. The *present/state next state map* and the *characteristic equation* are two additional ways of describing the operation of a latch or flip-flop. The same information is contained in all three.

From the next-state map in Figure 7.7 the characteristic equation of the S-R latch is seen to be $Q^+ = S + \overline{R}\cdot Q$ and $S\cdot R = 0$.

TABLE 7.2

	$\overline{S}$	$\overline{R}$	Q^+	Comment
	0	0	X	Not allowed
	0	1	1	Set latch
	1	0	0	Clear latch
	1	1	Q	No change

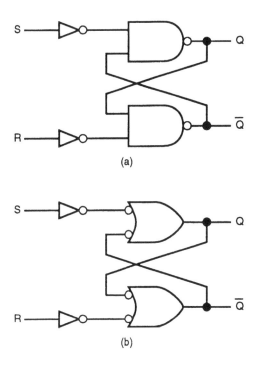

FIGURE 7.6. (a) The S-R latch constructed of two NAND gates plus two inverters, and (b) the latch with matched assertion levels.

TABLE 7.3

S	R	Q	Q^+
0	0	0	0
0	0	1	1
0	1	0	0
0	1	1	0
1	0	0	1
1	0	1	1
1	1	0	X
1	1	1	X

In English, the characteristic equation states that the next value of output Q will be 1 if either the latch is already set or if the present value of output Q is 1 and it is not cleared. The auxilliary equation $S \cdot R = 0$ guarantees that S and R are not both asserted true simultaneously.

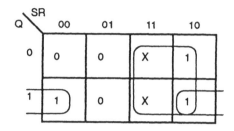

FIGURE 7.7. The next state map of an S-R latch.

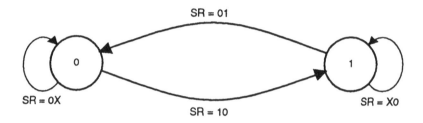

FIGURE 7.8. The state diagram of an S-R latch.

The *state diagram* provides a graphical representation of the characteristics of the latch by showing the possible transitions and is another method of displaying the same information. The state diagram of an S-R latch is shown in Figure 7.8. There are two states or output values of Q for the latch, 0 and 1. State 0 is indicated on the left, and state 1 is on the right. The directed line segments indicate transitions from the present state to the next state of the latch, based upon the inputs to the latch.

The transitions are labeled with the values of S and R that produce them. When the latch is in the 0 state and S = 0, the next state is 0 regardless of the value of R, and if the latch is in the 1 state and R = 0, the next state is 1 regardless of the value of S. These transitions are shown by the small loops in Figure 7.8. For the latch to change from the cleared or 0 state to the set or 1 state, the inputs must be SR = 10, and for the latch to change from the set or 1 state to the clear or 0 state, the inputs must be SR = 01.

7.3 THE CLOCKED OR GATED LATCH

For an asynchronous circuit to work properly, only one of the external input signals can be effective at any given time, and the circuit must have time to reach a steady-state value before the next input change occurs. Each external input signal becomes a timing event for that specific gate, and signals ripple through the

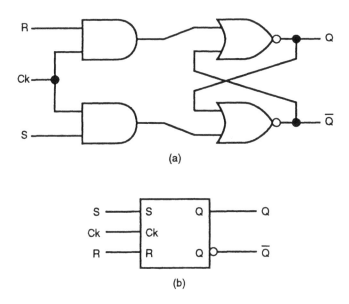

FIGURE 7.9. The AND-NOR gated S-R latch: (a) the circuit, and (b) the symbol.

circuit, making it very fast. Speed is the main advantage of asynchronous circuits. They are difficult to design properly and are only used when synchronous design will not work.[1]

Most digital systems operate in a *synchronous* mode, under the control of a *master clock signal.* The master clock generates a train of periodic pulses which are distributed throughout the system, allowing data to flow in a predetermined manner rather than ripple through the system. The clock usually generates a squarewave pulse train, although the actual triggering pulses seen by the circuit may be of very short duration. Such synchronous systems are referred to as *clocked systems* or *triggered systems.*

The three primary methods of clocking are level triggering, pulse triggering, and edge triggering. Level triggering is suitable for synchronous operation, but cannot always handle asynchronous input data. Storage devices which are level triggered are called latches, and storage devices that are pulse triggered or edge triggered are called flip-flops. Pulse triggering is satisfactory for asynchronous inputs, but is subject to noise problems. Edge triggering overcomes most timing problems and is the most reliable method of clocking flip-flops.

7.3.1 The Clocked S-R Latch

If the cross-connected NOR gates are augmented with two AND gates, the latch can be clocked as shown in Figure 7.9. The gated latch uses a level-sensitive gate or clock input which determines when the output data can change. The output

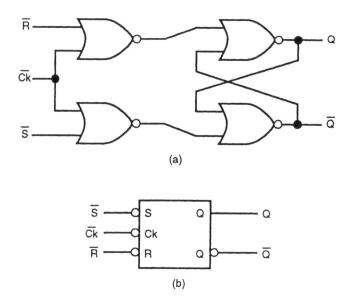

(a)

(b)

FIGURE 7.10. The NOR-NOR gated $\bar{S}$-$\bar{R}$ latch: (a) the circuit, and (b) the symbol.

is updated by the data inputs each time the clock is asserted, and the latch is said to be *synchronized* with the clock. The latch can also be said to be *gated* by the clock signal. The clock input is also called a *control* input since it controls the response from the latch.

The two AND gates in Figure 7.9 are referred to as *control* or *steering* gates since they steer the data to the latch and control the flow of data by forcing it to be synchronized with the clock. The clock signal enables the two control gates to respond to the data inputs. When the clock is asserted high, both AND gates are said to be *enabled*, and the latch follows the next-state table of an S-R latch. State changes occur when the clock is high for this circuit. When the clock is low, the steering gates are *disabled*, and the output of the latch remains unaffected by the data inputs S and R.

If the AND steering gates of Figure 7.9 are replaced by NOR steering gates, as shown in Figure 7.10, one has an active-low S-R latch which is enabled when the clock is low and is set or reset by the appropriate input going low. Changes in the latch output occur when the clock is low. This is shown in Figure 7.10(b) by the active-low bubble at the clock input. When the clock is high, the NOR gates are disabled and the latch retains its current value.

A clocked S-R latch can be constructed from four NAND gates also. The S and R inputs are reversed from those in Figure 7.9(a), but the circuit behaves the same as the circuit of Figure 7.9(a) and has the same symbol as in Figure 7.9(b). The NAND S-R latch is shown in Figure 7.11.

To see how the clocked latch works, consider the circuits shown in Figure 7.12. For the AND and NAND gates, an input which is low disables the gate, and the

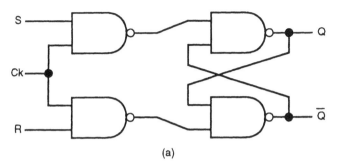

(a)

FIGURE 7.11. The NAND-NAND gated S-R latch.

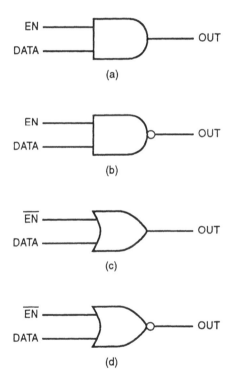

FIGURE 7.12. Enable inputs are active high for (a) AND and (b) NAND gates, and active low for (c) OR and (d) NOR gates.

output of the gate is low for the AND gate and high for the NAND gate, regardless of the other inputs. For the OR and NOR gates, an input which is high disables the gate, and the output is high for the OR gate and low for the NOR gate, regardless of the other inputs. The bar over EN indicates an active-low enable or active-high disable. A logic low to the AND and NAND gates, or a logic high to the OR and NOR gates, is said to *disable* the gates, rendering them unable to

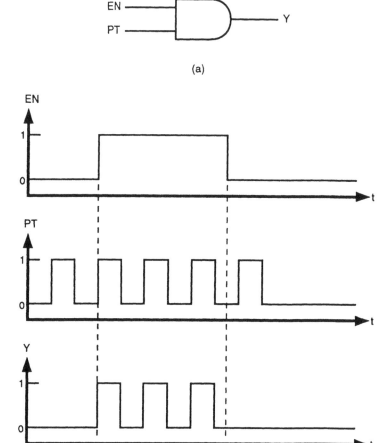

FIGURE 7.13. A pulse counter: (a) the circuit, and (b) the timing diagram.

respond to the other gate inputs. A logic high to the AND and NAND gates, or a logic low to the OR and NOR gates, is said to *enable* the gates, since they are now able to follow the other gate inputs.

EXAMPLE 7.1: Design a simple circuit to measure the pulse repetition rate (number of pulses per second) for a pulse-train input.

SOLUTION: Consider the AND gate of Figure 7.13(a), with inputs EN (enable) and PT (pulse train). When EN = 0, the AND gate is disabled and the output of the gate is low. When EN = 1, the AND gate is enabled and the output follows the PT input. The

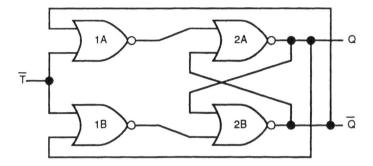

FIGURE 7.14. A NOR-NOR toggle latch.

output is a string of pulses of duration determined by the EN input. The timing diagram is shown in Figure 7.13(b). If EN is high for a known period of time and the gate output is fed to a pulse counter, the counter can determine the number of pulses per second generated.

There are restrictions on the timing of gated signals, regardless of the type latch used. The timing must always be such that changes in input signals occur when the clock is inactive, and the inputs must have reached their correct values a short time prior to the clock being asserted.

Setup time is defined as the time the input data must be in place before the clock is asserted. This is the minimum time required for data to precede the clock in order to reliably effect a state change at the output. In general, setup time is the time interval between the application of a signal that is maintained at one specified input terminal and a consecutive active transition at another specified input terminal.

Hold time is defined as the time required for the clock to effect a state change at the output. This is the minimum time during which the clock must enable the steering gates, and the inputs must remain fixed during this interval. In general, hold time is the interval during which a signal is retained at one specified input after an active transition occurs at another specified input terminal.

7.3.2   The Gated Toggle Latch

Connect the Q output of the NOR-NOR $\overline{S}$-$\overline{R}$ latch of Figure 7.10(a) to the $\overline{S}$ input and the $\overline{Q}$ output to the $\overline{R}$ input and a *toggle latch* is created, as shown in Figure 7.14. If the latch starts in the cleared state with Q = 0 and $\overline{Q}$ = 1, the signals at the inputs are $\overline{S}$ = 0 and $\overline{R}$ = 1. When the clock goes low the latch will toggle to the set state. The new outputs are Q = 1 and $\overline{Q}$ = 0, and the inputs are $\overline{S}$ = 1 and $\overline{R}$ = 0. If the clock is still low the latch will toggle again.

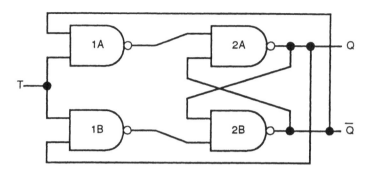

FIGURE 7.15.   A NAND-NAND toggle latch.

As long as the clock is low there is no stable state and the latch will *oscillate* from Q high to Q low, outputting a square wave signal which is high 50% of the time and low 50 percent of the time. Such a signal can be used to generate a clock waveform. The clock input is relabeled the T input since it causes the latch to toggle. The T input can be thought of as a trigger input also, since it triggers toggling, and a *trigger* or *toggle* latch has been created.

If one connects the Q output of the NAND-NAND S-R latch in Figure 7.11 to the R input and the $\overline{Q}$ output to the S input, a toggle or trigger latch is again created, as shown in Figure 7.15. This time the output toggles every time the clock goes high; and if the clock is held high, an astable oscillator or multivibrator is again created.

If short negative pulses are applied to the trigger input of the latch in Figure 7.14, it will toggle once on every pulse; and if short positive pulses are applied to the trigger input of the latch in Figure 7.15, it will toggle once on every pulse. Provided short pulses are periodically applied to the trigger input of a T-latch, the output will change at half the frequency of the input. However, if the input pulses are too long, the latch oscillates and is said to be *free running*. This situation is also referred to as a *race* condition, with the latch toggling an unknown number of times and ending up in an unknown state.

A race condition is usually unacceptable and to be avoided. As digital circuits become faster, it becomes more difficult to make the pulse width of the signal at the trigger input less than the propagation delays of the latch gates, and a different method of triggering is required.

There are two other timing problems with the toggle latch, due to the fact that it is a multiple feedback circuit. The basic assumption in using an S-R ($\overline{S}$-$\overline{R}$) latch is that the data inputs do not change while the clock is high (low). This condition cannot always be met by a toggle latch. If the toggle input signal is not held to a width of precisely two propagation delays ($2t_{pd}$), the output will oscillate.[2] Also, if the combined propagation delays of steering gates 1A and 1B are longer than the combined propagation delays of gates 2A and 2B, in either the circuit of Figure 7.14 or of Figure 7.15, the circuit will again oscillate.

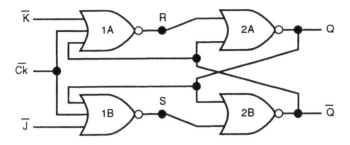

FIGURE 7.16. An active-low J-K latch created from an active-high S-R latch.

A second undesirable mode of operation occurs when the combined propagation delays of gates 1A and 1B are less than the combined delays of gates 2A and 2B. Under this condition, the latch of Figure 7.14 will "hang up" with the outputs of gates 1A and 1B at logic 1 and both Q and $\overline{Q}$ outputs at logic 0 until its clock goes high again; and the latch of Figure 7.15 will "hang up" with the outputs of gates 1A and 1B at logic 0 and both Q and $\overline{Q}$ outputs at logic 1 until its clock goes low again. See Problems 7.39 and 7.40. These problems will be overcome by the edge-triggered toggle flip-flop discussed in Section 7.5.

It is not always necessary to toggle on every trigger pulse. An enable input added to the circuit can be used to selectively trigger the latch. This idea will also be pursued further when edge-triggered toggle flip-flops are discussed.

7.3.3   The Gated J-K Latch

By combining the properties of an S-R latch and a T-latch one obtains a J-K latch. If the Q output of the $\overline{S}$-$\overline{R}$ latch of Figure 7.10 is fed back and NORed with the $\overline{S}$ input, and the $\overline{Q}$ output is fed back and NORed with the $\overline{R}$ input, an *active-low J-K latch* is created, as shown in Figure 7.16. The $\overline{S}$ data input is relabeled the $\overline{J}$ input, and the $\overline{R}$ data input is relabeled the $\overline{K}$ input. $\overline{J}$ is an active-low set input and $\overline{K}$ is an active-low clear input.

If the Q output of the S-R latch of Figure 7.11 is fed back and ANDed with the R input and the $\overline{Q}$ output is fed back and ANDed with the S input, an *active-high J-K latch* is created, as shown in Figure 7.17. The J and K inputs correspond to the active-high set and reset data inputs and perform in a similar manner to the S and R inputs of an S-R latch, except when J and K are both asserted.

When J = K = 1 or $\overline{J}$ = $\overline{K}$ = 0, the Q and $\overline{Q}$ signals cause the latch to toggle, and the J-K flip-flop eliminates the unallowed S = R = 1 input condition. The symbol, next-state table, next-state map, and characteristic equation of the J-K latch are given in Figure 7.18. In English, the characteristic equation states that the next state of the J-K is asserted if the present state is set and it is not cleared, or if the present state is clear and it is set. The relationship between the S-R and J-K inputs is S = $J\overline{Q}$ and R = KQ.

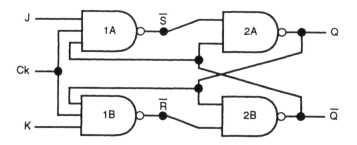

FIGURE 7.17. An active-high J-K latch created from an active-low S-R latch.

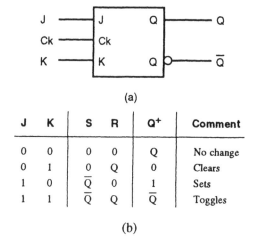

(a)

J	K	S	R	Q⁺	Comment
0	0	0	0	Q	No change
0	1	0	Q	0	Clears
1	0	$\overline{Q}$	0	1	Sets
1	1	$\overline{Q}$	Q	$\overline{Q}$	Toggles

(b)

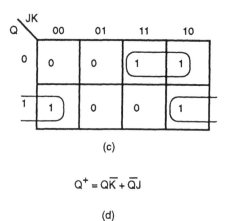

(c)

$$Q^+ = Q\overline{K} + \overline{Q}J$$

(d)

FIGURE 7.18. The (a) symbol, (b) next-state table, (c) next-state map, and (d) characteristic equation of a J-K latch.

The J-K latch still has timing problems. In particular, if J = K = 1 for a time interval longer than the propagation delay through the latch, the J-K will toggle an unknown number of times, and the latch is free running or astable. The master/slave J-K flip-flop can eliminate this problem. It is also possible for the J-K latch to "hang up" with both its outputs high if it is made of NAND steering gates and a NAND latch, or with both its outputs low if it is made up of NOR steering gates and a NOR latch. If this happens the latch will remain in this metastable condition as long as the clock is asserted, and a race occurs when the clock returns to the unasserted state.[3] (See Problems 7.41 and 7.42.)

7.3.4 The Gated Data Latch

S-R latches are useful in control applications requiring separate set and reset capability, but often it is necessary simply to store data temporarily. In storing and shifting data, inputs S = R = 0 are useless while inputs S = R = 1 are not allowed. The data latch eliminates these two input combinations, and if the enable input of the data latch is controlled by the system clock, the D latch can shift data on the clock pulses.

A *data latch,* also called a *delay latch, D latch,* or *transparent latch,* can be constructed by connecting an inverter from the set input to the reset input, as shown in Figure 7.19(a). This forces S = D and R = $\overline{D}$. To be useful, one needs to add an enable input, EN, to the latch. This can be accomplished by adding two NOR steering gates as was done in Figure 7.10(a). However, a little thought will obviate the need for an extra inverter, leading to the simplified circuit of Figure 7.19(b). When the clocked latch is enabled, the output of gate 1A is $\overline{D}$, which is steered through gate 1B to produce D, resulting in R = $\overline{D}$ and S = D as required.

When the circuit is enabled, the latch is "open" and the circuit is transparent to the presence of the latch. When the circuit is disabled, the latch is "closed" and the output retains its last value. The symbol for an asynchronous D latch with active-low enable is shown in Figure 7.19(c). The next-state table, next-state map, and characteristic equation are shown in Figure 7.20. When the circuit is enabled, the next state of the latch is the present value of D, or $Q^+ = D \cdot \overline{EN}$. If the data input in Figure 7.19(b) is connected to gate 1B and the output of gate 1B is connected to an input of gate 1A, R = D, S = $\overline{D}$, and one has an inverting delay latch with $Q^+ = \overline{D} \cdot \overline{EN}$.

The data latch eliminates the S = R = 1 problem, but there are timing restrictions on it. The data must be stable for a *setup time* before the enable signal appears, and it must not change until a *hold time* after the enable signal appears. This defines a "window" during which the data input must not change if the output is to be predictable.

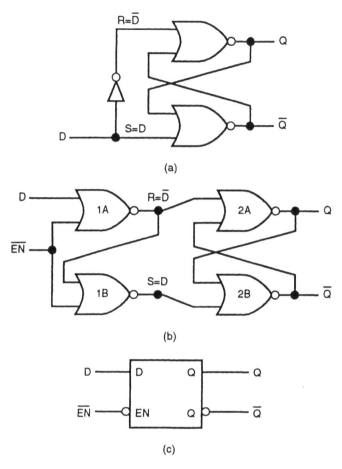

FIGURE 7.19. The asynchronous data latch: (a) the basic NOR circuit, (b) the NOR-NOR latch with enable input, and (c) the symbol.

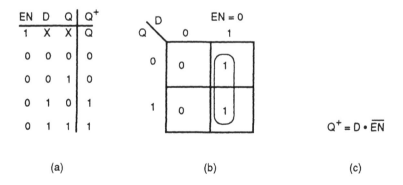

FIGURE 7.20. The (a) characteristic table, (b) next-state map, and (c) characteristic equation of the D latch.

7.4   THE FLIP-FLOP

A flip-flop uses the changing state of the clocking level to modify its output. There are three major catagories of flip-flop: the *pulse-triggered master/slave,* the *edge-triggered,* and the *data-lockout flip-flop.* The master/slave flip-flop is designed to interrupt the logic connection between its inputs and its outputs during the time the clocking signal is enabled. This removes the transparency of the device and eliminates a race condition while providing a memory device that can be used in synchronous sequential designs.

The edge-triggered flip-flop responds to a rising or falling edge of the clock signal, which is of very short duration and again avoids a race. The data-lockout flip-flop is a combination of an edge-triggered master flip-flop and a level-triggered slave latch. These will each be studied in detail.

7.4.1   The Master/Slave S-R Flip-Flop

A *master/slave S-R flip-flop* combines a pair of S-R latches, as shown in Figure 7.21(a). The first latch, called the *master,* accepts the data from the S and R inputs when the clock pulse is high and transfers the data to the second latch, called the *slave,* when the clock pulse goes low. The master is enabled and the slave is disabled when the clock is high; the master is disabled and the slave is enabled when the clock is low. The slave is enabled for the entire time the clock is low, but the master cannot change until the clock goes high again. Therefore, the output of the slave changes once at the negative-going transition of the clock and cannot change again until the next negative-edge of the clock. This eliminates the problem of race. If the inputs to an S-R master/slave flip-flop are both asserted, the operation of the flip-flop is again unpredictable and it might end up in a metastable condition.

Most master/slave flip-flops are *negative-edge pulse triggered,* meaning that the output data change on the clock transition from high to low. This is shown in the next-state table of Figure 7.21(b) by a positive clock pulse waveform.

The master/slave S-R flip-flop is not truly edge triggered. It behaves more like a latch in that the master follows the input while the clock is high, but the slave changes its output on the negative edge of the clock. To indicate this, a *postponed-output indicator* (an inverted L) is drawn at both outputs in Figure 7.21(c). This shows that the outputs change on the negative or falling edge of the clock pulse.

7.4.2   The Master/Slave J-K Flip-Flop

If feedback is employed from the $\overline{Q}$ output of the slave to the set input of the master and from the Q output of the slave to the clear input of the master of an S-R master/slave flip-flop, one has a master/slave J-K flip-flop. This is shown in Figure 7.22(a). The clocking is the same as that of the master/slave S-R flip-flop.

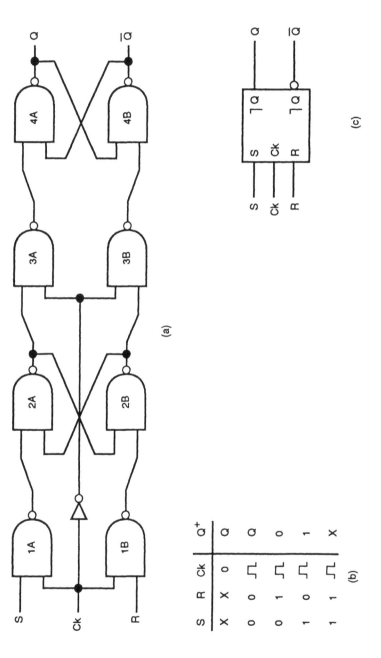

FIGURE 7.21. The (a) circuit, (b) next-state table, and (c) symbol of a master/slave S-R flip-flop.

Otherwise, the behavior is similar to that of a J-K latch. The master/slave J-K flip-flop is a pulse-triggered flip-flop also. This is shown in the next-state table of Figure 7.22(b) by a positive clock-pulse waveform, and in the circuit symbol of Figure 7.22(c) by the postponed-output indicators.

The master/slave clocking scheme is shown in Figure 7.23. The master NAND steering gates (gates 1A and 1B in Figures 7.21 or 7.22) are enabled when the clock signal is at least as high as the level defined by points b and c in Figure 7.23; and the slave-steering gates (gates 3A and 3B in Figures 7.21 or 7.22) are enabled when $\overline{Ck}$ is as high as points b and c, which occurs when Ck is lower than the level defined by points a and d.

Gates 3A and 3B are disabled when the clock rises above point a, and the slave is unable to respond to the master outputs. Gates 1A and 1B are enabled at point b on the clock pulse, and the master responds to the data inputs. The master stays enabled until the clock waveform reaches point c, at which time gates 1A and 1B are disabled and the master can no longer respond to the data inputs. The slave is disabled until the clock reaches point d, after which the slave is enabled and loads the data from the master. There is a very short period of time when both the master and the slave are disabled on the rising and on the falling edges of the clock, but they are never both enabled at the same time.

The master loads the data on the leading or rising edge of the clock signal, and the slave loads the data on the trailing or falling edge of the clock signal. The output changes on the trailing edge of the pulse when the slave responds to the master. Flip-flops which exhibit this type behavior are referred to as *pulse-triggered flip-flops.*

Since the output changes on the falling edge of the clock, the flip-flop symbol must contain an indication that the output changes after the clock makes a high-to-low transition. The state of the flip-flop (the state of the slave) depends upon the last state of the master prior to the falling edge of the clock signal. A master/slave device that changes on the rising edge of the clock pulse can be obtained by inverting the clock signal to the master and not to the slave. This requires moving the inverter, but adds no extra gates to the circuit.

The J-K master/slave flip-flop avoids the problem of both inputs being asserted simultaneously, and it avoids the race problem, but it does exhibit a false-triggering problem. Consider the case when the inputs to the flip-flop of Figure 7.22 are J = 0 and K = 1 and the flip-flop is in the cleared state with Q = 0. The inputs to the master latch are $J\overline{Q}$ and KQ. $\overline{Q}$ = 1 enables the J NAND gate, and Q = 0 disables the K NAND gate. If a noise spike produces a "1" at the J input while the clock is high, the master latch will go to the set state and the flip-flop is said to have "caught" a 1. The master cannot be cleared since the K input is disabled, and on the next clock transition, the output of the flip-flop will be 1.

Likewise, if J = 1 and K = 0 and the flip-flop is in the set state, then Q = 1 enables the K NAND gate and $\overline{Q}$ = 0 disables the J NAND gate. If a noise spike produces a "1" at the K input while the clock is high, the master latch will go to the clear state and the flip-flop is said to have "caught" a 0. The master cannot be

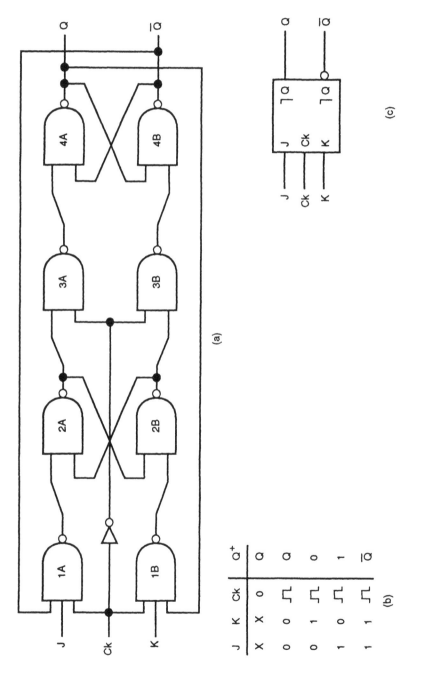

FIGURE 7.22.   The (a) circuit, (b) next-state table, and (c) symbol of a master/slave J-K flip-flop.

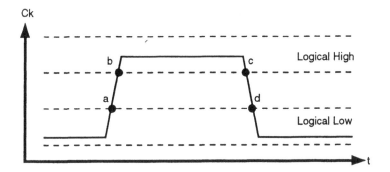

FIGURE 7.23. The pulse-clocking waveform.

set since the J input is disabled, and the output of the flip-flop will go to 0 on the next clock transition.

7.4.3   The Master/Slave Data Flip-Flop

A *master/slave data flip-flop* combines a pair of data latches, as shown in Figure 7.24(a). The master accepts the data D when the clock pulse is high and transfers the data to the slave when the clock pulse goes low; hence, this is also a pulse-triggered flip-flop, as shown in Figures 7.24(b) and (c).

The *pulse-triggered D flip-flop* still has a setup and hold time window, during which the D input must not change. If the setup and hold conditions are not met, the flip-flop usually will go to a stable but unpredictable state.

If a D latch has its active-low output connected to its input, then the output can oscillate from high to low at a frequency determined by the propagation delay of the signal from input to output. A master/slave D-type flip-flop cannot do this.

7.4.4   The Master/Slave Toggle Flip-Flop

Master/slave toggle flip-flops can be constructed in the same manner as the master/slave J-K flip-flop of Figure 7.22 by simply omitting the J and K inputs, or they can be designed from either D flip-flops or J-K flip-flops, as shown in Figure 7.25. They are ideal for use in counters and frequency dividers.

Toggle flip-flops need not be toggled on every clock pulse. If an enable input, EN, is added to the T flip-flop, it will change state on the triggering edge of the clock pulse only if the enable input is asserted. A toggle flip-flop can be created by exclusively ORing the Q output of a D flip-flop with the EN input, or by tying the inputs of a J-K flip-flop together and to the EN input. A *T flip-flop with enable* designed both ways is shown in Figure 7.26. In both cases, when EN = 0 there is no change, and when EN = 1 the flip-flop toggles on the clock pulse.

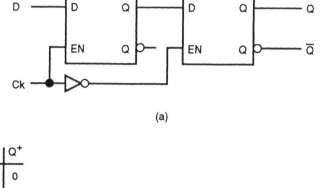

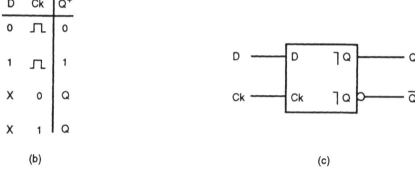

(a)

(b)   (c)

FIGURE 7.24.   The master/slave D flip-flop: (a) the circuit constructed of two D latches, (b) the next-state table, and (c) the logic symbol.

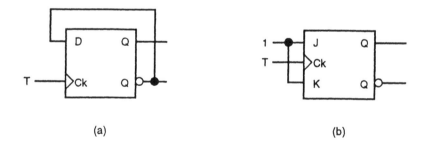

(a)   (b)

FIGURE 7.25.   A T flip-flop constructed from (a) a D flip-flop, and (b) a J-K flip-flop.

7.5   THE EDGE-TRIGGERED FLIP-FLOP

Edge triggering was invented to avoid the 1s and 0s catching problem of the master/slave J-K flip-flop. There are several forms of edge-triggered flip-flops. One method of designing toggle and delay flip-flops is as follows.[4]

A toggle latch constructed of four NOR gates is shown in Figure 7.14 and repeated in Figure 7.27 for convenience. Examination of Figure 7.27 shows that gates 3A and 3B form a conventional S-R latch as shown in Figure 7.4. This will be referred to as the *primary latch*. Gates 1 and 2 are used to feed the output

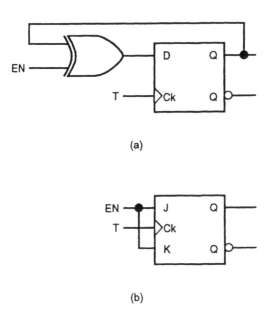

(a)

(b)

FIGURE 7.26. A gated T flip-flop constructed from (a) a gated D flip-flop, and (b) a gated J-K flip-flop.

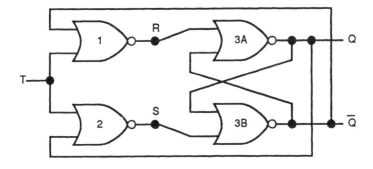

FIGURE 7.27. The basic NOR-NOR toggle latch.

signals back to the S-R latch on a toggle input. If one were to replace gates 1 and 2 with S-R latches, as shown in Figure 7.28, the circuit could latch the values of Q and $\overline{Q}$ and retain them. As long as T = 1, S and R are clamped at 0 and the output cannot change. When T = 0, S = $\overline{Q}$ and R = Q. If Q = 1, the primary latch receives a clear signal and toggles; if Q = 0, the primary latch receives a set signal and the primary latch again toggles.

The circuit must not toggle more than once on each clock transition. To prevent a race condition, feedback from R to an input of gate 2B prevents S from changing more than once following a toggle input; and feedback from S to an input of gate

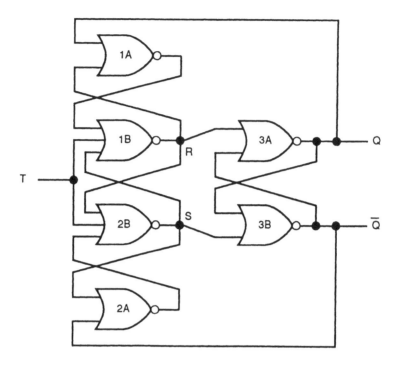

FIGURE 7.28.   The edge-triggered NOR-NOR toggle flip-flop.

1B prevents R from changing more than once following a toggle input. Whichever primary latch signal, R or S, goes to 1 first clamps the other signal at 0. Now, when T = 0 and Q = 1, R = 1 and S = 0 again and the primary latch toggles to the cleared state. However, R = 1 is fed back to an input to gate 2B. This keeps S = 0 and prevents additional toggling until after the trigger input returns to 1, which forces both S and R to 0. When T next goes from 1 to 0, R = 0 and S = 1 and the primary latch toggles back to the set state. Now the S = 1 signal is fed back to gate 1B and prevents R from changing from 0 to 1 again until after the trigger input returns to 1.

   The device is a flip-flop which toggles following an input signal change from 1 to 0 and is thus negative-edge triggered. The flip-flop has only six gates and a maximum of three propagation delays, whereas the master/slave requires eight gates and has a maximum of six propagation delays. It is thus 33% smaller than a master/slave flip-flop and 50% faster. The flip-flop shown in Figure 7.28 is not clocked. Clocking can be done by adding an AND gate to the toggle input and ANDing the toggle data signal with the enable or clock signal.

   The above circuit can be converted to a delay flip-flop by removing the Q and $\overline{Q}$ feedback lines and inputing D to gate 2A. The output of gate 2A is then fed back to an input of gate 1A as well as to an input of gate 2B. The feedback connection from S to an input of gate 1B is no longer needed, and the circuit of Figure 7.29 results.

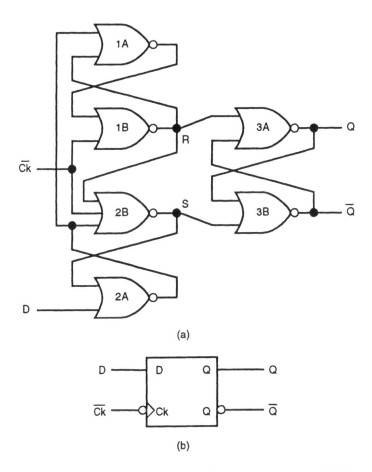

FIGURE 7.29. The edge-triggered NOR-NOR delay flip-flop: (a) the circuit, and (b) the symbol.

The circuit behaves as follows. When Ck = 1 and D = 1, R = S = 0 and $\overline{D}$ = 0. D = 1 is present at the output of gate 1A and $\overline{D}$ = 0 is at the output of gate 2A. The D = 1 input to gate 1B clamps R at 0 when the clock goes low. While the clock is low, R = 0 and S = 1, the basic latch is set and the next value of Q is 1. If D changes to 0 while the clock is low, the S = 1 signal keeps the output of gate 2A at 0, which keeps the output of gate 1A at 1, and R remains clamped to 0. The basic latch cannot change again while the clock line is at 0.

When Ck = 1, R = S = 0. If D = 0, $\overline{D}$ = 1 is present at the output of gate 2A and is fed back to an input of gate 1A, keeping its output at 0. When the clock goes low, both inputs to gate 1B are low and R changes to 1. This clears the primary latch while R = 1 is fed into gate 2B and keeps S at 0. If D now changes to 1, the output of gates 1A and 2B are both clamped at 0 by R, and the latch cannot respond.

This is true edge triggering, with transitions occurring on the falling edge of the clock. Edge triggering is shown by a triangle at the clock input in the symbol of

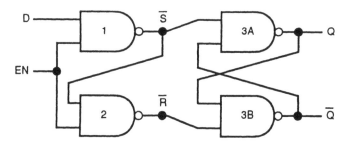

FIGURE 7.30.   The basic gated NAND-NAND delay latch.

an edge-triggered toggle or delay flip-flop, as shown in Figure 7.29(b). The invert bubble at the clock input indicates negative-edge triggering. In keeping with our notation, the clock input is a negative clock line and is labeled $\overline{Ck}$. Positive-edge triggering is indicated by the presence of a triangle and the absence of a bubble at the clock input, and the clock line is labeled Ck.

The circuit of Figure 7.28 requires additional gating before it can be used to construct an S-R or J-K flip-flop. If the Q and $\overline{Q}$ feedback lines are removed and their inputs replaced by set and reset inputs, one has an edge-triggered S-R flip-flop. If the Q and $\overline{Q}$ feedback lines are retained and set and reset inputs are added, one has an edge-triggered J-K flip-flop. The S-R and J-K flip-flops can have both data inputs at 0 prior to a clock transition to 0, i.e., S = R = 0 and J = K = 0 are permissible. When this occurs, either input changing to 1 while the clock is 0 can cause an output change. This is not true edge triggering, and additional gating is required to overcome this problem. The resulting circuit requires ten gates and is more complicated than the master/slave configuration. For this reason the circuit is not practical. See Mowle[4] for a derivation of the S-R and J-K edge-triggered flip-flops.

Edge-triggered delay and toggle flip-flops can be constructed from the active-low NAND S-R latch also. The basic gated delay NAND latch is shown in Figure 7.30, which is the dual circuit of Figure 7.19(b). By replacing gates 1 and 2 by latches again, as was done to the NOR toggle latch of Figure 7.27, one arrives at the edge-triggered delay flip-flop of Figure 7.31. This is a positive-edge-triggered D flip-flop. It can be converted into a positive-edge-triggered T flip-flop consisting of NAND gates by simply reversing the steps taken in going from Figure 7.28 to 7.29.

Commercial D flip-flops do not use a master slave latch design. A clocked master/slave D latch pair requires at least 8 NAND gates for the flip-flop and one inverter for the clock, i.e., two latches such as in Figures 7.19(b) or 7.30. With this construction, the longest path length of a signal is through six gates, requiring six propagation delays. The negative-edge-triggered D flip-flop of Figure 7.29 and the positive-edge-triggered D flip-flop of Figure 7.31 each require six gates and no inverters; and the longest signal path for this circuit is through three gates, corresponding to three propagation delays. This circuit is both much smaller and much faster than a master/slave D flip-flop.

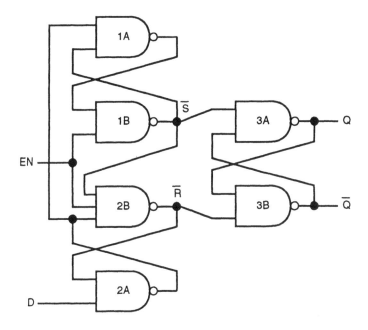

FIGURE 7.31. The edge-triggered NAND-NAND delay flip-flop.

An edge-triggered delay flip-flop can be converted into an edge-triggered J-$\overline{\text{K}}$ flip-flop by the addition of AND-OR logic, as shown in Figure 7.32. In this configuration, $D = J\overline{Q} + \overline{K}Q$. If the flip-flop is initially in the clear state, $\overline{Q} = 1$ and $Q = 0$. If $J = 1$, then $D = 1$ and the flip-flop sets. K is a don't care. If the flip-flop is initially set, $Q = 1$ and $\overline{Q} = 0$. If $\overline{K} = 0$, then $D = 0$ and the flip-flop will clear regardless of the value of J.

A J-K edge-triggered flip-flop can be constructed from the edge-triggered toggle flip-flop by adding an AND-OR gate also, as shown in Figure 7.33, such that $T = J\overline{Q} + KQ$. If the flip-flop is set and $K = 1$, it will toggle regardless of the value of J; if it is clear and $J = 1$, it will also toggle regardless of the value of K. Thus, it will toggle whenever $J = K = 1$. When $J = 1$ and $K = 0$, the flip-flop will end up set, regardless of its initial state, and when $J = 0$ and $K = 1$, the flip-flop will clear, again regardless of its initial state.

The edge-triggered J-K flip-flop inputs must still meet the setup and hold requirements in order to guarantee proper behavior, but unallowed inputs, race problems, and 1s and 0s catching due to noise are all avoided; and it is as close to foolproof as a flip-flop can be. The most common application of edge-triggered J-K flip-flops is in clocked synchronous-state machines. This device can be used in virtually every application where other flip-flops can be used. Because the edge-triggered flip-flop has a large immunity to noise, it is replacing the master/ slave flip-flop in many modern applications.

A novel way to get around the ten-gate requirement for an edge-triggered J-K flip flop is to construct a J-$\overline{\text{K}}$ flip-flop as follows. If the NAND D flip-flop of

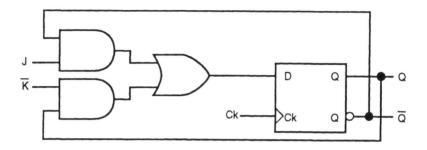

FIGURE 7.32.   An edge-triggered J-$\overline{\text{K}}$ flip-flop created from an edge-triggered D flip-flop.

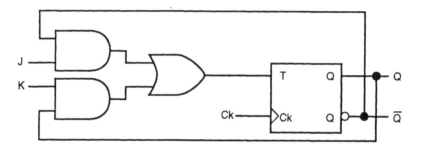

FIGURE 7.33.   A edge-triggered J-K flip-flop created from an edge-triggered T flip-flop.

Figure 7.31 is modified by changing gate 2A to a NOR gate and adding two AND gates and appropriate feedback from the Q and $\overline{\text{Q}}$ outputs, one has the positive-edge-triggered J-$\overline{\text{K}}$ flip-flop of Figure 7.34. This device requires only eight gates and is essentially the same circuit as in Figure 7.32.

An edge-triggered flip-flop can be constructed using propagation delay to generate very short triggering pulses also, although it is not as reliable as the latching edge-triggered flip-flop discussed above. Propagation delay can be used to obtain positive pulses, as shown in Figure 7.35, or negative pulses, as shown in Figure 7.36.

If input A in Figure 7.35 makes a low-to-high transition, the delay through gate 1 causes both inputs to gate 2 to be high for a period of time equal to the propagation delay of gate 1, t_{pd1}. After a delay t_{pd2} of gate 2, a positive pulse appears at the output Y. The width of the positive pulse is equal to t_{pd1}. High-to-low transitions of input A are ineffective and cause no pulses to be created.

If input A in Figure 7.36 makes a high-to-low transition, the delay through gate 1 causes both inputs to gate 2 to be low for a period of time equal to the propagation delay of gate 1, t_{pd1}. After a delay t_{pd2} of gate 2, a negative pulse appears at the output Y. The width of the negative pulse is again equal to t_{pd1}, and now a low-to-high transition is ineffective.

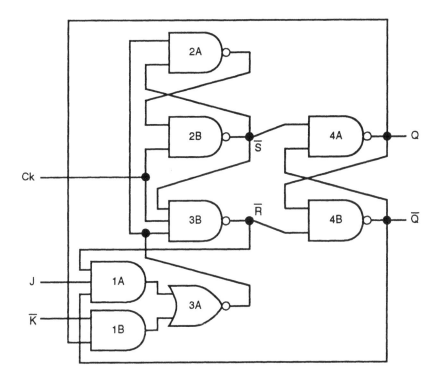

FIGURE 7.34.   A positive-edge triggered J-$\overline{\text{K}}$ flip-flop created from a modified edge-triggered D flip-flop of Figure 7.31.

To construct a propagation-delay edge-triggered J-K flip-flop, start with the basic gated AND-NOR latch of Figure 7.9(a). Split the AND gates in two as shown in Figure 7.37, with the clock enabling AND gates 2A and 3A and clock-BAR enabling AND gates 2B and 3B through gates 1A and 1B. When the clock is high, gates 2A and 3A close the feedback path and latch the outputs. When the clock is low, gates 2B and 3B close the feedback path and latch the outputs. A low-to-high transition is ineffective, but a high-to-low transition enables the basic latch. Figure 7.37 is a negative-edge-triggered J-K flip-flop. If Q is not fed back to the K input and $\overline{\text{Q}}$ is not fed back to the J input, one has a fast negative-edge-triggered S-R flip-flop.

7.6   THE J-K FLIP-FLOP WITH DATA LOCKOUT

Some flip-flops have a feature called *data lockout,* which allows the flip-flop to load next-state data on an edge of the clock, after which further input changes are locked out of the flip-flop until the next enabling clock edge. This is essentially an edge-triggered master and a level-triggered slave.

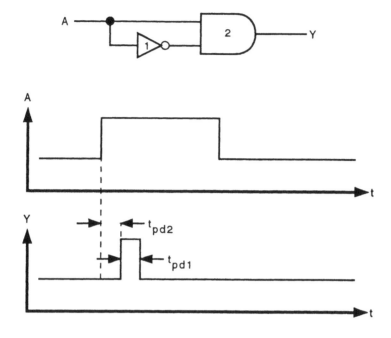

FIGURE 7.35.   A positive-pulse generating circuit and its timing diagram.

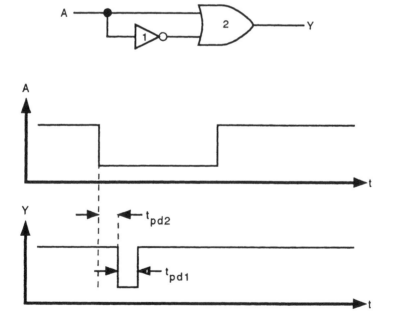

FIGURE 7.36.   A negative-pulse generating circuit and its timing diagram.

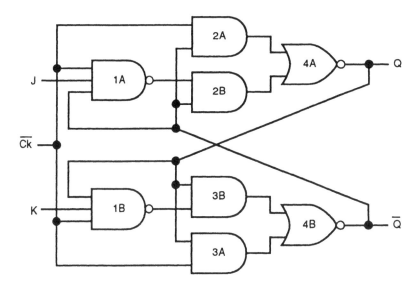

FIGURE 7.37. An edge-triggered J-K flip-flop requiring only eight gates.

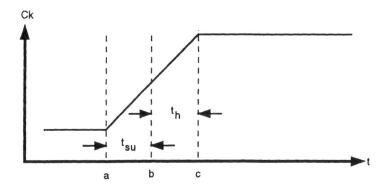

FIGURE 7.38. A typical clock waveform for a rising-edge data loading device, with setup and hold times indicated.

A typical clock waveform for a rising-edge-triggered device is shown in Figure 7.38. The inputs are allowed to change up to time a in Figure 7.38. The setup time is the time from point a to point b, and the hold time is the time from point b to point c. The inputs must not change during the setup or hold time. After point c, the input data are locked out and input changes are ineffective until the next clock pulse reaches point a.

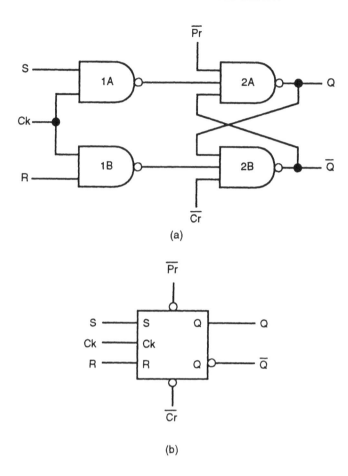

FIGURE 7.39.   An S-R latch with active-low direct set and reset: (a) the circuit, and (b) the symbol.

7.7   DIRECT SET AND RESET FLIP-FLOPS

Many flip-flops have *asynchronous inputs* that can set and clear the device by overriding the clock and synchronous data inputs. These inputs can be labeled Pr (preset) and Cr (clear) and behave the same as the set and reset inputs of an S-R flip-flop. Asynchronous inputs are usually used for initializing the flip-flop and for creating known states for testing purposes. The Pr and Cr inputs must be independent of the data inputs and the clock input.

A direct-reset capability is necessary whenever a counter must be set back to 0. For example, if a microwave oven is accidently set to one hour instead of one minute, a reset button will allow the timer to be cleared, and one can start over to program it. If the wrong number is loaded into the X register of a calculator, a CLX button will reset all the numbers in the X register to 0. The direct preset is

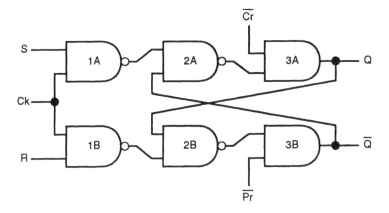

FIGURE 7.40. An S-R latch with correct active-low direct set and reset.

primarily useful for testing purposes, where one can set all flip-flops as well as clear all of them.

An S-R latch with direct set and reset (clear) is shown in Figure 7.39. NAND gates are disabled by inputting a 0 to them; hence, the preset and reset inputs are active low. If one started with cross-connected NOR gates, the preset and reset inputs would be active high. As with synchronous data inputs, one must not assert both the direct preset and the direct clear inputs simultaneously.

Consider the behavior of the circuit in Figure 7.39 when the data inputs are S $= 1$ and R $= 0$. If $\overline{Pr} = 1$ and $\overline{Cr} = 0$, the asynchronous inputs are clearing the latch while the synchronous data inputs are attempting to set the latch. When the clock input is low, there is no problem and the latch is cleared, but when the clock is asserted high, the output of gate 1A is 0 and the output of gate 1B is 1. Gates 2A and 2B both have one input low which forces both Q and $\overline{Q}$ high. As long as the direct clear signal is imposed upon the circuit, $\overline{Q}$ stays high, while Q follows the clock and reproduces the clock waveform.

The problem arises due to the fact that the data inputs and the asynchronous inputs are affecting the same gates. To override the clock completely, one must control the outputs of gates 2A and 2B, not some of their inputs. This is done by adding two AND gates behind gates 2A and 2B, as shown in Figure 7.40. This interchanges the asynchronous inputs, and the direct clear is now at gate 3A and the direct set is at gate 3B, but they are still active-low inputs, and the symbol remains the same as in Figure 7.39(b).

A master-slave S-R flip-flop with direct set and clear is shown in Figure 7.41. It is not sufficient to set or clear only the slave, since the master might be in the opposite state and reverse the operation after the direct input is removed. As before, care is required in presetting or clearing the master flip-flop. If the master latch is preset or reset by using an input to either gate 2A or 2B, the master will suffer the same problems discussed with respect to Figure 7.39. The addition of two AND gates (gates 3A and 3B in Figure 7.41) allows one to truly override the

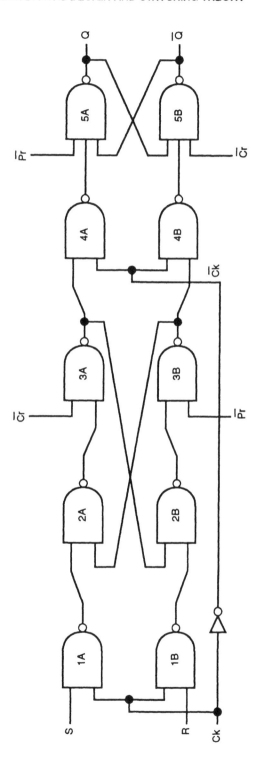

FIGURE 7.41. An S-R master/slave flip-flop with active-low direct set and reset.

synchronous data inputs and preset or clear the master. There is no corresponding problem with the slave latch, since simultaneously presetting or clearing the master guarantees that the synchronous and asynchronous inputs to the slave are both the same. As discussed before, adding the appropriate feedback converts the S-R master/slave to a J-K master/slave. This has no bearing on the direct inputs.

7.8 SUMMARY

Binary devices are divided into two major groups, latches and flip-flops. An unclocked latch monitors its inputs continuously and can change its output at any time. A clocked latch is level triggered, and a flip-flop is either edge or pulse triggered. They both sample their inputs and change their outputs only at times determined by the clocking signal.

Flip-flops are further divided into three major catagories: pulse triggered master/slave, with or without data lockout, and edge triggered. The master/slave flip-flop loads data into the master latch on the asserted clock level and transfers the data to the slave latch on the nonasserted clock level.

The master/slave flip-flop is a pulse-triggered device which eliminates race problems. The edge-triggered flip-flop is enabled by extremely narrow pulses and avoids the problem of 1s or 0s catching. The data lockout flip-flop is essentially a combination of an edge-triggered master and a pulse-triggered slave flip-flop. The memory or storage device used by all of these is the basic cross-coupled NAND or NOR latch.

The J-K flip-flop is the most versatile and can function as a clocked delay, toggle, or S-R flip-flop. It has no invalid input combinations. The edge-triggered J-K and the edge-triggered D flip-flop are the two most important types.

For reliable operation of latches and flip-flops, the data inputs must be stable for a period of time, called the setup time, before the effective clock signal and for an additional time period, called the hold time, after the effective clock signal.

Symbols for the basic latches and flip-flops are shown below along with truth tables. Figure 7.42 shows the S-R and J-K master/slave positive-pulse-triggered flip-flops and the next-state table, including clocking pulses. The symbols for pulse-triggered flip-flops whose outputs change on the falling edge of the clock pulse contain a *postponed-output indicator*, shaped as an inverted "L". The logic symbol for the master/slave flip-flop does not have a dynamic-input indicator (triangle) because the flip-flop is not truly edge triggered.

Figure 7.43 shows positive-edge-triggered delay and toggle flip-flops and negative-edge-triggered S-R and J-K flip-flops. Positive-edge-triggered flip-flops have a triangle-shaped indicator at the clock input, and negative-edge-triggered flip-flops have both a triangle-shaped indicator and an invert bubble at the clock input. The next-state tables, indicating the type clocking, are also shown.

Flip-flops which exhibit data lockout have both a triangle-shaped edge-triggering indicator and a postponed-output inverted "L" indicator. These are shown in Figure 7.44.

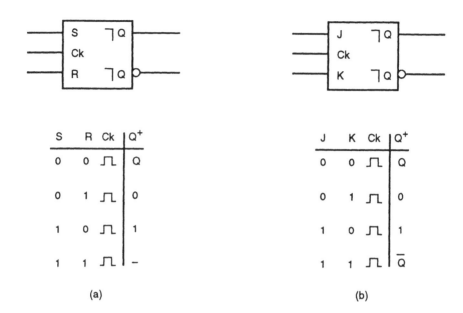

FIGURE 7.42. Master/slave pulse-triggered flip-flops, and their next-state tables. (a) The S-R, and (b) the J-K.

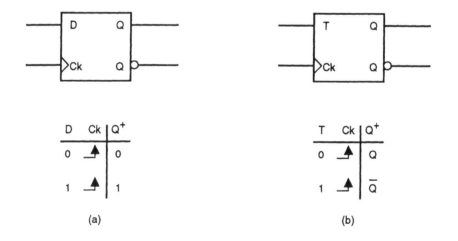

FIGURE 7.43. The symbols and next-state tables for four edge-triggered flip-flops. (a) and (b) The positive-edge-triggered D and T; (c) and (d) the negative-edge-triggered S-R and J-K.

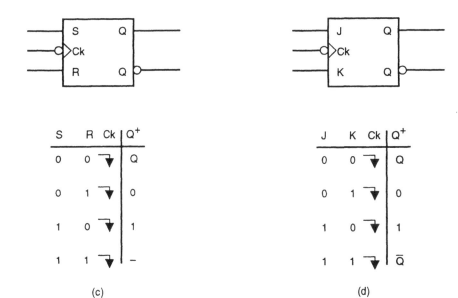

(c)   (d)

FIGURE 7.43. (continued).

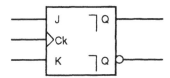

FIGURE 7.44. The symbol for a positive-edge-triggered J-K master/slave flip-flop with data lock-out.

PROBLEMS

7.1. (a) Design a gated set-dominant latch with a gated S-R latch and additional input circuitry to give the next-state table shown below. The latch inputs are A and B.

   (b) Obtain the characteristic equation of the latch.

TABLE 7.4

A	B	S	R	Q^+
0	0	0	0	Q
0	1	0	1	0
1	0	1	0	1
1	1	1	0	1

7.2. (a) Design a gated reset-dominant latch with a gated S-R latch and additional input circuitry to give the next-state table shown below. The latch inputs are C and D.

(b) Obtain the characteristic equation of the latch.

TABLE 7.5

C	D	S	R	Q^+
0	0	0	0	Q
0	1	0	1	0
1	0	1	0	1
1	1	0	1	0

7.3. (a) Design a gated E-F latch with a gated J-K latch and additional input circuitry to give the next-state table shown below. The latch inputs are E and F.

(b) Obtain the characteristic equation of the latch.

TABLE 7.6

E	F	J	K	Q^+
0	0	1	1	$\overline{Q}$
0	1	1	0	1
1	0	0	0	Q
1	1	0	1	0

7.4. (a) Design a gated G-H latch with a gated J-K latch and additional input circuitry to give the next-state table shown below. The latch inputs are G and H.

(b) Obtain the characteristic equation of the latch.

TABLE 7.7

G	H	J	K	Q^+
0	0	1	1	$\overline{Q}$
0	1	1	0	1
1	0	0	1	0
1	1	0	0	Q

7.5. (a) Design a gated L-M latch with a gated S-R latch and additional input circuitry to give the next-state table shown below.

(b) Obtain the characteristic equation of the latch.

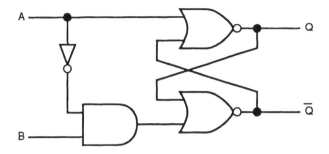

FIGURE 7.45.

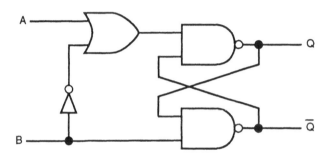

FIGURE 7.46.

TABLE 7.8

L	M	Q^+
0	x	Q
1	0	0
1	1	1

7.6. Repeat Problem 7.5 using a gated D latch.

7.7. Obtain the next-state table and the characteristic equation for the circuit shown in Figure 7.45.

7.8. Obtain the next-state table and the characteristic equation for the circuit shown in Figure 7.46.

7.9. (a) Obtain the next-state table and the characteristic equation for the latch shown in Figure 7.47.

   (b) Of what use is this latch?

7.10. (a) Obtain the next-state table and the characteristic equation for the latch shown in Figure 7.48.

   (b) Of what use is this latch?

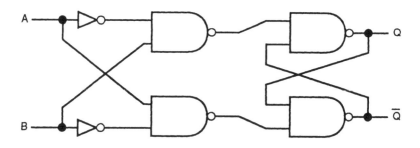

FIGURE 7.47.

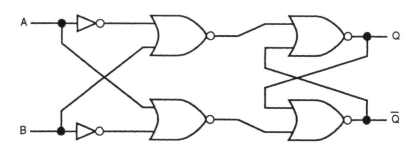

FIGURE 7.48.

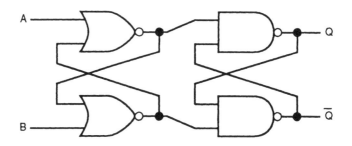

FIGURE 7.49.

7.11. For the latch shown in Figure 7.49, find the sequence of output states Q. The latch is initially cleared, and the inputs are in the sequence AB = 10, 00, 10, 01, 01.

7.12. If the NAND and NOR gates in Figure 7.49 are interchanged, find the sequence of output states Q. The latch is initially cleared, and the inputs are in the sequence AB = 10, 11, 10, 01, 01.

7.13. Repeat Problem 7.11 for the input sequence AB = 10, 00, 01, 10, 00.

7.14. Repeat Problem 7.12 for the input sequence AB = 10, 11, 01, 10, 11.

7.15. Repeat Problem 7.11 for the input sequence AB = 01, 01, 10, 10, 00.

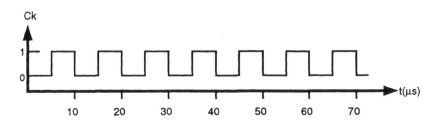

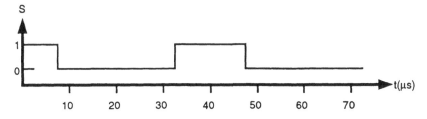

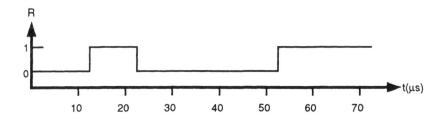

FIGURE 7.50.

7.16. Repeat Problem 7.12 for the input sequence AB = 01, 01, 10, 10, 11.

7.17. Complete the timing diagram of a positive-edge-triggered S-R flip-flop whose clock, S, and R inputs are shown in Figure 7.50. Neglect all propagation delays. The flip-flop is initially cleared.

7.18. Complete the timing diagram of the circuit shown in Figure 7.51. Neglect all propagation delays. Both flip-flops are initially cleared.

For Problems 7.19 through 7.38 inclusive, assume the clock signal is a square wave of period 200 ns, and the propagation delay of each gate is 10 ns, unless otherwise specified. Assume all setup and hold time requirements are met and that the bistables are initially cleared and work properly, unless otherwise specified.

7.19. (a) Show the timing diagram for the circuit of Figure 7.14 if the toggle input is a square wave of period 200 ns.

    (b) How many times does the latch toggle during an interval of 400 ns?

7.20. Repeat Problem 7.19 for the circuit of Figure 7.15.

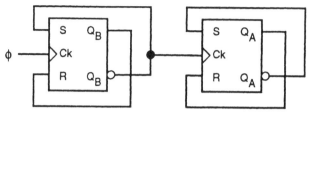

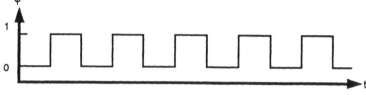

FIGURE 7.51.

7.21. (a) Show the timing diagram for the circuit of Figure 7.31 if D = 1
        and the enable signal is a square wave of period 200 ns. The flip-
        flop is initially in the cleared state.
     (b) Explain how the latch composed of gates 1A and 1B and the latch
        composed of gates 2A and 2B work to guarantee proper triggering.
7.22. Repeat Problem 7.21 for D = 0. The flip-flop is initially in the set state.
7.23. Show the timing diagram for the circuit of Figure 7.34 if J = 1, K = 0,
     and the flip-flop is initially in the cleared state.
7.24. Repeat Problem 7.23 for J = 0, K = 1, and the flip-flop is initially in
     the set state.
7.25. Show the timing diagram for the circuit of Figure 7.37 if J = 1, K = 0,
     and the flip-flop is initially in the cleared state.
7.26. Repeat Problem 7.25 for J = 0, K = 1, and the flip-flop is initially in
     the set state.

7.27. Show the Q and $\overline{Q}$ outputs of the circuit in Figure 7.39(a) for 500 ns if
     the clock period is 100 ns and all propagation delays are negligible.
     From 0 to 250 ns S = 1, R = 0, $\overline{Pr}$ = 1 and $\overline{Cr}$ = 0.
     From 250 to 500 ns S = 0, R = 1, $\overline{Pr}$ = 0, and $\overline{Cr}$ = 1.
7.28. Repeat Problem 7.27 for the circuit shown in Figure 7.40.
7.29. Show the timing diagram for ϕ_1 and ϕ_2 of the circuit in Figure 7.52 if
     the clock is a square wave of period 1 μs and both flip-flops are
     initially cleared. Neglect propagation delays at this clock speed.
7.30. Complete the timing diagram for signals B and C of the circuit in
     Figure 7.53 if each gate delay is 20 ns.

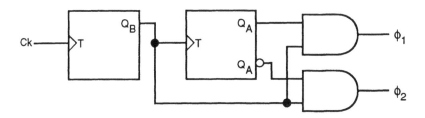

FIGURE 7.52.

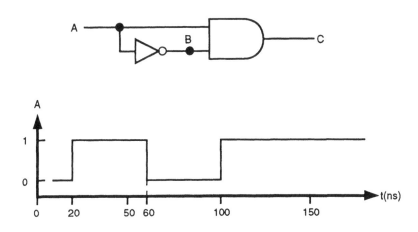

FIGURE 7.53.

7.31. Complete the timing diagram for signals B and C of the circuit in Figure 7.54 if each gate delay is 20 ns.

7.32. Complete the timing diagram for signals B and C of the circuit in Figure 7.55 if each gate delay is 20 ns.

7.33. Complete the timing diagram for signals B and C of the circuit in Figure 7.56 if each gate delay is 20 ns.

7.34. Show the timing diagram for outputs ϕ_1, $\overline{\phi}_1$, ϕ_2, and $\overline{\phi}_2$ of the circuit in Figure 7.57.

7.35. Show the timing diagram for the flip-flop of Figure 7.58. Initially, the flip-flop is cleared, and during the first effective clock transition the data inputs are $\overline{S} = 0$, $\overline{R} = 1$. During the second effective clock transition the data inputs are $\overline{S} = 1$ and $\overline{R} = 0$.

7.36. Repeat Problem 7.35 for the flip-flop in Figure 7.59, if JK = 10, then 01.

7.37. Show the timing diagram for T_M and T_S, the master and slave clock inputs of the master/slave J-K flip-flop created from two J-K latches, as shown in Figure 7.60.

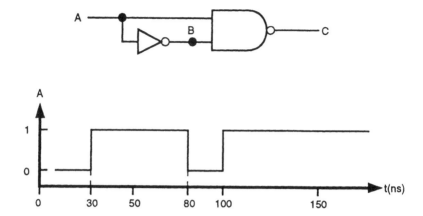

FIGURE 7.54.

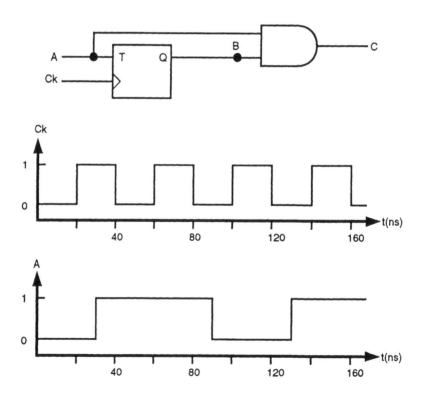

FIGURE 7.55.

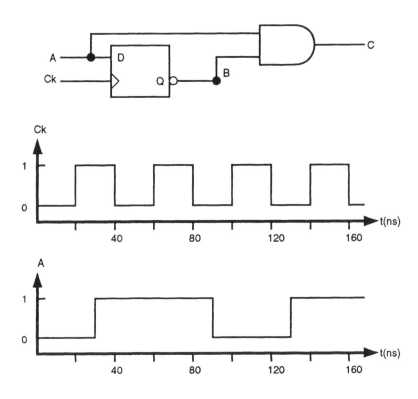

FIGURE 7.56.

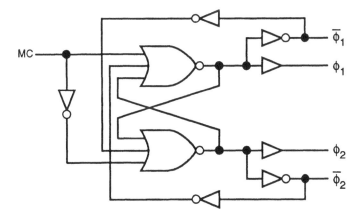

FIGURE 7.57.

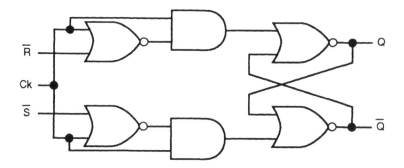

FIGURE 7.58.

7.38. Complete the timing diagram for the delay flip-flop shown in Figure 7.61. The flip-flop is initially in the cleared state.

7.39. Show that $Q = \bar{Q} = 0$ is a metastable state for the circuit of Figure 7.14, as long as input T is low, and that a race occurs when T makes a transition from low to high.

7.40. Show that $Q = \bar{Q} = 1$ is a metastable state for the circuit of Figure 7.15, as long as input T is high, and that a race occurs when T makes a transition from high to low.

7.41. Show that $Q = \bar{Q} = 0$ is a metastable state for the circuit of Figure 7.16, as long as the clock and the data inputs are low, and that a race occurs when the clock makes a transition from low to high while the data inputs are low.

7.42. Show that $Q = \bar{Q} = 1$ is a metastable state for the circuit of Figure 7.17, as long as the clock and the data inputs are high, and that a race occurs when the clock makes a transition from high to low while the data inputs are high.

7.43. Give the state diagram of a gated S-R latch.

7.44. Give the state diagram of a gated D latch.

7.45. Give the state diagram of a gated T latch.

7.46. Give the state diagram of a gated J-K latch.

7.47. (a) Sketch the setup and hold times of a master/slave J-K flip-flop with and without data lockout. Assume $t_{su} = t_h = 2$ ns, and the clock signal is a squarewave of frequency 100 MHz.

(b) How long must the S and R inputs be stable in each case?

SPECIAL PROBLEMS

7.48. Add appropriate asynchronous preset and clear inputs to:
(a) The positive-edge-triggered D flip-flop of Figure 7.31.
(b) The positive-edge-triggered J-$\bar{K}$ flip-flop of Figure 7.34.
(c) The negative-edge-triggered J-K flip-flop of Figure 7.37.

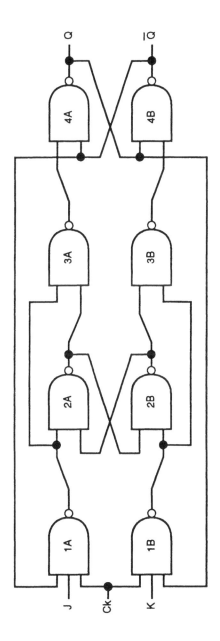

FIGURE 7.59.

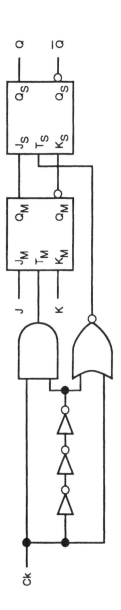

FIGURE 7.60.

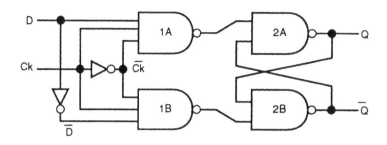

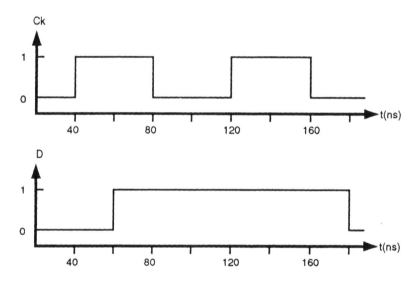

FIGURE 7.61.

Be sure they override any input signal and that the assertion levels are correctly labeled.

7.49. Design a simple sequence detector which will store three bits and output a logic high when either sequence 111 or sequence 000 is detected. Choose appropriate flip-flops and a minimum of supporting logic gates.

7.50. Design a simple comparator which will store two strings of three bits each and output a logic high when the two bit-strings are identical. Choose appropriate flip-flops and a minimum of additional supporting logic gates.

REFERENCES

1. **Breeding, K. J.,** *Digital Design Fundamentals,* Prentice-Hall, Englewood Cliffs, NJ, 1989, 111.
2. **Mowle, F. J.,** *A Systematic Approach to Digital Logic Design,* Addison-Wesley, Reading, MA, 1976, 341–342.
3. **Fletcher, W. I.,** *An Engineering Approach to Digital Design,* Prentice-Hall, Englewood Cliffs, NJ, 1980, 305, 307.
4. **Mowle, F. J.,** *A Systematic Approach to Digital Logic Design,* Addison-Wesley, Reading, MA, 1976, 356–367.

CHAPTER 8

Counters and Registers

8.1 INTRODUCTION

Two main applications of latches and flip-flops are in the construction of counters and registers. Methods of designing both will be studied in this chapter.

Binary counters are constructed with one flip-flop for each bit of the counter. A 4-bit or modulo-16 counter can count from 0000 to 1111 and requires four flip-flops to store the count. The output of each flip-flop of a counter changes at half the frequency of its input. Thus, a 4-bit counter will output square waves at 1/2, 1/4, 1/8, and 1/16 the frequency of the input signal. The counter is also called a divide-by-16 counter since it can divide by up to 16.

Counters are versatile and useful for many applications in digital test instrumentation, such as digital multimeters, pulse counters, and frequency division. An obvious application of a counter is in the construction of a digital clock. If the 60-Hz line frequency is converted into a 60-Hz square wave, this can be divided by 60 to output 1 pulse per second. If this signal is also divided by 60, an output of 1 pulse per minute is obtained, and this signal divided by 60 gives an output of 1 pulse per hour. Finally, a modulo-12 counter can be used to output the hour signals.

The counter output can be used to drive seven-segment displays, LEDs, or otherwise output the number stored in the counter. The binary ripple counter is the simplest to design and will be studied first. Feedback can be used to obtain counters of various moduli or of arbitrary count sequences, and this technique will also be studied.

Toggle flip-flops are the most useful for binary counting purposes, while delay flip-flops are the most useful for register applications. For counting in sequences other than up or down in straight binery, S-R and J-K flip-flops are more powerful and require less additional circuitry. To construct a general purpose counter/register, J-K flip-flops can be used and programmed to operate either as toggle or delay devices. When toggle or delay flip-flops are used in this chapter it is to be understood that J-K or S-R devices could be used.

Asynchronous counters are easy to design, but propagation delays are additive, and a large asynchronous counter becomes very slow. The synchronous counter applies the input signal to all stages simultaneously and is usually much faster, but the penalty for this speed is more complicated control circuitry.

It is sometimes desirable to have a down-count capability. A down counter can be set to some initial value, usually all 1s, after which it counts down until it reaches a count of 0. Down counters have many applications ranging from microwave ovens and VCRs to countdowns for launch of a rocket. Counters can be constructed with the capability of counting either up or down, depending upon a control signal. Circuitry for doing this will be discussed.

A register is a collection of flip-flops designed to store binary data. A single flip-flop used as a register is often called a *flag*. A register of more than one stage is constructed by interconnecting flip-flops such that each flip-flop can either store or shift its data bit, depending upon the control signal.

Registers are used to temporarily store data in digital systems, and shift registers are used to store and to shift data. There are two methods of entering data into or extracting data from shift registers. In a serial shift register, data are shifted in and out of the register one bit at a time in a serial manner, while in a parallel shift register all the data bits are shifted in and out of the register simultaneously. Serial-to-parallel converters load data in a serial form and output the data in a parallel form, while parallel-to-serial converters load data in parallel and output the data in series.

Registers can be used to shift data left to higher bit weights or right to lower bit weights. The number 20_{10} stored in a six-bit register is 010100_2. If this number is shifted left one position it becomes $101000_2 = 40_{10}$, and if it is shifted right one position it becomes $001010_2 = 10_{10}$. Shifting a binary number left n bit positions is equivalent to doubling its value n times, while shifting it right n bits is equivalent to halving its value n times.

8.2 BINARY COUNTERS

Any clocked sequential circuit whose state diagram contains a single cycle is called a counter. Counters are classified as either synchronous or asynchronous. In synchronous binary counters, the system clock drives the control inputs to all the latches or flip-flops in the counter, and they all change state at the same time. There is no system clock in an asynchronous binary counter, and changes ripple through the counter from each flip-flop to the next.

A counter can be described by the number of bits or latches it contains or by the number of states in the primary counting sequence. A 4-bit counter can have 16 states and be a modulo-16 or divide-by-16 counter. It could also have any number of states less than 16. A modulo-12 or divide-by-12 counter would be useful for counting hours, while a modulo-10 or divide-by-10 counter would be preferable when outputing decimal numbers. In general, a counter that has n latches or flip-flops has m £ 2^n discrete states and is called a *modulo-m counter,* or a *divide-by-m counter.*

Counters with a high modulus can be designed directly or can be constructed by cascading a combination of lower-modulus counters. Thus, a modulo-60 counter could be constructed directly using six flip-flops and feedback to stop the count at $60_{10} = 111100_2$, or a modulo-15 counter and a modulo-4 counter can be cascaded to achieve the desired result.

8.2.1   Asynchronous Binary Counters

An *asynchronous,* or *ripple,* n-bit binary counter can be constructed with n flip-flops and no additional components. A four-bit binary ripple counter, constructed with negative-edge-triggered toggle flip-flops, is shown in Figure 8.1. The clock drives the first flip-flop of the counter. Each remaining flip-flop is driven by the output of the preceding flip-flop, and the counter is triggered in a ripple manner. The T flip-flop changes state on every falling edge of its clock input, and each bit of the counter toggles if, and only if, the immediately preceding bit changes from 1 to 0. This corresponds to a normal binary counting sequence, with each bit change from 1 to 0 generating a carry to the next most significant bit (MSB) position, causing that flip-flop to toggle.

The next state of the counter is achieved after all the flip-flops reach a stable state. The ripple effect can cause momentary "false" states to occur, which can cause *glitches* at the output. Momentary erroneous pulses, called *hazards* or *glitches,* are seen in the timing diagram of the three-bit ripple counter which is shown in Figure 8.2. The output B cannot change until one propagation delay, t_{pd}, after C changes, and the output of A cannot change until one propagation delay after B changes. Glitches occur on clock pulses 2, 4, and 6, and the counting sequence is 000, 001, (000) 010, 011, (010, 000) 100, 101, (100) 110, 111. The states in parantheses are due to the undesired glitches. This is likely to cause false triggering of any device being driven by the counter output and is a major reason for designing synchronous circuits.

The ripple counter uses less components than an equivalent synchronous counter, but it is slower. The worst-case time delay occurs when the most significant bit must change. An n-bit counter requires n propagation delays of a flip-flop before the signal ripples through to the output of the MSB. For n large, the counter becomes unacceptably slow. An 8-bit counter with 20-ns propagation delays per flip-flop requires at least 160 ns to change the MSB.

Asynchronous binary counters that count in straight binary sequence are relatively easy to design. For most other counting sequences synchronous counters are easier to design.

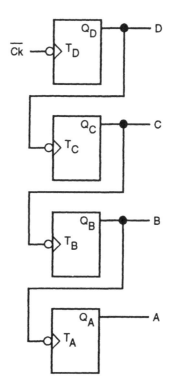

FIGURE 8.1.   An asynchronous 4-bit binary counter using T flip-flops.

8.2.2   Synchronous Binary Counters

A *synchronous binary counter* has all of its flip-flop clock inputs connected to the same common clock signal so that all its flip-flops change state at the same time, with a propagation delay of one flip-flop. A synchronous binary counter stores a binary code or number and increments the counter with each clock event.

Combinational logic is used to control when any given flip-flop will toggle. A synchronous counter can be built with either serial enable logic or with parallel enable logic. A four-bit synchronous counter with serial enable logic is shown in Figure 8.3, and a four-bit synchronous counter with parallel enable logic is shown in Figure 8.4. The circuit with parallel enable is the fastest binary counter structure.

8.3   DESIGNING SYNCHRONOUS COUNTERS

The binary sequence to be realized by the counter determines the control circuitry required to build the counter. The design starts with a statement of the

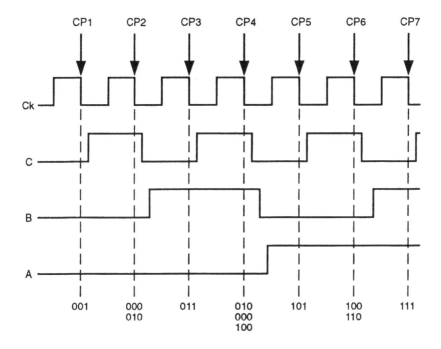

FIGURE 8.2. The timing diagram of a 3-bit asynchronous counter using negative-edge-triggered T flip-flops.

counting sequence, either in the form of a *state graph*, a *present-state/next-state table*, or *present-state/next-state maps*. A set of flip-flop input tables is generated next, which specifies the behavior of the flip-flop. This is then mapped and simplified to give the equations controlling the flip-flop inputs.

Example counter designs will be done using three-bit, modulo-8 counters, since nothing is gained pedagogically by designing with more than three flip-flops. The simplest counting sequence is straight binary, and either D or T flip-flops are the simplest to use since they have only one input. In designing with T flip-flops, the flip-flop must be toggled whenever the present state and the next state of that bit of the counter differ, i.e., $T = Q \oplus Q^+$. When using D flip-flops the next state is the present value of the input, or $D = Q^+$. The first design example will employ T flip-flops.

EXAMPLE 8.1: Design a binary modulo-8 counter which counts from 0 to 7, using positive-edge-triggered T flip-flops.

SOLUTION: The state table is shown below, with the flip-flop inputs added to the state table, and the state graph is shown in Figure 8.5(a). Flip-flop C toggles every time; hence, $T_C = 1$. The other T

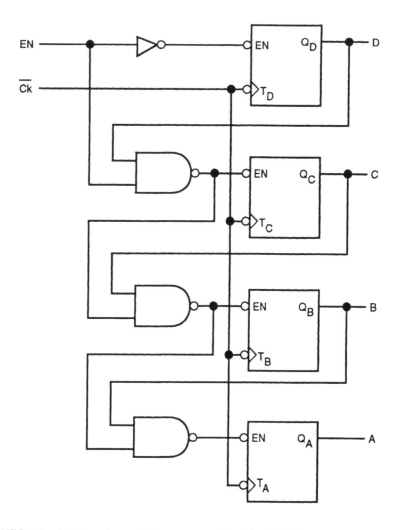

FIGURE 8.3.   A 4-bit synchronous binary counter with serial enable logic.

inputs are mapped and simplified as shown in Figure 8.5(b).
The input toggle equations, called *excitation equations,* are T_A
$= BC$, $T_B = C$, and $T_C = 1$, and the counter is realized as
shown in Figure 8.5(c).

TABLE 8.1

Present state			Next state			Flip-flop inputs		
A	B	C	A^+	B^+	C^+	T_A	T_B	T_C
0	0	0	0	0	1	0	0	1
0	0	1	0	1	0	0	1	1
0	1	0	0	1	1	0	0	1
0	1	1	1	0	0	1	1	1
1	0	0	1	0	1	0	0	1
1	0	1	1	1	0	0	1	1
1	1	0	1	1	1	0	0	1
1	1	1	0	0	0	1	1	1

The binary counter of Example 8.1 used feedforward only. The LSB flip-flop toggled on each clock pulse, and the remainder were toggled by signals from flip-flops in lower bit positions. If an n-bit counter is desired that counts to less than the maximum possible modulus of 2^n, it must be reset when it reaches its maximum count state. It requires feedback to do this.

For instance, a modulo-5 binary counter can count from ABC = 000 to 100, after which it must be reset to 000. If the counter has an active-low reset capability, the count of 101 can be used to reset the counter. To do this, outputs A and C are used as inputs to a NAND gate which is connected to the common reset line, so that when the counter attempts to count 101 it resets itself.

Counters are often needed which cycle in a sequence other than straight binary. When this occurs, feedback is used to cause the counter to omit some states. The techniques discussed in Example 8.1 can be extended to include both these cases of feedback.

The states that are skipped are determined by the feedback connections, and it is desired to minimize the feedback circuitry required. When the counter does not use all possible states, the missing states become don't cares in the next-state table, and they become don't cares when mapping the flip-flop inputs. This is shown in Example 8.2.

EXAMPLE 8.2: Design a three-bit, five-state counter which counts in the sequence 000, 100, 111, 010, 011, 000. Use positive-edge-triggered T flip-flops.

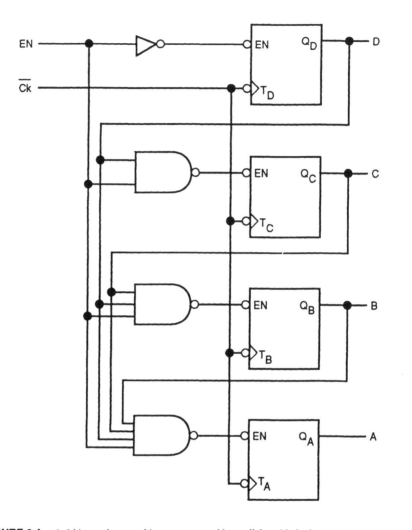

FIGURE 8.4.   A 4-bit synchronous binary counter with parallel enable logic.

SOLUTION: The state table is shown below and the state graph, input maps, and the counter are shown in Figure 8.6. The toggle inputs (excitation equations) are read from the maps as

$$T_A = \overline{A}\,\overline{B} + AB$$
$$T_B = \overline{A}\,C + A\overline{C}$$
$$T_C = A + B$$

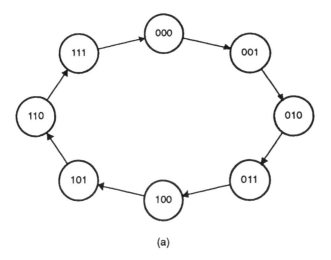

(a)

FIGURE 8.5.   A modulo-8 synchronous binary up-counter designed with T flip-flops; (a) the state graph, (b) the excitation maps, and (c) the circuit.

TABLE 8.2

Present state			Next state			Flip-flop inputs		
A	B	C	A^+	B^+	C^+	T_A	T_B	T_C
0	0	0	1	0	0	1	0	0
0	0	1	x	x	x	x	x	x
0	1	0	0	1	1	0	0	1
0	1	1	0	0	0	0	1	1
1	0	0	1	1	1	0	1	1
1	0	1	x	x	x	x	x	x
1	1	0	x	x	x	x	x	x
1	1	1	0	1	0	1	0	1

The counter of Figure 8.6 is designed with AND-OR control circuitry, done by obtaining a sum-of-products minimization of the input excitation maps. Sometimes it is simpler to design with OR-AND control circuitry. This can be done by looping the 0s of the input excitation maps and obtaining a product-of-sums minimization, as shown in Example 8.3.

EXAMPLE 8.3:   Realize the control circuitry of the five-state counter of Example 8.2 in OR-AND form.

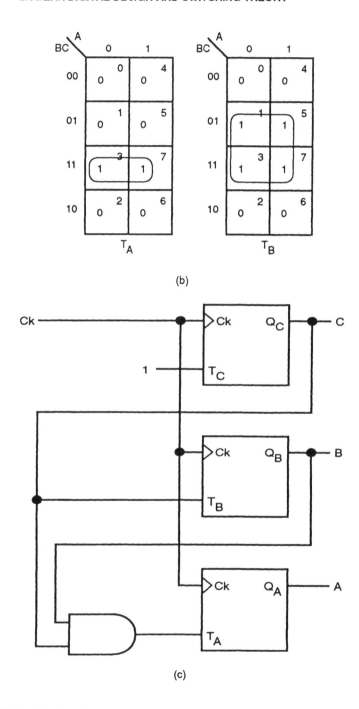

(b)

(c)

FIGURE 8.5.   (continued).

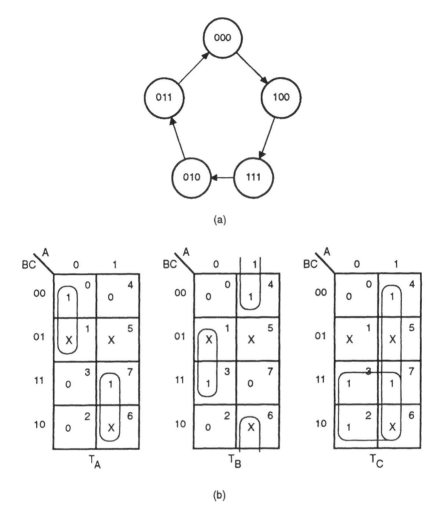

(a)

(b)

FIGURE 8.6. A 5-state synchronous counter with counting sequence 0, 4, 7, 2, 3, 0. (a) The state graph, (b) the excitation maps, and (c) the circuit. The connections from outputs to inputs are omitted.

SOLUTION: The counter excitation maps are redrawn in Figure 8.7(a) and simplified by looping the 0s in each case, to give toggle inputs

$$T_A = (A + \overline{B})(\overline{A} + B)$$
$$T_B = (A + C)(\overline{A} + \overline{B})$$
$$T_C = A + B$$

The counter circuit is shown in Figure 8.7(b).

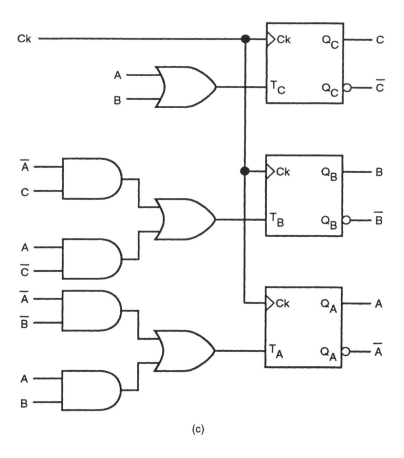

(c)

FIGURE 8.6. (continued).

Designing counters with S-R and J-K flip-flops is similar to designing with T flip-flops, but more work is required because there are now two inputs to each flip-flop. On the other hand, the J-K flip-flop is more powerful than the S-R, which is more powerful than the T, and the control circuitry should get simpler as one goes to more powerful flip-flops. To design with S-R and J-K flip-flops, one must determine the input signals required to achieve the desired outputs. Examination of the S-R and J-K next-state tables leads to the following excitation tables.

	TABLE 8.3					TABLE 8.4			
Q	Q⁺	S	R	Comment	Q	Q⁺	J	K	Comment
0	0	0	x	Don't set	0	0	0	x	Don't set
0	1	1	0	Set	0	1	1	x	Set or toggle
1	0	0	1	Clear	1	0	x	1	Clear or toggle
1	1	x	0	Don't clear	1	1	x	0	Don't clear

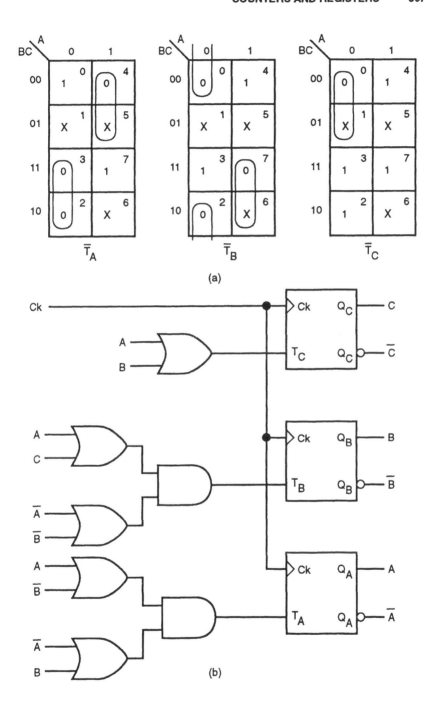

FIGURE 8.7. An alternate design of the 5-state synchronous counter with sequence 0, 4, 7, 2, 3, 0. (a) The excitation maps and (b) the circuit. Feedback connections again omitted.

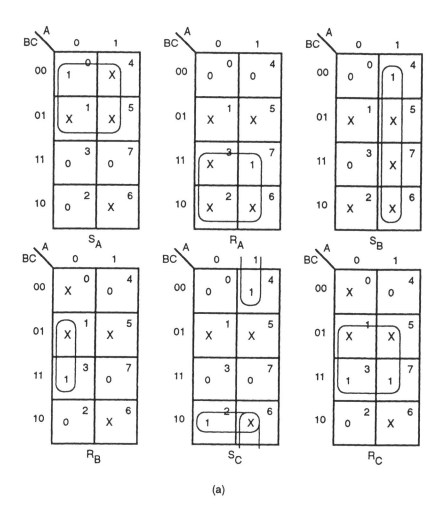

(a)

FIGURE 8.8.   The 5-state synchronous counter with sequence 0, 4, 7, 2, 3, 0 constructed with S-R flip-flops. (a) The excitation tables and (b) the circuit.

EXAMPLE 8.4:   Design the previous counter (sequence 0, 4, 7, 2, 3, 0) with positive-edge-triggered S-R flip-flops.

SOLUTION: The next-state and S-R excitation table are shown in Table 8.5. The excitation maps are shown in Figure 8.8(a). In simplifying them, care must be taken that SR = 11 does not occur. Simplifying as shown will avoid this, and the inputs are

$$S_A = \overline{B}, R_A = B$$
$$S_B = A, R_B = \overline{A}\,C$$

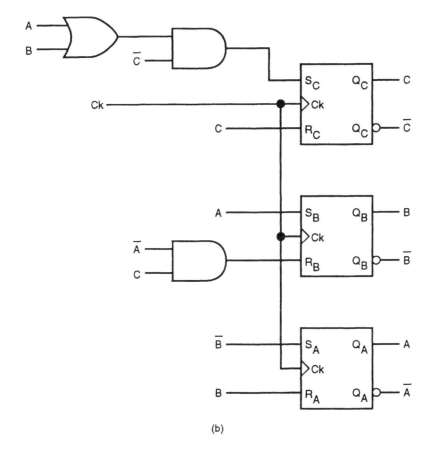

(b)

FIGURE 8.8. (continued).

$$S_C = A\overline{C} + B\overline{C} = (A + B)\overline{C}, R_C = C$$

The counter is shown in Figure 8.8(b).

TABLE 8.5

A	B	C	A⁺	B⁺	C⁺	S_A	R_A	S_B	R_B	S_C	R_C
0	0	0	1	0	0	1	0	0	x	0	x
0	0	1	x	x	x	x	x	x	x	x	x
0	1	0	0	1	1	0	x	x	0	1	0
0	1	1	0	0	0	0	x	0	1	0	1
1	0	0	1	1	1	x	0	1	0	1	0
1	0	1	x	x	x	x	x	x	x	x	x
1	1	0	x	x	x	x	x	x	x	x	x
1	1	1	0	1	0	0	1	x	0	0	1

EXAMPLE 8.5: Design the above five-state counter with positive-edge-triggered J-K flip-flops.

SOLUTION: The next-state and J-K excitation table are shown in Table 8.6.

TABLE 8.6

A	B	C	A^+	B^+	C^+	J_A	K_A	J_B	K_B	J_C	K_C
0	0	0	1	0	0	1	x	0	x	0	x
0	0	1	x	x	x	x	x	x	x	x	x
0	1	0	0	1	1	0	x	x	0	1	x
0	1	1	0	0	0	0	x	x	1	x	1
1	0	0	1	1	1	x	0	1	x	1	x
1	0	1	x	x	x	x	x	x	x	x	x
1	1	0	x	x	x	x	x	x	x	x	x
1	1	1	0	1	0	x	1	x	0	x	1

The J-K excitation maps of Figure 8.9(a) are obtained from Table 8.6, and the inputs are found to be

$$J_A = \overline{B}, \, K_A = B$$
$$J_B = A, \, K_B = \overline{A}\,C$$
$$J_C = A + B, \, K_C = 1$$

The counter is implemented as shown in Figure 8.9(b).

8.4 LOCKOUT

There is still a serious problem that can arise in the designs of Examples 8.2 through 8.5, a problem which is inherent in every synchronous sequential circuit that does not use all its possible states. If, for any reason, the counter gets into a state which is not one of the states of the primary counting sequence, it is possible that the counter has no path from this state to the primary counting sequence. This situation is called *lockout* and could occur due to any noise signal which erroneously triggers one of the flip-flops and throws the counter out of the primary sequence.

To avoid lockout there must be a path from any unused states back to the primary sequence. Avoiding lockout in this manner is called *illegal state recovery*, and a counter that avoids lockout is referred to as *self-starting*. To check the counter of Example 8.2 for illegal state recovery, one must determine which states the counter will go to if it gets into one of the three unused states. If, for instance, the counter is found to go from state 5 to state 6 to state 5, or from state 5 to state 6 to state 1 to state 5, the counter exhibits potential lockout. A counter could

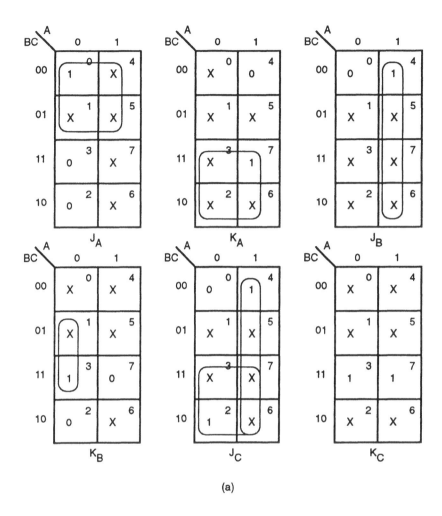

FIGURE 8.9.   The 5-state synchronous counter of Figure 8.10, implemented with J-K flip-flops. (a) The excitation tables and (b) the circuit.

conceivably go to a state for which all the toggle inputs are 0 and nothing further happens also.

Examination of the excitation maps for the three flip-flops in Figure 8.6(b) shows that the don't care inputs were treated as 1s if they were looped and treated as 0s if they were not looped. This is shown in the three maps of Figure 8.10, from which it can be seen that if the counter gets into state 001, the toggle signals received by the counter are $T_A T_B T_C = 110$. Flip-flops A and B will toggle, while flip-flop C will not. The counter will therefore change from state 001 to state 111, which is a valid state of the primary counter sequence.

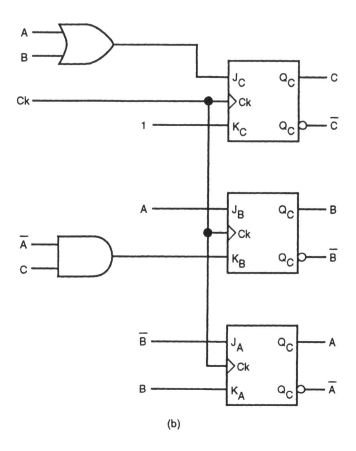

(b)

FIGURE 8.9.   (continued).

If the counter gets into state 101, the toggle signals are $T_A T_B T_C = 001$, and the counter goes from state 101 to state 100, which is in the primary sequence. Starting in state 110, the toggle signals are $T_A T_B T_C = 111$, sending the counter from state 110 to state 001, from which it will go to state 111, which is a valid state in the primary sequence. The complete counter state graph is as shown in Figure 8.11, and the counter does not have the potential for lockout.

Of course, the counter could be designed with every state specified. One solution would be to have the counter go from any unused state to the 0 state. This means replacing the don't cares in the next-state graph by 0s. If this is done to the counter of Example 8.2, one obtains input excitation controls

$$T_A = AB + AC + \overline{A}\,\overline{B}\,\overline{C}$$
$$T_B = A\overline{C} + \overline{A}BC$$
$$T_C = C + A\overline{B} + A\overline{B}$$

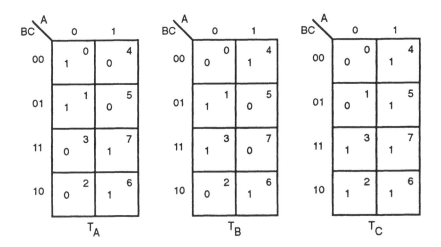

FIGURE 8.10.   The 1 and 0 assignments of the three toggle maps of Figure 8.6.

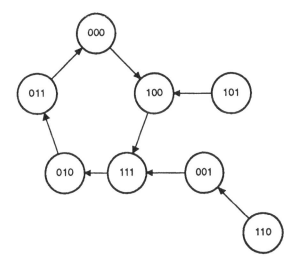

FIGURE 8.11.   The complete state graph of the counter of Figure 8.6.

The counter designed in this manner requires considerably more complicated control circuitry.

Another solution would be to design the counter with asynchronous-clear flip-flops that would automatically reset the counter when it enters an illegal state. For the counter in Example 8.2, it should reset if states 1, 5, or 6 occur. The signal to clear the counter is $CR = \overline{A}\,\overline{B}C + A\overline{B}C + AB\overline{C} = \overline{B}C + AB\overline{C}$. Two AND gates and an OR gate are required to reset the counter when it strays from the primary sequence.

8.5   SERIAL AND PARALLEL CARRY SYNCHRONOUS COUNTERS

The counter of Example 8.1 is called a binary up-counter, since it counts up from 0 to 7 in binary. The toggle inputs were found to be $T_A = BC$, $T_B = C$, and $T_C = 1$. Flip-flop C toggles every time, flip-flop B toggles whenever $C = 1$, and flip-flop A toggles when B and C are both 1. For any flip-flop to toggle, the outputs of all the lower flip-flops must be 1.

If the counter is expanded to a modulo-16 counter it requires four flip-flops, and the toggle excitations become $T_A = BCD$, $T_B = CD$, $T_C = D$, and $T_D = 1$. This can be generalized to give two representations of the toggle excitation signal, T_n, of flip-flop n as

$$T_n = Q_{n-1} \cdot Q_{n-2} \cdot Q_{n-3} \cdot Q_{n-4} \cdot \ \cdots \ \cdot Q_1 \tag{8.1}$$

$$T_n = Q_{n-1} \cdot T_{n-1} \tag{8.2}$$

where Q_{n-i} is the output of flip-flop $n - i$, and $T_0 = 1$.

The carries required to transfer a signal from the lower bit positions to the flip-flop in the nth bit position can be realized in two ways. The realization of Equation 8.1 gives a parallel carry, also called a *look-ahead carry,* and the realization of Equation 8.2 gives a serial carry. The parallel carry is the faster since there are fewer propagation delays, but the fan-out requirement of the LSB stage and the fan-in requirement of the MSB stage both quickly becomes excessive as the number of stages increases. A four-bit binary up-counter with parallel carry is shown in Figure 8.12.

The serial carry of Equation 8.2 requires a two-input AND gate for each stage of the counter and has the same fan-in and fan-out for every stage. It is slower than the parallel-carry counter due to the accumulative propagation delays of the carry chain. A four-stage binary up-counter with serial carry is shown in Figure 8.13.

A binary down-counter is one which starts from all 1s and counts down to all 0s. A modulo-8 down counter counts from 111 to 000. The next-state tables of both a modulo-8 binary up-counter and a modulo-8 binary down-counter are shown in Table 8.7.

It can be seen from Table 8.7 that if the active-high outputs of a binary counter are counting up, the complementary outputs are counting down. If the true outputs of a binary counter are interchanged for the complemented outputs, one has a binary down-counter. The counter can be connected as shown in Figure 8.12 or 8.13, except that the active-low outputs of each flip-flop are used. If the toggle inputs of the 3-bit counter are $T_A = \overline{B}\,\overline{C}$, $T_B = \overline{C}$, and $T_C = 1$, one has a binary modulo-8 down-counter. The toggle inputs to a modulo-16 binary down-counter are $T_A = \overline{B}\,\overline{C}\,\overline{D}$, $T_B = \overline{C}\,\overline{D}$, $T_C = \overline{D}$, and $T_D = 1$.

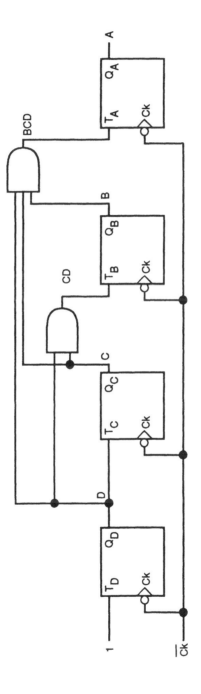

FIGURE 8.12. A 4-bit parallel-carry synchronous up-counter.

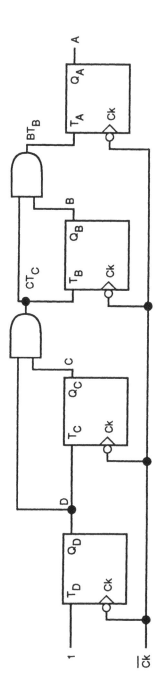

FIGURE 8.13. A 4-bit serial-carry synchronous up-counter.

TABLE 8.7

Up-counter						Down-counter					
A	B	C	A⁺	B⁺	C⁺	$\overline{A}$	$\overline{B}$	$\overline{C}$	$\overline{A}{}^+$	$\overline{B}{}^+$	$\overline{C}{}^+$
0	0	0	0	0	1	1	1	1	1	1	0
0	0	1	0	1	0	1	1	0	1	0	1
0	1	0	0	1	1	1	0	1	1	0	0
0	1	1	1	0	0	1	0	0	0	1	1
1	0	0	1	0	1	0	1	1	0	1	0
1	0	1	1	1	0	0	1	0	0	0	1
1	1	0	1	1	1	0	0	1	0	0	0
1	1	1	0	0	0	0	0	0	1	1	1

If a simple method of interchanging the outputs of each flip-flop can be found, one could design an up/down-counter. Alternatively, if an EOR gate is inserted in the output circuit of each flip-flop as shown in Figure 8.14, one can selectively complement or not complement each flip-flop output, depending upon the control signal. This same scheme also works for the serial carry binary counter. If the control line is low ($\overline{UP}$ = 0 or active), one has an up-counter, and if the control line is high (DOWN = 1 or active), one has a down-counter.

Asynchronous binary up/down-counters can also be designed, but they have a unique problem associated with them. When the direction of counting changes, clock pulses from the EOR gates can affect the count. One way to avoid this problem is to connect the toggle inputs to an enable input which can be used to force each T input to 0 before the direction of counting changes, so that undesired clock pulses will not change the counter.[1]

8.6 PROGRAMMABLE RIPPLE COUNTERS

Modulo-n or divide-by-n ripple counters are commercially available which can be user programmed to obtain any desired modulus. Asynchronous clear inputs can be used to reset the counter after the desired number of states. Thus, a modulo-5 counter counts up from 0 to 4 and must reset itself on $5_{10} = 101_2$. To achieve this behavior, connect outputs A and C to the reset through a NAND gate so that when the counter reaches 101, it is reset. Figure 8.15 shows this. The clear inputs are active low and the NAND gate outputs a low when both A and C are high.

When the circuit of Figure 8.15 is reset there is a glitch at the output. This is because the counter output does reach 101 for a brief period of time equal to the propagation delay of the NAND gate before the clear signal resets the counter. In this case the glitch occurs when the counter counts from ABC = 100 to 101 to 000, and a positive pulse appears at output A that should not occur. This is a fundamental problem with asynchronous counters.

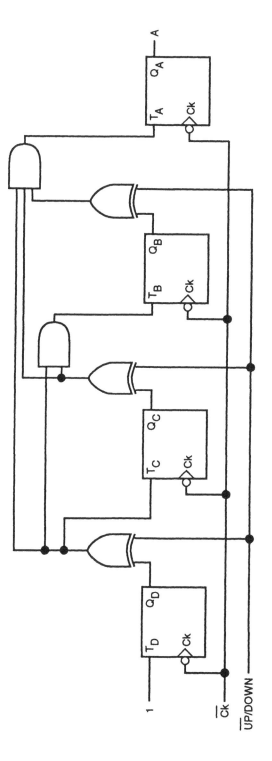

FIGURE 8.14. A 4-bit parallel-carry synchronous up/down counter.

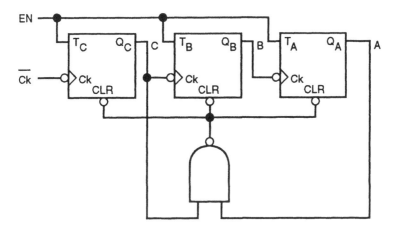

FIGURE 8.15.   A 3-bit modulo-5 asynchronous counter.

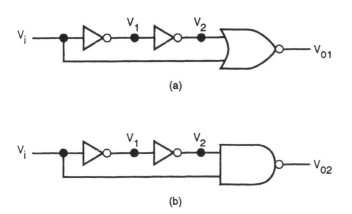

FIGURE 8.16.   A glitch filter or mask for (a) negative pulses, (b) positive pulses, and (c) the timing diagrams.

If the glitch can be either suppressed or ignored, the ripple counter with self-reset can be used. Since it can be established whether the glitch is a positive or negative pulse, it can be filtered out. The circuit in Figure 8.16(a) can be used to *mask* or filter a negative glitch, and the circuit in Figure 8.16(b) can be used to *mask* or filter a positive glitch. As shown in Figure 8.16(c), the positive-glitch filter produces two negative glitches, and the negative-glitch filter produces two positive glitches. If this is not satisfactory, a synchronous counter must be used.

A simple programmable ripple binary counter, which can divide by any number up to 16, is shown in Figure 8.17(a). The counter consists of a divide-by-2 stage (one flip-flop) and a divide-by-8 stage (three flip-flops). When EN = 1, the

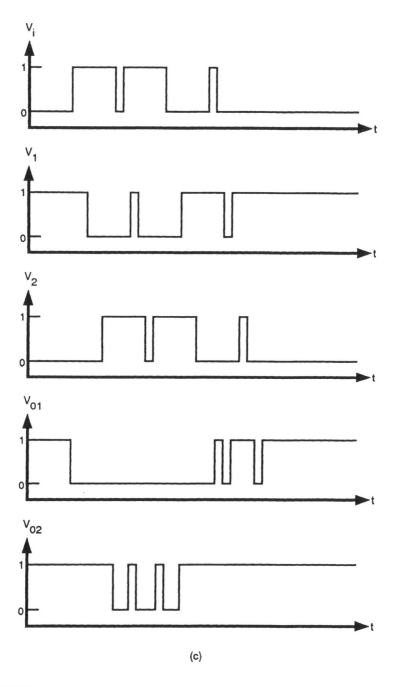

(c)

FIGURE 8.16.   (continued).

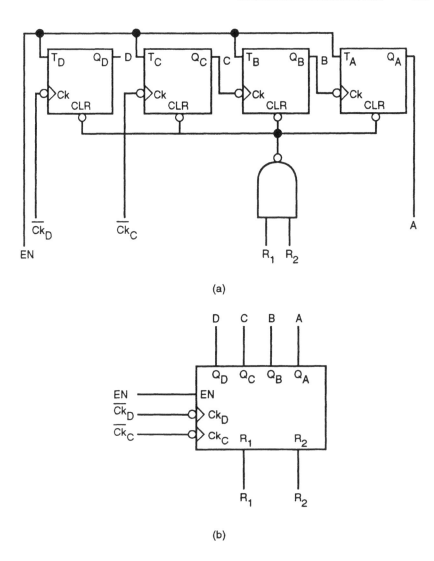

FIGURE 8.17. A programmable modulo-2/modulo-8 asynchronous binary counter. (a) The circuit and (b) a symbol.

flip-flops will toggle on the negative edge of the clock signal. A two-input NAND gate is connected to all the reset (CR) inputs. A suitable symbol for this counter is shown in Figure 8.17(b). This is essentially a 7493 MSI counter.

The circuit can be used as both a divide-by-2 counter and a divide-by-8 counter as is. To obtain a divide-by-16 counter, connect output Q_D to clock input Ck_C, as shown in Figure 8.18(a). To divide by any n other than 2, 8, or 16, one must reset the clock at count n. To construct a counter which counts to less than 8, one can use the divide-by-8 circuit. To divide by a number greater than 7, one can first

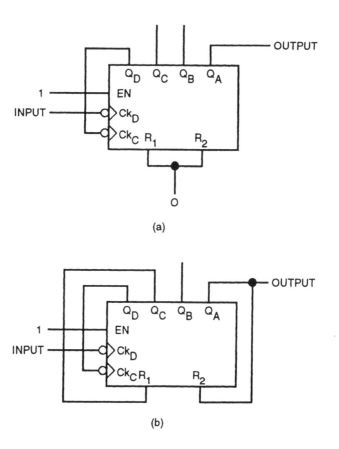

FIGURE 8.18. The ripple counter of Figure 8.17 configured as (a) a divide-by-16 and (b) a divide-by-10 counter.

convert the counter to a divide-by-16 counter and then program a reset signal to clear the counter at count n.

To construct a modulo-10 or divide-by-10 counter, the counter must count up to 9 and reset on the next clock pulse. To be able to count past 7, first connect the counter as a modulo-16 counter. This is done as before by connecting output Q_D to clock input C. The number $10_{10} = 1010_2$; hence, the counter is to reset when $A = 1$ and $C = 1$. If output A is connected to one of the two reset inputs and output C is connected to the other reset input, as shown in Figure 8.18(b), the two reset signals will be high when the counter reaches 1010 and the output of the NAND gate will go low and clear all the flip-flops. The counter will thus count up to 9 and reset.

When counters are cascaded their moduli are multiplied. To count modulo-17 up to modulo-256, one can cascade two modulo-16 counters. To count modulo-64, one can simply connect the two divide-by-8 circuits in cascade. Every 8 counts

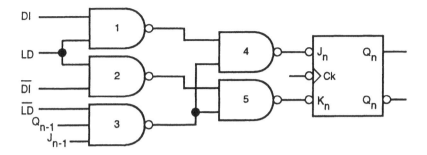

FIGURE 8.19. One stage of a programmable counter/storage register.

of the first counter will send a pulse to the second counter until the count reaches 63. To count to 79 (divide by 80), one counter can be configured to count modulo-16 and feed a second counter which counts modulo-5.

The 7493 counter requires an external AND gate to count to any number whose vector magnitude is greater than 2. A modulo-7 counter has to reset on count 0111, and a modulo-11 counter has to reset on count 1011. Both require ANDing three input signals to the reset. Since the counter comes with a two-input reset NAND gate, an additional AND gate is required in order to reset the counter on any binary string with three or more 1s.

Sometimes a counter is needed whose modulus can be changed either by clearing at the appropriate time or by presetting the counter to a given value. If the counter could be further generalized to store data it would be quite general. Counting is done easily by toggle flip-flops, and data storage is done easily by delay flip-flops. The J-K flip-flop can be made to toggle if its inputs are both the same or to store data if its inputs are complementary. Stage n of a circuit that can do this is shown in Figure 8.19.[2] DI is the data input signal, Q_{n-1} is the output of the next lower stage, J_{n-1} is the input of the next lower stage, and LD is the load signal. When the LD signal is high, gate 3 is disabled and it outputs a 1. DI is then fed to the J_n input through gates 1 and 4, and $\overline{DI}$ is fed to the K_n input through gates 2 and 5. When LD is low, gates 1 and 2 are disabled and output 1s, while gate 3 is enabled and ANDs Q_{n-1} and J_{n-1} and feeds the product to the J and K inputs via gates 4 and 5. Breeding[2] gives a four-bit presettable binary counter with parallel carry. For a thorough treatment of counters, see Kostopoulos[3] or Williams.[4]

8.7 BINARY SHIFT REGISTERS

Another major application of latches and flip-flops is in the implementation of registers, which are locations for the temporary storage of data. The data can be incoming information or the intermediate results of a computation. Registers must

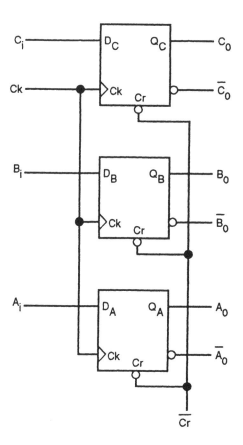

FIGURE 8.20. A 3-bit PIPO register composed of edge-triggered D flip-flops.

be clocked, and the data stored in them can only change on a clock pulse. During the time the data is stored, it is available for use by other processing elements of the digital system.

A *register* is a collection of latches or flip-flops which are clocked on a common signal. It is used to store 1s and 0s, and can be constructed with edge-triggered D flip-flops, as shown in Figure 8.20, where the subscripts i and o refer to the input and output data bits, respectively. Edge-triggered and master/slave flip-flops require only one clock to strobe them. Registers normally have direct clear inputs and both active-high and active-low outputs.

S-R and J-K flip-flops can be used as registers also, but $SR = 00$ or $JK = 00$ accomplishes nothing, and $SR = 11$ is not allowed, while $JK = 11$ toggles the flip-flop and changes the stored bit. In a gated or strobed register composed of S-R or J-K flip-flops, the S or J input can be used as the data input and the R or K input can be used for direct reset or clear.

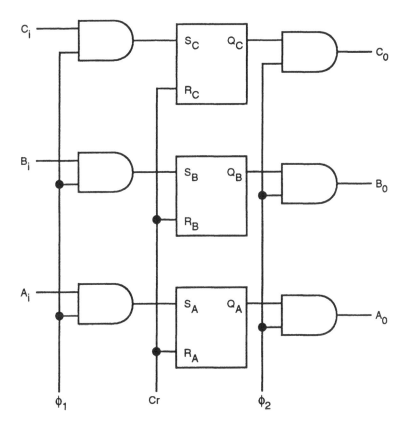

FIGURE 8.21. A 3-bit PIPO register composed of S-R latches.

A register composed of S-R latches and using two-phase clocking is shown in Figure 8.21. The reset line is activated to clear the register asynchronously before data are loaded. The data are loaded into the register on clock phase ϕ_1, the input data strobe, and the data will appear on the output lines when clock phase ϕ_2, the output data strobe, is activated. If gated S-R latches are used the clock input can be used as an enable input. To avoid the SR = 11 condition, the data strobe and reset signals must not both be asserted at the same time.

A *shift register* is an n-bit registor with a provision for shifting its stored data by one bit position on each clock pulse. A unidirectional shift register can shift its data in only one direction, while a bidirectional shift register can shift its data in either of two directions.

Data can be loaded into a shift register in either a serial or a parallel manner, and the contents of the register can be output in either a serial or a parallel manner. Registers are labeled according to how they are loaded and read, and can thus be classified into four kinds: serial in/serial out (SISO), parallel in/parallel out

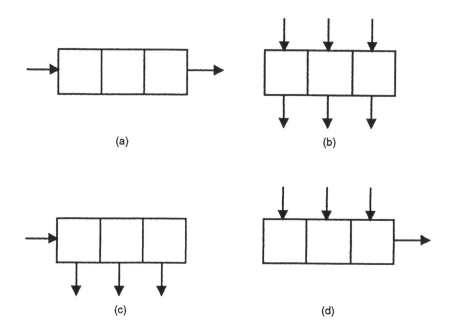

FIGURE 8.22. The four basic types of shift register: (a) SISO, (b) PIPO, (c) SIPO, and (d) PISO.

(PIPO), serial in/parallel out (SIPO), and parallel in/serial out (PISO). These four types of shift register are shown symbolically in Figure 8.22.

A *serial-in register* loads the input data into the D input of the first flip-flop one bit at a time on each clock pulse. A strobe enable signal can be used to disable the shift operation if it is desired to hold the data during one or more clock pulses. If the flip-flops in the register do not have an enable input, a 2-to-1-line MUX, controlled by an enable signal, would be needed to drive the input of each flip-flop.

A *parallel-in register* loads all the bits into the register on the same clock pulse, and it can either load new data or shift the stored data. To do this, it requires either an enable input or a 2-to-1-line MUX, controlled by a LOAD/SHIFT signal, to drive the input of each flip-flop. The MUX is needed to guarantee that LOAD and SHIFT are disjoint and are never both high at the same time.

A *serial-out register* outputs its contents one bit at a time, while a *parallel-out register* outputs its contents on one strobe signal. In a serial-out register, the output of each flip-flop is connected to the D input of the next flip-flop, and the output is read at the last stage. For the parallel-out register, the output of each flip-flop is available on the same strobe pulse.

Figure 8.23 shows a SISO shift register and Figure 8.24 shows a PIPO shift register. Again, LOAD and SHIFT must be disjoint, and the subscripts i and o refer to the input and output data, respectively.

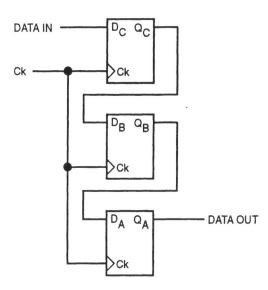

FIGURE 8.23. A 3-bit SISO shift register.

Sometimes a shift register is used to receive serial information and output parallel information, or the converse. A SIPO register can perform serial-to-parallel conversion, and a PISO register can perform parallel-to-serial conversion.

Registers can be designed to shift data in either direction. The two directions are called left and right. A *shift-left* operation moves the data bits toward higher bit positions, and a *shift-right* operation moves the data bits toward lower bit positions. If the register is storing binary numbers, a left shift corresponds to doubling the stored number, and a right shift corresponds to halving the stored number.

A *unidirectional shift register* can only shift data in one direction, while a *bidirectional shift register* can shift data both left and right, with the addition of proper gating. A bidirectional shift register that can input and output data either serially or in parallel is called a *universal shift register* because it can perform all four possible combinations of shifting or storing data.

A universal shift register is shown in Figure 8.25. The 4-to-1-line MUXs are controlled by a shift-left/shift-right ($L/\overline{R}$) signal and a load (LD) signal and follow the truth table shown in Table 8.8. On a shift-left signal, a new bit of data is loaded into the register at the LSB position by the LEFT IN line, and each remaining flip-flop loads a bit from the immediately lower bit position, Q_{n-1}. On a shift-right signal, a new data bit is loaded into the register at the MSB position by the RIGHT IN line, and each remaining flip-flop loads a bit from the immediately higher bit position, Q_{n+1}.

On a load data signal, new data are loaded into the register. The fourth control signal of the MUX is a hold or no-operation signal, and the register retains the data

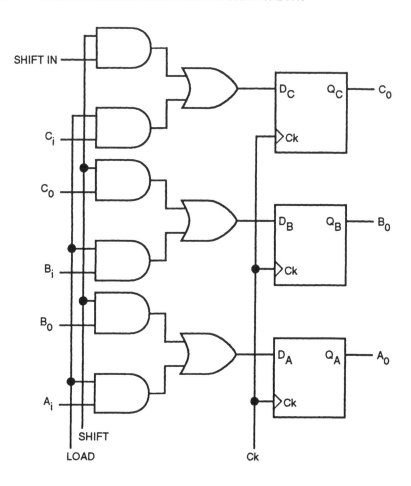

FIGURE 8.24. A 3-bit PIPO shift register, output connections omitted.

stored in it. This can be accomplished either by making no connection to the MUX, or each datum can be fed back to the input of its flip-flop to refresh itself.

TABLE 8.8

	Inputs		Next state			
Function	S_1	S_0	A_0	B_0	C_0	D_0
SHR	0	0	R_i	A_0	B_0	C_0
Load	0	1	A_i	B_i	C_i	D_i
SHL	1	0	B_0	C_0	D_0	L_i
Hold	1	1	A_0	B_0	C_0	D_0

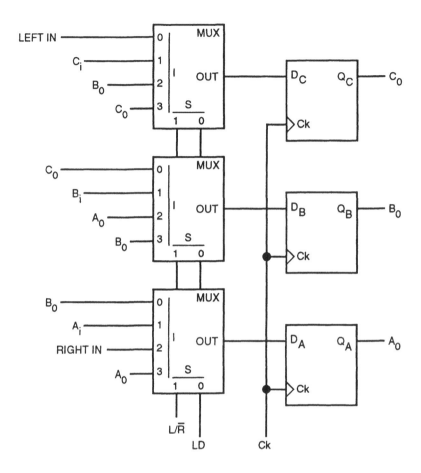

FIGURE 8.25. A universal shift register constructed with D flip-flops.

8.8 SUMMARY

Binary counters can be classified as synchronous or asynchronous, and as up- or down-counters. Clocked counters are synchronized with a master clock and are designed such that all the flip-flops are triggered simultaneously by the clock. They are faster than asynchronous counters and can count in any desired sequence. An unclocked counter is called a ripple counter since the data ripples through the counter in an asynchronous manner. An n-bit ripple counter has a delay n times that of a single flip-flop, and the clock frequency must be such that the counter has time for all its outputs to change. This requires a clock period of more than n delays, and as n becomes large, the maximum clock frequency is decreased.

Counters can also be specified by the counting sequence and can be classified by their modulus. Counters that either skip states or count in nonbinary sequences employ feedback as well as feedforward. The feedback paths can be complex, and it is necessary to design for minimum circuitry to achieve the desired count sequence. This is done by examining a present-state/next-state table to see when each flip-flop input must be changed, then mapping the information and simplifying the circuitry. Toggle and delay flip-flops are the easiest to design with, but they require the most complicated feedback circuitry. S-R flip-flops are more powerful and require less feedback circuitry, and J-K flip-flops require the least.

An n-bit counter has 2^n potential states. If they are not all used, the counter must be designed to avoid lockout by having a path from any unused state to the primary sequence. This can be done in one of several ways. The unused states can be checked to assure that lockout cannot occur. If the potential for lockout exists, the counter can be designed to go from any unused state to a suitable state in the main sequence or to go to state 0. If the next state of any unused states is state 0, the counter will automatically reset if it gets into an illegal state. Circuitry can also be provided that will not affect the primary sequence but will asynchronously reset the counter if it enters an illegal state.

Registers are useful for the temporary storage of data. They can be classified according to whether they load their data in serial or parallel form and whether they output their data in serial or parallel form. They also can be classified by whether they shift data to the left or to the right.

SISO registers are slow. An n-stage SISO register can be used to delay data by $n-1$ clock pulses. PIPO shift registers are faster. Sometimes it is desirable to use serial and parallel operation. Parallel registers are generally used in high-speed computers, and serial registers are needed for modems that transmit signals on a single telephone line. In this case, a PISO register could be used by the sender and an SIPO register by the receiver.

A parallel loading register can shift data either to the left or right, and registers that can shift both ways are called bidirectional or shift-right/shift-left registers. A universal shift register has serial inputs to allow it to shift left or right, and it can load data or output data in parallel. It can also store data.

Shift registers are useful in some counter circuits. Two of the most common shift-register counters are the ring counter and the Johnson counter. These counters are discussed in the assignments.

PROBLEMS

8.1. How many flip-flops are required to construct binary counters modulo-4, -6, -9, and -60?

8.2. What is the highest state of a ripple counter if it has (a) five stages, and (b) nine stages?

8.3. How many unused states are there in a 7-bit modulo-100 binary counter?

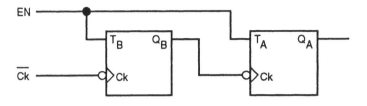

FIGURE 8.26.

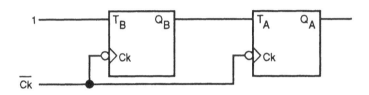

FIGURE 8.27.

8.4. What is the delay through an 8-stage serial shift-register which is clocked at 10 MHz?

8.5. An 8-stage bidirectional shift-register contains the number N.
(a) How many shifts in what direction are required to divide the number N by 16?
(b) How many shifts in what direction are required to divide 256 by 8?

8.6. If each flip-flop of a modulo-64 binary ripple counter has a 20-ns propagation delay, what is the maximum clock frequency at which the counter can operate correctly?

8.7. A clock frequency of 20 MHz is used as input to a ripple counter, and the output is taken from the last flip-flop.
(a) What is the frequency of the output of a divide-by-12 counter?
(b) What is the frequency of the output of a divide-by-16 counter?

8.8. A 6-stage bidirectional shift register contains the number 010101_2. What number, base 10, is stored in the register after (a) four shifts left, and (b) four shifts right? Zeros are loaded into the register as needed in both cases.

8.9. Sketch the timing diagram of the 2-bit asynchronous counter shown in Figure 8.26. The clock is a square wave of period 200 ns, and the propagation delay of each flip-flop is 20 ns.

8.10. Repeat Problem 8.9 for the 2-bit synchronous circuit shown in Figure 8.27.

8.11. Sketch the timing diagram of a modulo-5 binary ripple counter and show what, if any, glitches occur. Assume a clock period of 200 ns and glitches of width 20 ns.

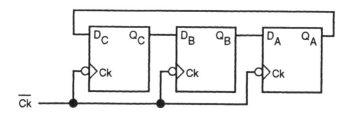

FIGURE 8.28.

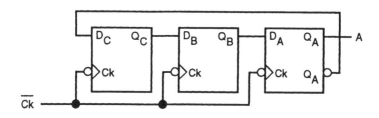

FIGURE 8.29.

8.12. Modify the circuit of Figure 8.12 to include an enable input.

8.13. Design a military clock which counts from 0 to 24 hours. Binary counters modulo-2, -4, -6, -8, and -10 are available and a 60-Hz square wave is available for an input signal.

8.14. A ring counter consisting of a 3-stage shift register is shown in Figure 8.28. Sketch the timing diagram of this counter and give the output sequence
  (a) if the counter starts from ABC = 001.
  (b) if the counter starts from ABC = 011.
  (c) Name one advantage and one disadvantage of the ring counter vs. a binary ripple counter.

8.15. Give the state diagram of a 5-stage ring counter
  (a) if the starting sequence is 00001.
  (b) if the starting sequence is 01001.

8.16. A Johnson or twisted-ring counter is obtained by interchanging an input as shown in Figure 8.29.
  (a) Give the count sequence ABC for this counter if it starts in state 000.
  (b) Give one advantage of the Johnson counter over a similar ring counter.

8.17. Design a divide-by-9 counter from a 7493, shown in Figure 8.17.

8.18. Design a divide-by-7 counter from a 7493, shown in Figure 8.17.

8.19. Design a modulo-4 Gray-code counter using J-K flip-flops.

8.20. The counter shown in Figure 8.30 starts in state 00000. Give the state table or state diagram of this counter.

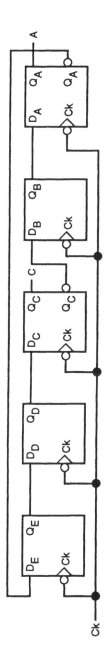

FIGURE 8.30.

A	B	C	A^+	B^+	C^+
0	0	0	1	0	1
0	0	1	1	1	1
0	1	0	0	1	1
0	1	1	1	1	0
1	0	0	0	0	0
1	0	1	1	0	0
1	1	0	0	1	0
1	1	1	0	0	1

FIGURE 8.31.

A	B	C	A^+	B^+	C^+
0	0	0	0	0	1
0	0	1	0	1	1
0	1	0	1	1	1
0	1	1	1	1	0
1	0	0	0	0	0
1	0	1	1	0	0
1	1	0	0	1	0
1	1	1	1	0	1

FIGURE 8.32.

8.21. Complete the state table or state diagram for a counter that counts in the sequence 0, 3, 1, 5, 4, 6, 2, 7, 0.

8.22. Complete the state table or state diagram for a counter that counts in the sequence 0, 4, 1, 5, 2, 6, 3, 7, 0.

8.23. Obtain the toggle maps for flip-flops A, B, and C of the counter, whose next-state table is given in Figure 8.31.

8.24. Obtain the toggle maps for flip-flops A, B, and C of the counter whose next-state table is given in Figure 8.32.

8.25. Design the counter whose excitation maps are shown in Figure 8.33.

8.26. Design the counter whose excitation maps are shown in Figure 8.34.

8.27. A counter that counts by 3s, modulo-8 is to be designed using clocked toggle flip-flops and AND and OR gates. The count sequence is 0, 3, 6, 1, 4, 7, 2, 5, 0.

8.28. Repeat Problem 8.27 using clocked S-R flip-flops.

8.29. Repeat Problem 8.27 using clocked J-K flip-flops.

8.30. Design a synchronous counter which counts in the sequence 0, 2, 3, 4, 1, 0. The next state of any unused states is to be 000. Flip-flop A is a toggle, flip-flop B is an S-R, and flip-flop C is a J-K. AND and OR gates are available as needed.

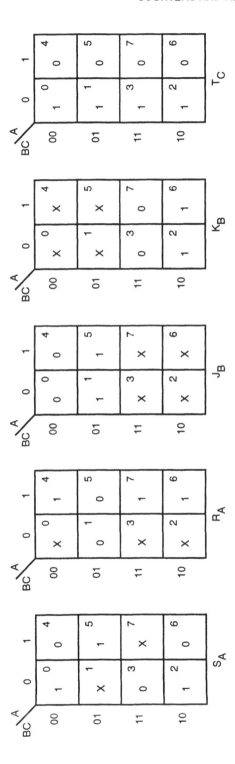

FIGURE 8.33.

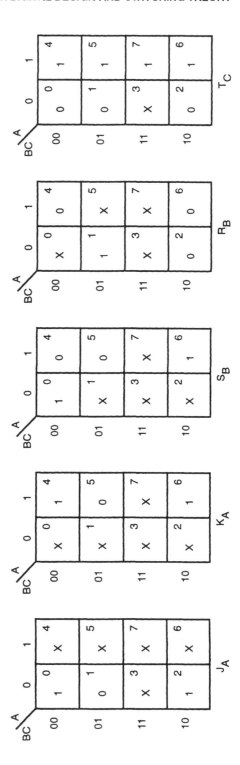

FIGURE 8.34.

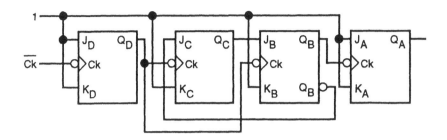

FIGURE 8.35.

8.31. Design a synchronous counter which counts in the sequence 0, 2, 7, 4, 6, 3, 1, 0. State 5 is a don't care. Flip-flop A is a toggle, flip-flop B is an S-R, and flip-flop C is a J-K. AND and OR gates are available as needed.

8.32. Design a synchronous counter which counts in the sequence 0, 1, 4, 3, 5, 2, 6, 7, 0. Flip-flop A is a J-K, flip-flop B is an S-R, and flip-flop C is a toggle. AND and OR gates are available as needed.

8.33. Design a synchronous counter that uses S-R flip-flops and counts in the sequence 7, 6, 5, 4, 3, 2, 1, 0, 7. AND, OR, EOR, and ENOR gates are available.

8.34. Design a synchronous counter that uses S-R flip-flops and counts in the sequence 0, 1, 2, 3, 5, 7, 0. The unused inputs are don't cares. AND, OR, EOR, and ENOR gates are available. Show that lockout does not occur for your design.

8.35. Repeat Problem 8.34 using J-K flip-flops.

8.36. Design a synchronous counter that uses S-R flip-flops and counts in the sequence 0, 1, 3, 2, 6, 7, 5, 4, 0. AND, OR, EOR, and ENOR gates are available. Check SOP and POS solutions and obtain the simpler solution.

8.37. Repeat Problem 8.36 using J-K flip-flops.

8.38. Design a synchronous modulo-16 counter which counts in the sequence 0, 8, 12, 10, 14, 1, 9, 13, 11, 15, 0. Use toggle flip-flops, AND and OR gates.

8.39. Repeat Problem 8.38 using S-R flip-flops, AND and OR gates.

8.40. Repeat Problem 8.38 using J-K flip-flops, AND and OR gates.

8.41. Repeat Problem 8.38 using D flip-flops, AND and OR gates.

8.42. Design a synchronous decade counter that counts in the sequence 0, 1, 2, 3, 4, 5, 6, 7, 8, 9, 0. Use toggle flip-flops, AND and OR gates.

8.43. Repeat Problem 8.42 using S-R flip-flops, AND and OR gates.

8.44. Repeat Problem 8.42 using J-K flip-flops, AND and OR gates.

8.45. Repeat Problem 8.42 using D flip-flops, AND and OR gates.

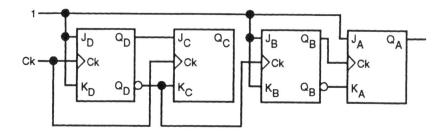

FIGURE 8.36.

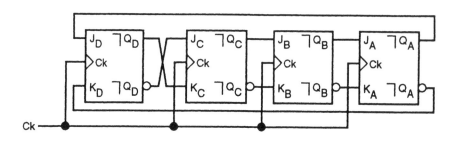

FIGURE 8.37.

8.46. An X-Y flip-flop clears when $XY = 00$, sets when $XY = 11$, toggles when $XY = 10$, and idles (doesn't change) when $XY = 01$. Design a synchronous 3-bit counter that uses three X-Y flip-flops, AND and OR gates to count in the sequence 0, 1, 3, 7, 5, 4, 0. The unused inputs are don't cares.

8.47. The counter shown in Figure 8.35 starts in the cleared state with all outputs equal to 0.
  (a) Tabulate the outputs for the first six clock pulses.
  (b) Give the timing diagram for the first six effective clock transitions. Neglect propagation delays.

8.48. Repeat Problem 8.47 for the counter shown in Figure 8.36.

8.49. (a) Determine the sequence of states in the counter of Figure 8.37. All flip-flops are master/slave with data lockout. Initially each flip-flop is in the cleared state.
  (b) Give the timing diagram for the first six effective clock transitions. Neglect propagation delays.

8.50. (a) Design a circuit which adds 5 to the contents of a 4-bit register. Use J-K flip-flops and AND and OR gates.
  (b) Show that the circuit works properly when the counter initially contains 0, 5, and 7.

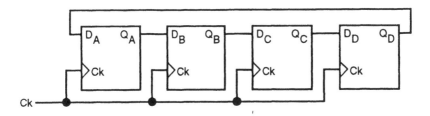

FIGURE 8.38.

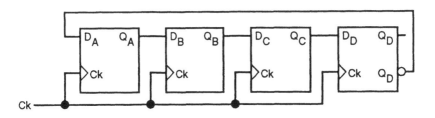

FIGURE 8.39.

SPECIAL PROBLEMS

8.51. A 4-bit ring counter is shown in Figure 8.38.
   (a) Give the complete state diagram for this circuit.
   (b) Sketch the timing diagram for a square wave clock signal if the counter starts in the state ABCD = 1000.
   (c) Sketch the timing diagram for a square wave clock signal if the counter starts in the state ABCD = 1100.
   (d) Sketch the timing diagram for a square wave clock signal if the counter starts in the state ABCD = 1110.
   (e) Show that the counter can be used to generate two nonoverlapping clock phases, each with 25% duty cycle. (25% of the time the clock signals are high, and 75% of the time they are low.)
   (f) What happens if the clock gets into state 0000?
   (g) What happens if the clock gets into state 0101?
8.52. A 4-bit Johnson counter is shown in Figure 8.39.
   (a) Give the counting sequence for this circuit, starting from state 0000. What is the modulus of the counter?
   (b) Sketch the timing diagram of the counter, starting from state 0000.
   (c) How is the output frequency related to the clock frequency?
   (d) What happens if the counter gets into state ABCD = 0010?
8.53. Design a counter which uses S-R flip-flops to count to 10 by 2s and then reset. Use clocked flip-flops with a direct-reset capability. AND

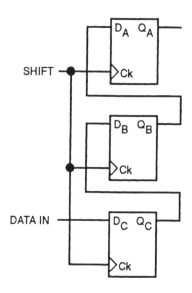

FIGURE 8.40.

and OR gates are available. Give the next-state diagram, the excitation maps, and the circuit realization. Design an auxiliary circuit that will reset the counter if it gets into an unused state.

8.54. Repeat Problem 8.53 using J-K flip-flops.

8.55. A 3-bit shift-left register is shown in Figure 8.40. Sketch the complete state diagram of the register.

8.56. Design a modulo-8 Gray-code counter using three flip-flops. The flip-flops available are clocked toggle, delay, S-R, and J-K, and AND, OR, NAND, NOR, EOR, and ENOR gates are available.

(a) Specify your reasons for choosing the type flip-flop used.

(b) Specify the states in such a manner that lockout is avoided.

8.57. Repeat Problem 8.56 for a modulo-8 XS3 counter.

REFERENCES

1. **Forbes, M. and Brey, B. B.,** *Digital Electronics,* Bobbs-Merrill Educational Publishing, Indianapolis, 1985, 192–193.

2. **Breeding, K. J.,** *Digital Design Fundamentals,* Prentice-Hall, Englewood Cliffs, NJ, 1989, 288–291.

3. **Kostopoulos, G. K.,** *Digital Engineering,* John Wiley & Sons, New York, 1975, chap. 6.

4. **Williams, G. E.,** *Digital Technology,* 2nd ed., *Science Research Associates,* Chicago, 1982, chap. 7.

CHAPTER 9

Application-Specific Integrated Circuits

9.1 INTRODUCTION

Chips designed for a particular limited product or application are called *application-specific integrated circuits (ASICs)*, and usually fall into one of five groups of semicustom components: programmable logic devices (PLDs), erasable programmable logic devices (EPLDs), field programmable gate arrays (FPGAs), gate arrays, and standard cells. When compared to other circuit realizations, ASICs generally require less components. This leads to smaller, faster designs which consume less power.

PLDs are composed of programmable logic array (PLA) devices, programmable read-only memory (PROM) devices, programmable array logic (PAL) devices, and dynamic logic array (DLA) devices. They are the logical choice with which to implement two-level logic when a finished product is required as soon as possible or when frequent circuit modifications are anticipated. They are ideal for introducing students to ASIC technology due to their low cost, flexibility, and rapid turnaround time.

The AND-OR structure of the PLA is the core of all PLDs since this structure can be used to implement any two-level logic functions. PLAs are also attractive to the custom VLSI logic designer because their regular structure requires a minimum number of separate cell designs and allows for ease in testing, while offering the opportunity for simple, rapid expandability.

Most commercial PLDs contain flip-flops since many digital designs require them. Such PLDs are called registered PLDs, because they generally have a single common clock signal to all their flip-flops. Recall that a register is defined as a collection of two or more flip-flops having a common clock signal. Shift registers can be implemented with registered PLDs; hence, any synchronous counters which combine shift registers and combinational logic can be easily implemented with registered PLDs.

Adding feedback to a PLD converts it to a finite-state machine (FSM). The finite-state machine under discussion is a synchronous system, with data moving between registers that are alternately clocked on a two-phase scheme so that there will be no races. The result of this approach is a finite-state machine that is simple and easy to lay out using established design rules. It is easy to test each FSM module, and there is less chance of error in designing the system. This FSM is the basic pipeline structure, consisting of register-to-register transfer of data, with local feedback added.

The next level of complexity is the folded-NAND or the folded-NOR multi-level programmable array structure. Each NAND of the folded-NAND structure or NOR of the folded-NOR structure generates one level of logic. A single level of logic can be implemented with these devices more efficiently than with the two-level PLD structure, and multilevel logic is realizable with the folded-NAND and folded-NOR by cascading multiple terms. A single programmable array can generate N levels of logic with N passes through the array. These structures can realize two-level SOP and POS solutions with two passes through the array, or they can implement higher complexity multilevel logic functions with more passes through the array.

Gate arrays (also called uncommitted logic arrays or ULAs) incorporate up to several thousand NOR and/or NAND gate cells on a single chip, arranged in rows and columns, with routing channels running vertically and horizontally between the cells. There is no opportunity to create custom cells in gate-array architecture; only the routing can be changed.

A recent newcomer to the family of programmable devices is the FPGA, which consists of regular arrays of programmable logic blocks called macrocells that can be interconnected by a programmable routing network. These are more general than PLA structures, and will be considered separately.

9.2 GENERAL PLA CONSIDERATIONS

PLDs are small-scale application-specific integrated circuits that can be programmed to realize one or more specific logic functions. PROM was the first user-programmable device, followed by the FPLA and PAL devices. Conventional FPLAs consist of programmable AND/OR arrays, and can include registered or latched inputs and outputs. Some FPLAs can be programmed only once, by blowing internal links called fuses to customize the interconnection of the internal gates. Others can be either electrically or optically erasable, and can be reprogrammed many times. The PLA is also a digital, application-specific integrated circuit that is widely used in custom VLSI design. Both types of ASIC will be covered in this chapter.

The most general form of PLD is the PLA, which has a programmable AND array and a programmable OR array. Physically, the OR-plane matrix is identical in form to the AND-plane matrix, but its layout is rotated 90° with respect to the AND plane. The input and output registers need not be identical, but they are also repetitive structures.

The size of a PLA is determined by the number of inputs, including signals that are fed back to the input, the number of outputs, and the number of product terms. A simple criterion of PLA size is the total number of crosspoints required. The number of crosspoints is the product of the number of rows and the number of columns of both the AND plane and the OR plane. A transistor placed at a crosspoint can connect a row to a column, in which case the crosspoint becomes an electrical node. Each crosspoint can be thought of as a potential node.

Let R, C, and N represent rows, columns, and crosspoints (potential nodes), respectively. A PLA requires two inputs per input variable, one for the true and one for the complemented variable. A PLA with three input variables, nine product terms, and four outputs requires an AND plane of $9R \times 6C = 54N$, and an OR plane of $9R \times 4C = 36N$. The total size of the PLA is thus $9R \times 10C = 90N$. In general, R rows by C columns produce $N = R \times C$ crosspoints.

Area, speed, and power are all critical parameters as PLA size increases. Large PLAs take up more area and are slower due to the higher fan-in and fan-out. Large PLAs require both inverting and noninverting input buffers[*] to drive the large fan-in of the AND array, and the output often requires large buffers to drive the PLA load. For these reasons, it is necessary to reduce the size of each PLA as much as possible.

The number of input and output lines of a PLA is determined by the system requirements. This fixes the number of columns of the PLA, and the design engineer usually has control over the number of product lines only. Thus, to create the smallest possible PLA for any specific application, the designer must be able to minimize the number of product lines needed.

It is advantageous if the designer can rearrange the functions in a manner so as to reduce the number of input lines at the expense of added product lines. Since each additional input line increases the size of the input register, it adds more to the overall size of the PLA than does a product line, even though the number of crosspoints may not be reduced.

PLA design is an exercise in multi-output gate minimization, as was discussed in Chapter 5, Section 5.4. In particular, a minimization of three functions was performed in Chapter 5, Section 5.4.2, and shown in Chapter 5, Figure 5.15. The original three functions were $X(A,B,C,D) = \sum m\,(2,4,10,11,12,13)$, $Y(A,B,C,D) = \sum m\,(4,5,10,11,13)$, and $Z(A,B,C,D) = \sum m\,(2,10,11,12) + d(3)$. The minimized three functions were found to be

$$X = AB\overline{C}D + \overline{B}C\overline{D} + B\overline{C}\,\overline{D} + A\overline{B}C \qquad (9.1)$$

$$Y = AB\overline{C}D + \overline{A}\,B\overline{C} + A\overline{B}C \qquad (9.2)$$

$$Z = AB\overline{C}\,\overline{D} + \overline{B}C\overline{D} + A\overline{B}C \qquad (9.3)$$

---

[*] A buffer is a gate which can source and sink large currents and can drive a large fan-out with a minimum propagation delay.

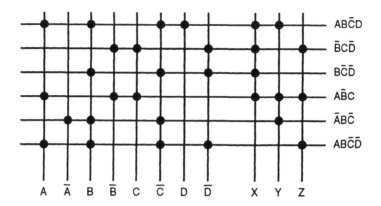

FIGURE 9.1. The circuit of Figure 5.15, Chapter 5, realized as a minimized PLA.

The minimized PLA is shown in Figure 9.1. It consists of 6 rows, 11 columns, 66 crosspoints or nodes where product lines and either input or output lines cross, and 30 connected crosspoints.

9.2.1   PLA Input Decoding

Frequently, two input signals occur in minterm or maxterm pairs more often than they occur independently. If pairs of input signals are decoded prior to their entering the AND plane, it might reduce the number of product lines required as well as the fan-in of the AND plane, thereby compacting the design and reducing the size and the delay of the PLA.

If two input signals can be ANDed or ORed and then steered into the AND plane, this is referred to as *input decoding* or *partitioning*. The input can be decoded into either the minterms or maxterms of the input literals. The cost penalty for this is small since no extra inputs and little extra area are required. If $Y = \overline{A}\,B{\cdot}CD + A\overline{B}{\cdot}C\overline{D} + A\overline{B}{\cdot}CD$ and $Z = AB{\cdot}\overline{C}\,\overline{D} + AB{\cdot}\overline{C}D$, then A and B, as well as C and D, can be partitioned by 2-to-4-line decoders.

If two input variables appear suitable for decoding, except that they do not always occur in decodable products, partitioning can still be useful. If A and B as well as C and D are suitable for partitioning, except for terms $\overline{A}\,C\overline{D}$ and BCD, the terms can be obtained as follows: $\overline{A}\,C\overline{D} = (\overline{A}\,\overline{B})(C\overline{D}) + (\overline{A}\,B)(C\overline{D})$ and BCD $= (\overline{A}\,B)(CD) + (AB)(CD)$. Each term can now be realized with decoded input minterms, and no additional input lines are required.

Properly partitioned inputs can implement the 14 useful combinations of 2 variables, whereas nonpartitioned inputs can only implement 8 useful combinations of 2 variables. The 16 combinations of inputs A and B, both direct and decoded, along with the corresponding outputs, are given in Tables 9.1 and 9.2.

A 1 means the input is connected to the product line, and a 0 means it is not connected to the product line. If the AND plane is realized with NOR gates, the inputs need to be partitioned into minterms, as shown in Table 9.1, and if the AND plane is realized with NAND gates, the inputs need to be partitioned into maxterms, as shown in Table 9.2.

For a NOR-NOR PLA implementation, the 14 nontrivial minterm decoded combinations of 2 variables and the 8 nondecoded combinations of variables are all the combinations other than 0 and 1. Since the four partitioned inputs are disjoint, all combinations of one, two, or three inputs are useful, whereas the unpartitioned inputs are trivial when A and $\overline{A}$ or when B and $\overline{B}$ are both inputs to a NOR gate.

For a NAND-NAND PLA implementation, the 14 nontrivial maxterm decoded combinations of 2 variables are shown in Table 9.2 along with the 8 nondecoded combinations of variables. Again, the four decoded inputs are disjoint and all combinations of one, two, or three inputs are useful, whereas the same undecoded inputs are trivial, as in the NOR-NOR PLA implementation.

TABLE 9.1   NOR-NOR PLA

Inputs				NOR output	Inputs				NOR output
$\overline{A}$	A	$\overline{B}$	B	Function	m_0	m_1	m_2	m_3	Function
0	0	0	0	1	0	0	0	0	1
0	0	0	1	$\overline{B}$	0	0	0	1	$\overline{AB}$ NAND
0	0	1	0	B	0	0	1	0	$\overline{A}$ + B
0	0	1	1	0	0	0	1	1	$\overline{A}$
0	1	0	0	$\overline{A}$	0	1	0	0	A + $\overline{B}$
0	1	0	1	$\overline{A+B}$ NOR	0	1	0	1	$\overline{B}$
0	1	1	0	$\overline{A}$ B	0	1	1	0	$\overline{A \oplus B}$ ENOR
0	1	1	1	0	0	1	1	1	$\overline{A+B}$ NOR
1	0	0	0	A	1	0	0	0	A + B OR
1	0	0	1	A$\overline{B}$	1	0	0	1	A $\oplus$ B EOR
1	0	1	0	AB AND	1	0	1	0	B
1	0	1	1	0	1	0	1	1	$\overline{A}$ B
1	1	0	0	0	1	1	0	0	A
1	1	0	1	0	1	1	0	1	A$\overline{B}$
1	1	1	0	0	1	1	1	0	AB AND
1	1	1	1	0	1	1	1	1	0

It is usually not obvious what the best input decoding is, and it may be necessary to try several combinations of input variables in order to determine which arrangement gives a minimum-size PLA. It is very difficult to decode three or more variables satisfactorily and it is not normally attempted.

EXAMPLE 9.1: Realize the four functions $AC\overline{D}$, $\overline{B}\,\overline{C}D$, $A \oplus B$, and $\overline{A}B +$ $\overline{A}\,\overline{C}D$ with a PLA whose inputs A, B and inputs C, D are partitioned.

SOLUTION: $AC\overline{D} = (A\overline{B})(C\overline{D}) + (AB)(C\overline{D})$, $\overline{B}\,\overline{C}D = (\overline{A}\,\overline{B})(\overline{C}D) +$ $(A\overline{B})(\overline{C}D)$, $A \oplus B = \overline{A}B + A\overline{B}$, and $\overline{A}B + \overline{A}\,\overline{C}D = \overline{A}B +$ $\overline{A}\,\overline{B}\,\overline{C}D$. The four functions are realized in the NAND-NAND PLA of Figure 9.2.

TABLE 9.2   NAND-NAND PLA

Inputs				NAND output	Inputs				NAND output
$\overline{A}$	A	$\overline{B}$	B	Function	M_0	M_1	M_2	M_3	Function
0	0	0	0	1	0	0	0	0	1
0	0	0	1	$\overline{B}$	0	0	0	1	AB AND
0	0	1	0	B	0	0	1	0	$A\overline{B}$
0	0	1	1	1	0	0	1	1	A
0	1	0	0	$\overline{A}$	0	1	0	0	$\overline{A}B$
0	1	0	1	$\overline{A}\overline{B}$ NAND	0	1	0	1	B
0	1	1	0	$\overline{A} + B$	0	1	1	0	$A \oplus B$ EOR
0	1	1	1	1	0	1	1	1	$A + B$ OR
1	0	0	0	A	1	0	0	0	$\overline{A+B}$ NOR
1	0	0	1	$A + \overline{B}$	1	0	0	1	$\overline{A \oplus B}$ ENOR
1	0	1	0	$A + B$ OR	1	0	1	0	$\overline{B}$
1	0	1	1	1	1	0	1	1	$A + \overline{B}$
1	1	0	0	1	1	1	0	0	$\overline{A}$
1	1	0	1	1	1	1	0	1	$\overline{A} + B$
1	1	1	0	1	1	1	1	0	$\overline{AB}$ NAND
1	1	1	1	1	1	1	1	1	1

9.2.2 Registered PLDs

Many digital systems such as finite-state machines, shift registers, and counters can be designed using registered PLDs and feedback. Some commercial registered PLDs are constructed with D flip-flops and some with J-K flip-flops. A D flip-flop can be converted into a J-K and vice versa, as discussed in Chapter 7.

Two OR-plane outputs are necessary for each J-K flip-flop, and it can drive a buffer which must be able to source sufficient current to drive the load. If the load fan-out is very large, it may be necessary to take the PLA output directly from buffered product lines, leaving the flip-flop outputs to drive only the AND plane lines.

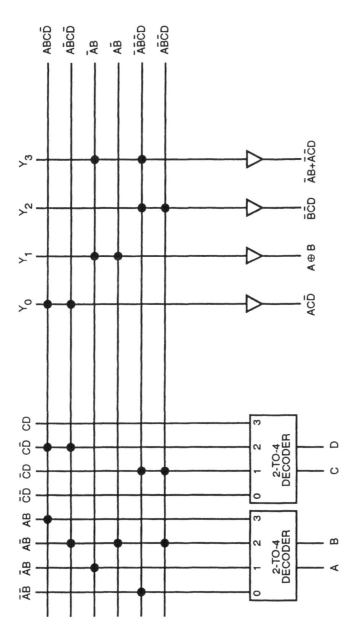

FIGURE 9.2. A PLA realization with decoded inputs to the AND plane.

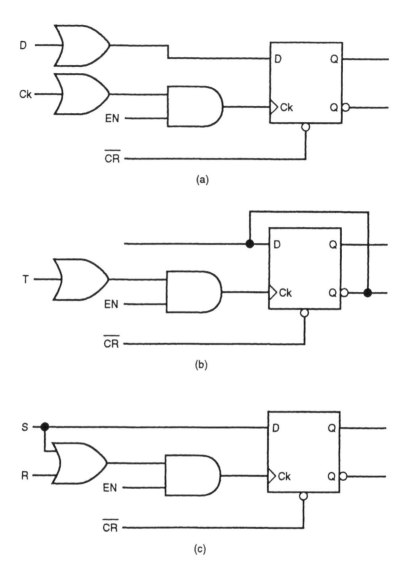

FIGURE 9.3.   A D flip-flop configured as (a) a clocked or asynchronous D flip-flop, (b) a T flip-flop, and (c) an S-R flip-flop.

A gated D flip-flop can be *programmably reconfigured,* as shown in Figure 9.3. The OR gates in the figure are part of the OR plane of the PLA. The flip-flop and the AND gate form a simple *macrocell.* In Figure 9.3(a), a synchronously clocked D flip-flop is obtained using EN as a clock enable input to the flip-flop. This is useful for temporarily storing data. An asynchronous D flip-flop is realized when the enable input is set to logical 1, allowing the PLA to clock the flip-flop. In both cases the clock and data inputs are driven by outputs from the OR plane.

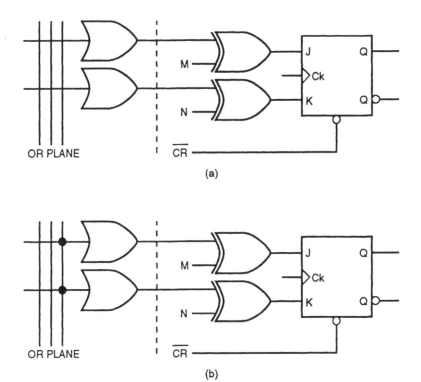

FIGURE 9.4. A J-K flip-flop which can be reconfigured as (a) a normal or inverted J-K flip-flop and (b) as a normal or inverted D or T flip-flop.

When the active-low output $\overline{Q}$ is fed back to the D input and there is no connection to the D input from the OR plane, as shown in Figure 9.3(b), one input of the AND gate is a toggle input to a T flip-flop. If an OR gate from the OR plane is connected, as shown in Figure 9.3(c), an S-R flip-flop results. If these changes are software driven, the D mode can be used to load or store data into the PLA, and the S-R mode can be used during normal operation. If a 2-to-1-line multiplexer is added to the circuit of Figure 9.3, one has the option of selectively bypassing the flip-flop entirely.

A programmably reconfigurable J-K flip-flop is shown in Figure 9.4(a).[1] The macrocell consists of the flip-flop and two EOR gates. If inputs M and N are set to logical 0, one has a simple J-K flip-flop triggered by two OR-plane output lines, and if inputs M and N are both set to logical 1, the EOR gates invert the signals from the OR plane. This inverted J-K flip-flop reverses the polarity of the output of the macrocell, and, in this mode, the clear input becomes an asynchronous preset input.

If input M = 0 and input N = 1, the K input is complemented by the EOR gate and the J input is not. If the NOR gate inputs are now tied together, the configuration is that of a D flip-flop, as shown in Figure 9.4(b). If M = 1, N = 0, and the

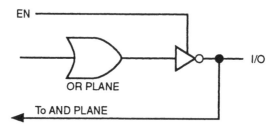

FIGURE 9.5. A programmable I/O cell utilizing a three-state buffer.

NOR gate inputs are tied together, the macrocell emulates an inverting D flip-flop or $\overline{D}$ flip-flop. Finally, if M and N are both low and the NOR gate inputs are tied together, as shown in Figure 9.4(b), a toggle or T flip-flop is created; and if M and N are both high and the NOR inputs are tied together, an inverting toggle flip-flop is created.

A programmable I/O cell can be created by using a 3-state gate, as shown in Figure 9.5. A 3-state buffer has a control input that can disable it. In the disabled state the buffer looks like an open circuit or open switch to a high or to a low logic signal. In this mode of operation, the I/O pin can be used as an input pin provided there is a path from the 3-state buffer to the AND plane. If the 3-state buffer is enabled and the path to the AND plane is left unconnected, the signal from the OR plane is sent to the output pin. Thus, the output node of the 3-state buffer can be used as an input pin or an output pin, and it can be software programmable to change from one to another.

A *registered PLD* has output flip-flops and can have input flip-flops also. Both PALs and PLAs can be designed with clocked flip-flops in the output circuitry. Figure 9.6 shows a general form of PLA with seven primary inputs, two feedback inputs, seven regular outputs, and four registered outputs. Inputs I_0 through I_3 are normal PLA inputs clocked through an input buffer, and inputs I_4 through I_6 are registered inputs which are clocked through D flip-flops. All six of these inputs are clocked on the same phase ϕ_1. The two feedback inputs I_7 and I_8 are taken from the asserted-low outputs of J-K flip-flops at the output of the PLA, clocked on phase ϕ_2, and fed back to the AND plane as inputs. Y_7 and Y_8 can be available as outputs or these flip-flops can be *buried registers,* whose output states are unavailable to the circuit designer. Whether or not some registers are buried, there can be registered outputs, such as Y_9 and $\overline{Y}_9$, and normally buffered outputs such as Y_0 through Y_6.

Field-programmable PLAs often come with output MUXs and an output *polarity circuit,* consisting of an EOR gate and a *polarity fuse,* as shown in Figure 9.7. With the fuse intact, input $X = 0$, output $Z = Y$, and the EOR gate acts as a noninverting output buffer. When the fuse is blown, input $X = 1$, $Z = \overline{Y}$, and the EOR gate acts as an inverting output buffer.

An ENOR gate can be used also, in which case an intact fuse gives an inverted

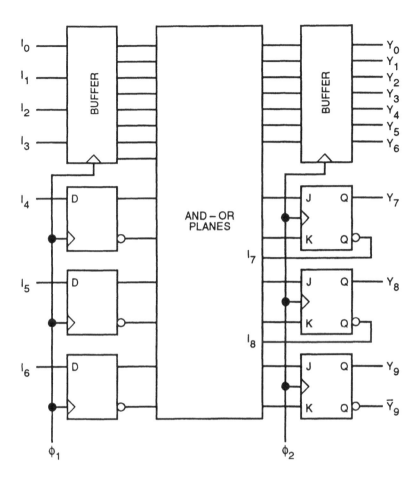

FIGURE 9.6. A PLA with buffered and registered inputs and outputs, including feedback registered outputs.

output and a blown fuse gives a noninverted output. In either case, one can design either for an SOP or a POS solution and choose the appropriate output polarity. A PLA with input decoding, output polarity circuitry, and both flip-flops and multiplexers in the output circuitry is an extremely flexible device.

9.3 PROGRAMMING THE PLA

Every PLA must be programmed by connecting appropriate crosspoints. This can be done by blowing fuses or by connecting diodes or transistors at the correct locations in the AND and OR arrays. It is necessary to specify the location of these connections, and the personality matrix is used to do this. The *personality matrix,*

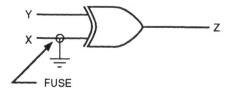

FIGURE 9.7. A typical PLA output polarity circuit.

Q, of a PLA is defined as follows.[2] In the PLA matrix, the presence of a connection at a crosspoint site will be indicated by a 1, and the absence of a connection will be indicated by a 0. The matrix so formed will consist of 1s at every crosspoint site in both the AND and OR planes for which a fuse must be blown or a transistor must be located.

The personality matrix can be defined as follows. In the AND plane, element $q_{ij} = 1$ if product line P_i is to be connected to input line I_j, and $q_{ij} = 0$ if product line P_i is not to be connected to input line I_j. In the OR plane, $q_{ij} = 1$ if product line P_i connects to output Y_j and $q_{ij} = 0$ if product line P_i does not connect to output line Y_j.

Figure 9.8 shows a PLA representation which implements the functions $Y_2 = \overline{A}\,\overline{B} + A\overline{C} + AB$, $Y_1 = \overline{A}\,\overline{B}$, and $Y_0 = AB + \overline{A}C$. The personality matrix of this PLA is

$$Q = \begin{bmatrix} 0\,1\,0\,1\,0\,0 & 1\,1\,0 \\ 1\,0\,0\,0\,0\,1 & 1\,0\,0 \\ 1\,0\,1\,0\,0\,0 & 1\,0\,1 \\ 0\,1\,0\,0\,1\,0 & 0\,0\,1 \end{bmatrix} \tag{9.4}$$

The PLA has product terms $P_0 = \overline{A}\,\overline{B}$, $P_1 = A\overline{C}$, $P_2 = AB$, and $P_3 = \overline{A}C$ and output terms $Y_2 = P_0 + P_1 + P_2$, $Y_1 = P_0$, $Y_0 = P_2 + P_3$.

9.4 DYNAMIC LOGIC ARRAYS

The *dynamic logic array (DLA)* resembles a PLA with the AND and OR planes merged such that the OR logic is performed in the AND plane. The DLA is ideally suited to implementations with few crosspoint connections in the OR plane. It can realize an SOP solution with less area than a PLA and can be clocked the same as a PLA structure. The DLA is less programmable than an equivalent PLA, but it combines the speed, simplicity, and small size of dynamic logic with the programmability of a PLA.

Figure 9.9(a) shows an implementation of three output functions using AND and OR gates. The first output function consists of the OR of two product terms,

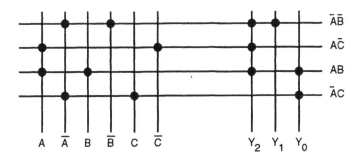

FIGURE 9.8. A PLA representation of the personality matrix of Equation 9.4.

whereas the second output function is the OR of three inputs, and the third output function consists of only one product term. The outputs are

$$Y_0 = \overline{A}\,\overline{C} + BC, \ Y_1 = A + B + \overline{C}, \ \text{and} \ Y_2 = ABC$$

They can be realized by three AND and two OR gates, as shown in Figure 9.9(a). The equivalent DLA is shown in Figure 9.9(b), and the personality matrix of this DLA is

$$Q = \begin{bmatrix} 0\,1\,0\,0\,0\,1 & 1\,0\,0 \\ 0\,0\,1\,0\,1\,0 & 1\,0\,0 \\ 1\,0\,0\,0\,0\,0 & 0\,1\,0 \\ 0\,0\,1\,0\,0\,0 & 0\,1\,0 \\ 0\,0\,0\,0\,0\,1 & 0\,1\,0 \\ 1\,0\,1\,0\,1\,0 & 0\,0\,1 \end{bmatrix}$$

9.5 THE FINITE-STATE MACHINE AS A PLA STRUCTURE

The PLA can be used to store information related to the past history of its inputs. This is memory and causes the same inputs to combine with different feedback signals to produce different outputs. When feedback is added to the AND-OR PLA structure, the PLA becomes a *finite-state machine (FSM)*. If clocked input and output buffers are added, the structure is as shown in Figure 9.10.

An FSM can be designed as a Mealy machine or a Moore machine. The Mealy machine shown in Figure 9.11 has outputs which may change with input changes in an asynchronous manner and cause erroneous behavior as signals ripple downstream. The Moore machine shown in Figure 9.12 has outputs which depend upon and change only with state changes, since all the outputs of the boolean logic

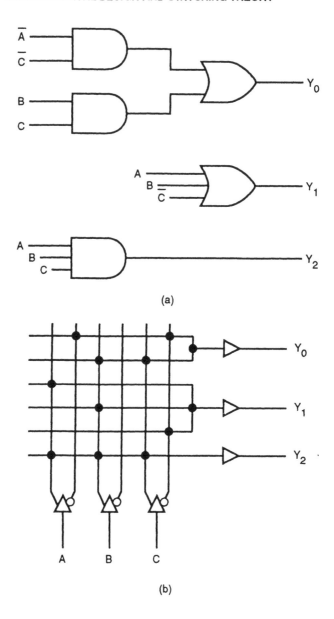

FIGURE 9.9. (a) A circuit representation and (b) a DLA realization of three functions.

block go through a state register and are synchronously clocked. It is easier to design a reliable Moore machine than an equivalent Mealy machine, just as it is easier to design a properly working synchronous counter than an asynchronous counter.

In designing FSMs, one should always be able to reset from any state. For N feedback loops there are 2N states, since each flip-flop can be in one of two states.

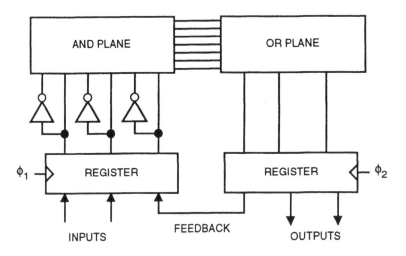

FIGURE 9.10. An FSM constructed from a PLA with feedback and two-phase nonoverlapping clocking.

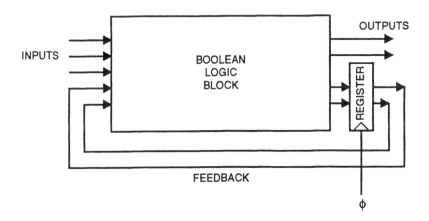

FIGURE 9.11. A Mealy realization of an FSM; some outputs are asynchronous.

Lockout can be avoided by designing for all 2N states. If this is done properly, the additional size cost can be minimized. For functions of many variables this does become a serious problem.

Any large sequential system can be constructed from a series of FSMs, as shown in Figure 9.13. Then, everything goes through only one combinational-logic block between clock pulses. There is still the basic clocking problem which led to the design of master/slave and edge-triggered flip-flops; the clock pulses must be very short to avoid rippling in a Moore machine. This can be accomplished by using two clock phases, just as the master/slave flip-flop uses and edge-

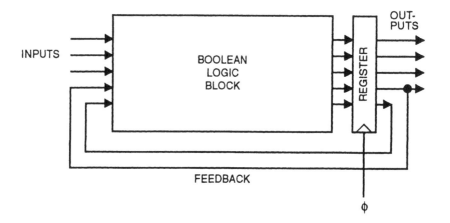

FIGURE 9.12.   A Moore realization of an FSM; all outputs are clocked.

triggering of the input and output registers will protect against asynchronous noise spikes.

EXAMPLE 9.2:   Find the personality matrix, Q, for a four-bit modulo-10 counter with delay flip-flops in the feedback loop. Give an implementation of the FSM.

SOLUTION: Let a plus superscript refer to the next state of a variable, whose present state has no superscript. The minimum SOP functions required as flip-flop inputs are:

$$A^+ = A\overline{D} + BCD$$
$$B^+ = B\overline{C} + B\overline{D} + \overline{B}CD \qquad\qquad (9.6)$$
$$C^+ = \overline{A}\,\overline{C}D + C\overline{D}$$
$$D^+ = \overline{D}$$

The OR plane matrix drives the four flip-flops, labeled D, C, B, and A, whose outputs are fed back to the AND plane. Either true or complemented latch outputs can be fed back. The PLA outputs could be taken directly from these four flip-flops, and a PLA of size 8R × 12C = 96N would result, but this loads the flip-flop outputs. If it is desired to obtain the outputs directly from the OR-matrix, the PLA is shown in Figure 9.14. The PLA inputs and outputs must be clocked by two disjoint clock phases, as discussed before, and as shown in Figure 9.10. The gated input and output buffers are omitted in Figure 9.14 to

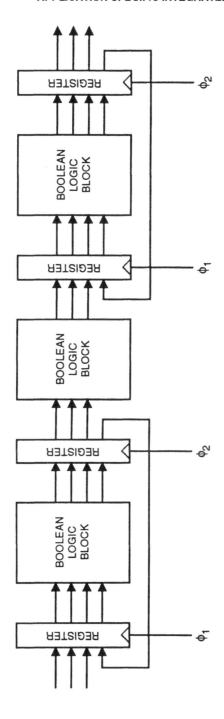

FIGURE 9.13. A sequential system consisting of cascaded finite-state machines.

simplify the figure. The PLA proper has dimensions of 12R × 16C = 192N, and a sparsity factor of 34/192 = 0.177.

$$
Q =
\begin{bmatrix}
1\,0\,0\,0\,0\,0\,0\,1 & 0\,0\,0\,1\,0\,0\,0\,0 \\
0\,0\,1\,0\,1\,0\,1\,0 & 0\,0\,0\,1\,0\,0\,0\,0 \\
0\,0\,1\,0\,0\,1\,0\,0 & 0\,0\,1\,0\,0\,0\,0\,0 \\
0\,0\,1\,0\,0\,0\,0\,1 & 0\,0\,1\,0\,0\,0\,0\,0 \\
0\,0\,0\,1\,1\,0\,1\,0 & 0\,0\,1\,0\,0\,0\,0\,0 \\
0\,1\,0\,0\,0\,1\,1\,0 & 0\,1\,0\,0\,0\,0\,0\,0 \\
0\,0\,0\,0\,1\,0\,0\,1 & 0\,1\,0\,0\,0\,0\,0\,0 \\
0\,0\,0\,0\,0\,0\,0\,1 & 1\,0\,0\,0\,0\,0\,0\,0 \\
1\,0\,0\,0\,0\,0\,0\,0 & 0\,0\,0\,0\,1\,0\,0\,0 \\
0\,0\,1\,0\,0\,0\,0\,0 & 0\,0\,0\,0\,0\,1\,0\,0 \\
0\,0\,0\,0\,1\,0\,0\,0 & 0\,0\,0\,0\,0\,0\,1\,0 \\
0\,0\,0\,0\,0\,0\,1\,0 & 0\,0\,0\,0\,0\,0\,0\,1
\end{bmatrix}
\tag{9.7}
$$

The personality matrix for the counter is given above. The inputs are taken to be A, B, C, and D, respectively. The outputs are taken directly in the order A, B, C, and D, and are fed back in the order D, C, B, A, respectively, as shown in Figure 9.14.

A PLA can implement a variety of fixed- or variable-modulo counters. The inputs to each flip-flop are obtained from the OR plane of the PLA. The inputs to the AND plane are the external controls and the outputs of the flip-flops. The AND plane is used to detect the present state and send the appropriate control signals to the flip-flop array that will determine the next state of the system.

In general, one should check various designs to determine the smallest PLA. An excellent discussion of PLA design with flip-flops in the feedback paths can be found in Carr and Mize.[3]

9.6 REDUCTION OF CUSTOMIZED PLAs

Often the active devices in a PLA are sparsely distributed and require a large area, a typical sparse PLA using about 10 to 20% of the AND and OR plane nodes. This results in unnecessarily long lines which require large drivers and have poor performance. Techniques that reduce the size of a PLA will also yield a faster PLA.

One should never become so absorbed in high-powered reduction techniques as to overlook simple approaches. The four-bit modulo-10 counter shown in Figure 9.14 can be designed with J-K flip-flops. The required boolean expressions for the flip-flop input excitations are

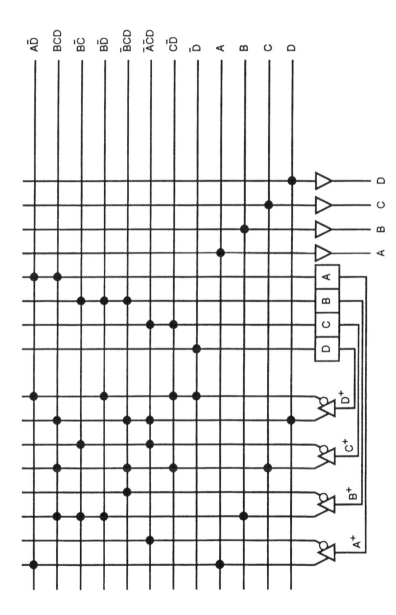

FIGURE 9.14. The PLA realization of a 4-bit module-10 counter realized with delay flip-flops; clocking omitted for clarity.

$$J_A = BCD, \; K_A = D, \; J_B = CD, \; K_B = CD,$$
$$J_C = \overline{A}D, \; K_C = \overline{A}D, \; J_D = 1, \; K_D = 1 \tag{9.8}$$

$K_C = CD$ is also correct and gives an equivalent realization with J-K flip-flops. Direct implementation yields a PLA of size 8R × 21C = 168N, and a sparsity factor of S = 24/168 = 0.143. This is shown in Figure 9.15(a).

The PLA can be reduced by first noting that, provided $K_C = \overline{A}D$, flip-flops B, C, and D have both inputs identical. They are operating in the toggle mode and can be replaced by toggle flip-flops. This immediately reduces the number of columns of the PLA by three, giving dimensions of 8R × 18C = 144N, and S = 21/144 = 0.146. This has not improved the space utilization of the PLA (the sparsity factor), but the size has been reduced to 85.7% of the original size.

$\overline{A}$ is the only complemented variable present in this PLA. If one can omit the inverting AND plane drivers for inputs B, C, and D, three more columns can be deleted. The PLA now has dimensions 8R × 15C = 120N and a sparsity factor of 21/120 = 0.175. This is still a relatively sparse PLA, but it has 71.4% of the area of the original design. The reduced PLA is shown in Figure 9.15(b). Again, the clocking is omitted.

9.7   PLA FOLDING

Row and/or column folding is a technique for reducing the physical size of a PLA by merging rows and/or columns and reducing the sparsity of the matrix. The maximum reduction in area is achieved if individual inputs and outputs are custom folded.

Once a boolean expression is simplified into a suitable form for PLA implementation, the AND plane inputs and the product lines should be ordered such that pairs of columns in the AND plane allow for input signals to enter from the top and bottom of the physical PLA. Pairs of columns in the OR plane can also be folded to allow output signals from the top and bottom of the PLA.

An example of a column-folded PLA is shown in Figure 9.16. The functions realized are $Y_1 = AB + \overline{A}\,\overline{B}$, $Y_2 = \overline{A}B + BC$, $Y_3 = \overline{C}$, and $Y_4 = C\overline{D} + \overline{C}D$. Inputs A, $\overline{A}$, B, and $\overline{B}$ enter from the top of the PLA, and Y_1 and Y_2 exit from the top of the PLA. Inputs C, $\overline{C}$, D, and $\overline{D}$ enter from the bottom of the PLA, and Y_3 and Y_4 exit from the bottom of the PLA. Generally, it is desirable to keep input I_i and input $\overline{I}_i$ on the same side of the PLA, as they usually come from the same physical location in the circuit and can be easily routed together to the input site of the PLA.

Row folding can reduce the number of product lines by splitting either the AND plane or the OR plane. Since the OR plane is usually much smaller than the AND plane, it is the logical one to split. One then has a structure with the AND plane between two segments of the OR plane. If two product lines are not both required by an output, they may be folded because the two rows or product lines

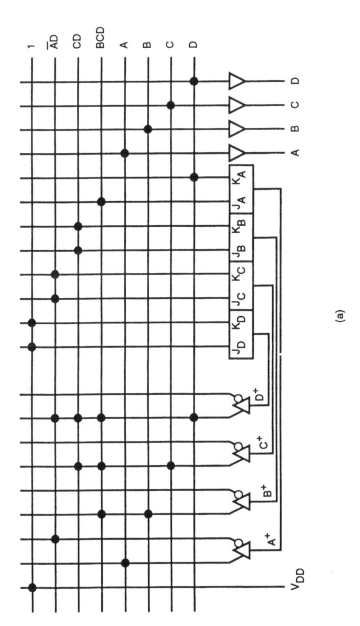

(a)

FIGURE 9.15.   The PLA implementation of a 4-bit modulo-10 counter realization (a) with J-K flip-Flops and (b) reduced to minimum size; clocking omitted for clarity.

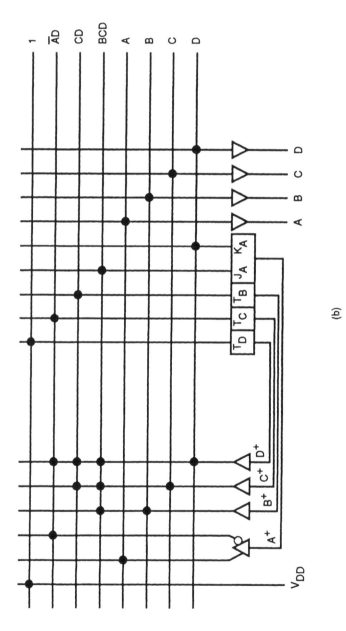

FIGURE 9.15. (continued).

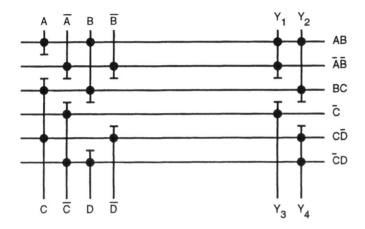

FIGURE 9.16. A PLA reduced by folding the input and output columns.

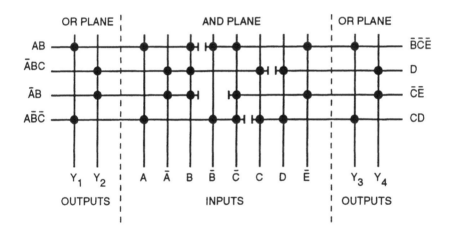

FIGURE 9.17. A PLA reduced by splitting the OR plane and folding pairs of product lines or rows.

do not have connections located at the same crosspoint sites. Figure 9.17 shows a PLA with a split OR plane that realizes the functions $Y_1 = AB + A\overline{B}\,\overline{C}$, $Y_2 = \overline{A}BC + AB$, $Y_3 = \overline{B}\,\overline{C}\,\overline{E} + CD$, and $Y_4 = D + \overline{C}\,\overline{E}$.

It is necessary to keep track of the crosspoints which are connected in order to fold the PLA. In the following discussion, rows will be ordered from top to bottom and columns will be ordered from left to right. Each input line will be treated separately, so that it can be folded independently of any other column.

The function must first be reduced to a minimum SOP form. A basic criterion for folding a row or column is that the pairs of rows or columns merged must not contain 1s at the same crosspoint locations. Next, pairs of columns of the PLA are

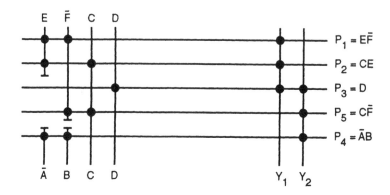

FIGURE 9.18. A PLA minimized by column folding.

folded in the AND arrays (OR arrays) to allow signals to be input (output) at both the top and bottom of the physical array.

EXAMPLE 9.3: Design a minimum-size PLA to realize the functions
$$Y_1 = CE + D + E\bar{F}$$
$$Y_2 = \bar{A}B + C\bar{F}$$

SOLUTION: A direct implementation requires 12 input lines for the 6 variables and their complements, plus 2 output columns and 5 product lines, for a total size of $5R \times 14C = 70N$ and a sparsity of $S = 15/70 = 0.214$.

A customized PLA requires only six input lines for the six variables actually present in the functions. Further examination shows that if the last two rows of the PLA are interchanged, then input pairs $\bar{A}$, E and B, $\bar{F}$ can be merged to occupy one column per pair. The reduced PLA has dimensions $5R \times 6C = 30N$, which is 43% of the original size, and sparsity $15/30 = 0.50$, which is a considerable improvement. The reduced PLA is shown in Figure 9.18.

With large PLAs, a systematic approach to PLA folding is required in order to successfully minimize the chip area and maximize the PLA speed.[4,5] Two columns of a PLA, c_i and c_j, are said to be disjoint if they do not have 1s at the same row in the personality matrix of a PLA. In the AND plane, two input lines are disjoint if they do not drive the same product line. In the OR plane, two output lines are disjoint if they are not driven by the same product line. Two rows (product lines) are disjoint if they do not have 1s at the same column in the PLA matrix, i.e., if they are not both driven by the same input lines and do not drive the same output lines.

EXAMPLE 9.4:   Three rows of a PLA are r_1 = 000110, r_2 = 011001, and r_3 = 101001. Determine which rows can be folded together.

SOLUTION: Rows r_1 and r_2, as well as rows r_1 and r_3, are disjoint and can be folded. Rows r_2 and r_3 cannot be folded since they both require connections at two column sites.

Given the personality matrix, Q, of a PLA, define R_i as the set of rows with 1s at the crosspoints of column c_i, and define C_i as the set of columns with 1s at the crosspoints of row r_i. Columns i and j are disjoint if $C_i \cap C_j = 0$, and rows i and j are disjoint if $R_i \cap R_j = 0$. Interchanging r and c, R and C, has no effect on the discussion, and the approach is the same for both rows and columns.

EXAMPLE 9.5:   Show mathematically which of the rows r_1 = 000110, r_2 = 011001, and r_3 = 101001 of Q can be folded as pairs.

SOLUTION: $C_1 = \{c_4, c_5\}$, $C_2 = \{c_2, c_3, c_6\}$, and $C_3 = \{c_1, c_3, c_6\}$. $C_1 \cap C_2 = 0$, and rows 1 and 2 can be paired. $C_1 \cap C_3 = 0$, and rows 1 and 3 can be paired. $C_2 \cap C_3 = \{c_3, c_6\}$. Rows 2 and 3 have overlapping connections at two column sites and they cannot be paired.

An ordered folding pair $<c_i, c_j>$ specifies two columns that can be folded, with column c_i above column c_j in the same physical column of the folded PLA. The column disjoint graph is an undirected graph, whose vertex set, V, represents the columns of the personality matrix and whose edge set, E, is defined as E = {c_i, $c_j | R_i \cap R_j = 0$}. The edge set consists of pairs of columns which are disjoint; hence, they can be folded.

The procedure for column/row folding consists of constructing the column/row disjoint graph from the personality matrix, finding the connected components of the disjoint graph, and constructing a path for each ordered pair of vertices on the graph.[6]

Next, a path with the maximum length among those constructed is taken in order to extract the best ordered folding set from the selected path. If the generated partial ordering is found to be cyclic, the path can be deleted and the next longest path chosen. A linear ordering is obtained from the generated partial ordering, and the folded PLA is generated.

EXAMPLE 9.6:   Give the edge set and the row disjoint graph for the three rows of the previous example.

SOLUTION: Rows 1 and 2 can be folded as a pair, and they form an edge in the disjoint graph, as do rows 1 and 3. The edges are E = {{r_1, r_2}, {r_1, r_3}}, and the graph is a line from row 2 to row 1 to row 3, as shown in Figure 9.19.

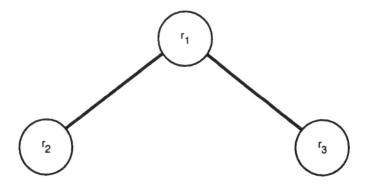

FIGURE 9.19. The row disjoint graph, showing which rows can be folded or merged together.

EXAMPLE 9.7: Column fold the PLA whose personality matrix, Q, is

$$Q = \begin{bmatrix} 1 & 1 & 0 & 0 & 0 & 0 & 1 & 0 & & 0 & 0 & 1 \\ 0 & 1 & 0 & 0 & 1 & 1 & 0 & 0 & & 0 & 1 & 0 \\ 1 & 1 & 1 & 0 & 1 & 0 & 0 & 0 & & 0 & 1 & 1 \\ 1 & 0 & 0 & 1 & 0 & 1 & 1 & 0 & & 1 & 0 & 1 \\ 0 & 0 & 0 & 1 & 0 & 1 & 0 & 1 & & 1 & 0 & 0 \\ 1 & 0 & 0 & 0 & 0 & 0 & 1 & 0 & & 1 & 0 & 0 \end{bmatrix} \tag{9.8}$$

SOLUTION: First, compute the row sets, R_i. For the AND plane

$$\begin{aligned}
R_1 &= \{r_1, r_3, r_4, r_6\} \\
R_2 &= \{r_1, r_2, r_3\} \\
R_3 &= \{r_3\} \\
R_4 &= \{r_4, r_5\} \\
R_5 &= \{r_2, r_3\} \\
R_6 &= \{r_2, r_4, r_5\} \\
R_7 &= \{r_1, r_4, r_6\} \\
R_8 &= \{r_5\}
\end{aligned} \tag{9.9}$$

To determine which columns can be paired, construct the column disjoint table for the AND plane as shown below to determine the disjoint of R_i and R_j for the AND plane. Zero entries in the column disjoint table indicate pairs in that row and column of the table which can be merged or folded together.

	R_2	R_3	R_4	R_5	R_6	R_7	R_8
R_1	r_1,r_3	r_3	r_4	r_3	r_4	r_1,r_4,r_6	0
R_2		r_3	0	r_2,r_3	r_2	r_1	0
R_3			0	r_3	0	0	0
R_4				0	r_4,r_5	r_4	r_5
R_5					r_2	0	0
R_6						r_4	r_5
R_7							0

For the OR plane.

$$R_9 = \{r_4, r_5, r_6\}$$
$$R_{10} = \{r_2, r_3\}$$
$$R_{11} = \{r_1, r_3, r_4\}$$

The OR plane can be done by inspection, and

$$R_9 \cap R_{10} = \varnothing$$
$$R_9 \cap R_{11} = \{r_4\}$$
$$R_{10} \cap R_{11} = \{r_3\}$$

Columns 9 and 10 can be folded; column 11 cannot be folded with either column 9 or 10.

From the table, it is seen that columns 1 and 8, 2 and 4, 2 and 8, 3 and 4, 3 and 6, 3 and 7, 3 and 8, 4 and 5, 5 and 7, 5 and 8, and 7 and 8 can be paired. The column disjoint graph is shown in Figure 9.20. From the graph it is seen that the longest path merges columns 1 and 8, 2 and 4, 5 and 7, and 3 and 6. The path is traced on Figure 9.20 with dark lines linking columns to be folded. The pairs form ordered sets of two columns each, and the first numbered columns in each set must be on the same edge of the AND plane. They will be chosen to be located above the second column of the pair.

To merge columns 1 and 8, row 5 must be below row 6. To merge columns 3 and 6, row 3 must be above row 2, and to merge columns 5 and 7, row 1 must be below row 2. The original rows are then in the sequence 3, 2, 1, 6, 4, 5. Columns 9 and 10 of the OR plane can be folded, while column 11 cannot. The folded PLA is shown in Figure 9.21. The original PLA consisted of 6 rows and 11 columns, or 66 crosspoints. The folded PLA consists of 6 rows and 6 columns, or 36 crosspoints, for a net area savings of 45.5%.

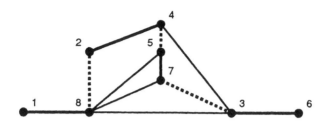

FIGURE 9.20. The column disjoint graph for the PLA of Example 9.7. The largest path length gives the maximum number of columns that can be folded.

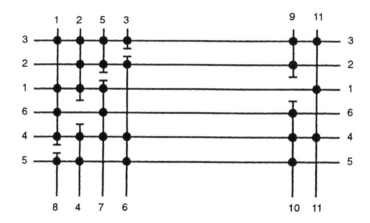

FIGURE 9.21. The minimized column-folded PLA of Example 9.7.

9.8 MULTILEVEL LOGIC STRUCTURES

Often it is desirable to add additional logic between the AND plane and the OR plane of a PLA. A channel of columns between the two planes allows dot-ANDing[*] between product lines, as shown in Figure 9.22. Note the similarity of this structure to the DLA structure.

The PLAs studied so far realize only two levels of logic. Multilevel logic requires a PLA with logic levels internally fed back into the array. Consider the carry-out of an adder stage, $C_o = AB + A\overline{B}C_i + \overline{A}BC_i$. This can be factored to

---

[*] Certain logic gates allow one to connect or tie their outputs together and attach the output to a pull-up resistor. The result is the same as if an AND gate had been used to connect these outputs; and this is referred to as a dot-AND structure, also called a wired-AND structure. Gates whose outputs can be connected together and attached to a pull-down resistor can simulate an OR connection of their outputs; and this is referred to as a wired-OR or dot-OR structure.

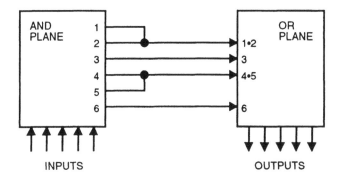

FIGURE 9.22.   Dot-ANDing between the AND plane and the OR plane of a PLA.

obtain $C_o = AB + C_i(A\overline{B} + \overline{A}B)$. If kernal $A\overline{B} + \overline{A}B$ is fed back to the AND plane, as shown in Figure 8.23(a), the two-level SOP form of the function has been converted to a factored form with four levels of logic. A PLA which has internal feedback paths to the AND plane, as shown in Figure 8.23(b), can realize the factored form directly with no additional external inputs and input registers.

The feedback paths do not actually form a loop in the circuit, but rather a "coil." The PLA can often be reduced greatly in size by factoring, but the circuit is much slower due to the multiple loops.

In general, outputs can be clocked through either buffers or flip-flops and fed back to either the AND plane or to the OR plane. When registered outputs are fed back to the OR plane, as shown in Figure 9.24, multilevel logic is realizable, and the PLA now has one AND plane and more than one OR plane. The utility of a PLA can be greatly increased if the OR-plane gates are fed into EOR gates which drive a D flip-flop. The registered output can be fed back to the OR plane, as shown in Figure 9.25, to obtain a powerful multilevel PLD structure.

The *folded-NOR* and *folded-NAND* array permit successive levels of NOR or NAND gates to be fed back and cascaded to implement multilevel logic designs.[1] A folded-NOR array is shown in Figure 9.26. The device consists of a single NOR plane, but by feeding back successive levels of NOR gates, one can construct a NOR array of arbitrary depth. This is very useful in implementing multilevel logic designs with NOR logic. The folded-back NOR terms perform asynchronous feedback. If the J-K or D flip-flops of Figures 9.3 and 9.4 are added to the folded-NOR array, one has a very powerful multilevel logic circuit.

A folded-NAND array is shown in Figure 9.27. This device consists of a single NAND plane with feedback to produce a NAND array of arbitrary depth. It can be used with J-K or D flip-flops also to realize complex multilevel logic with NAND gates.

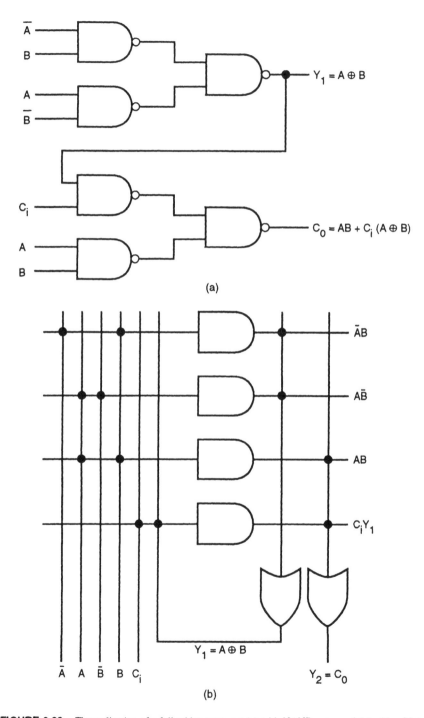

FIGURE 9.23. The realization of a full-adder carry-out (a) with NAND gates and (b) with a PLA utilizing internal feedback.

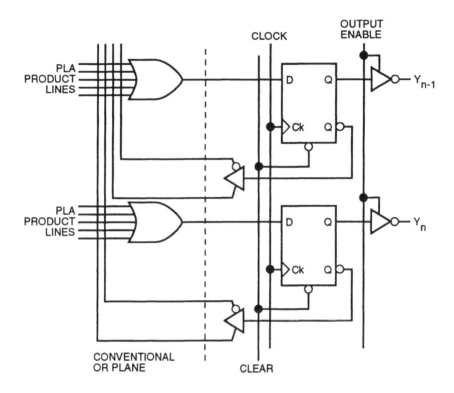

FIGURE 9.24. Feedback of registered outputs to form a folded-OR plane.

For a more thorough discussion of programmable logic devices and applications of the ABEL* language in programming PLAs see Wakerly.[7] For a comprehensive coverage of PLDs and FPGAs see Pellerin and Holley.[8]

9.9   FIELD PROGRAMMABLE GATE ARRAYS

Field programmable gate arrays (*FPGAs*) are the latest addition to the field of user programmable circuits. They are designed to provide greater freedom in the selection and interconnection of devices. A major difference between PLDs and FPGAs is that a PLD is programmed by modifying the logic function of a circuit with fixed interconnections, whereas a FPGA is programmed by modifying the routing of the interconnections of one or more separate fixed function blocks called macrocells. As a result, FPGAs are not based on the PLA architecture as were all the previously studied programmable devices. FPGAs are similar to

---

* Advanced Boolean Expression Language or ABEL *is* a trademark of Data I/O Corporation, Redmond, Washington.

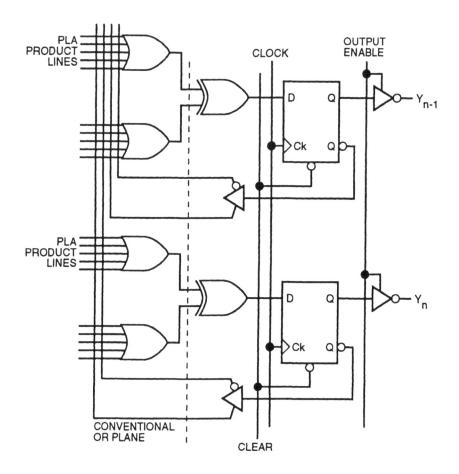

FIGURE 9.25. The folded-OR plane with EOR gates added.

PLDs in their user programmability and their rapid design and debug capability. This makes them extremely versatile choices for rapid turn-around prototyping. FPGAs are better suited than PLDs to the implementation of multilevel logic functions.

FPGA architecture can be divided into logic cells or *macros* and *programmable interconnects*, as shown in Figure 9.28. An FPGA may contain static memory cells that allow the interconnection pattern to be loaded and changed after the device is manufactured, giving the circuit programmable reconfigurability, as was the case of the D and J-K flip-flops discussed earlier.

9.9.1 Table-Look-Up Architecture

Because FPGAs have evolved from both *mask programmable gate arrays*

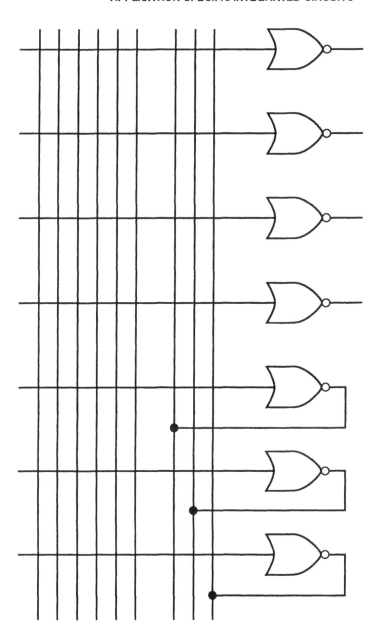

FIGURE 9.26. The folded-NOR multilevel-logic structure.

(MPGAs) and PLDs, there are two main catagories of logic block structures: *table-look-up* and *MUX-based*.[9] This gives rise to two architectures: channeled gate arrays with programmable interconnects and fixed interconnect routing with arrays of PLD-like programmable logic blocks.

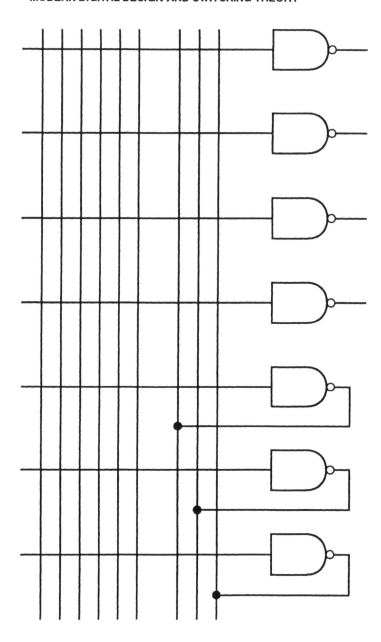

FIGURE 9.27. The folded-NAND multilevel-logic structure.

A basic table-look-up macrocell implements any function with up to m inputs, where m is a fixed number, greater than 2, for a given table-look-up architecture. There are five general-purpose inputs (m = 5) in the Xilinx architecture shown in Figure 9.29.[10] There are 2 to the 2^5 possible functions of 5 variables. This is over

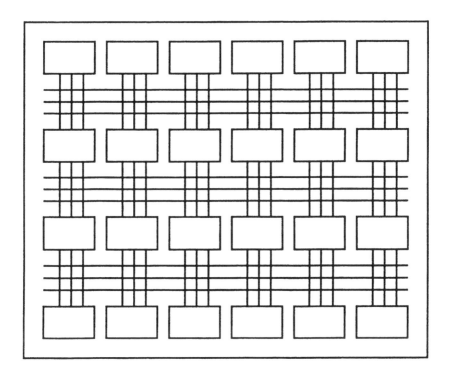

FIGURE 9.28.   The basic architecture of an FPGA, consisting of cells and programmable interconnects laid out in channels between the cells.

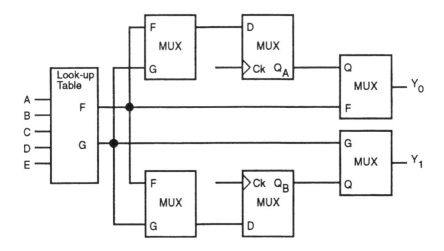

FIGURE 9.29.   The basic Xilinx table-look-up macrocell.

4.25 billion functions. The logic functions and the interconnections are determined by a "configuration program" stored in the internal static memory cells.

In table-look-up architecture, the number of variables is an indicator of the number of logic blocks required to implement it. There may be no correlation with the number of literals in the function. For example, the function Y_1 = ABCDEF has 6 literals and requires 2 5-input logic blocks, while the function Y_2 = ABC + BD$\overline{E}$ + $\overline{A}$CE + $\overline{C}$DE has 12 literals but only requires 1 logic block since it consists of only 5 variables.

One can define a *feasible function* as any function with no more variables than the number of general-purpose inputs m, and a *feasible network* as one for which every intermediate node of the network realizes a feasible function.[9] The best FPGA solution is then a feasible boolean network with a minimum number of nodes.

The desired solution is the one that reduces the number of look-up tables required, and the problem of obtaining a best utilization of look-up tables is mathematically equivalent to a bin-packing problem.[11] A two-level implementation realizes an SOP or POS solution and consists of one high-level node and more than one lower-level nodes. Some simple examples of the bin-packing approach follow.

EXAMPLE 9.9: Realize the two functions X = AB + CDE and Y = CDE + FG using five-input function blocks.

SOLUTION: The simplest SOP solution requires three AND gates and two OR gates, as shown in Figure 9.30(a). This requires a seven-input function block. The functions can be realized separately with two five-input function blocks, as shown in Figure 9.30(b).

EXAMPLE 9.10: Realize the function Z = AB + CD + DEF + GH with the least number of five-input function blocks.

SOLUTION: The simplest SOP solution requires four AND gates and one OR gate, as shown in Figure 9.31(a), and it could fit into an eight-input function block. If the function is partitioned as Z = (AB + GH) + (CD + DEF), it can be realized using one five-input function block and one four-input function block, as shown in Figure 9.31(b).

If the two-level decomposition is still too large to fit the function blocks, it must be decomposed into a multilevel function that will fit. This is equivalent to factoring the function into subfunctions which require no more than the available number of table-look-up inputs. An example of factoring a function follows.

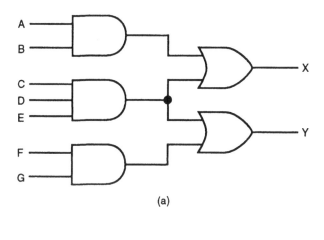

(a)

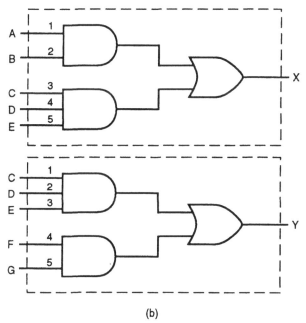

(b)

FIGURE 9.30. Realization of the functions X = AB + CDE and Y = CDE + FG (a) in AND-OR logic and (b) using two five-input function blocks.

EXAMPLE 9.11: Realize the function Z = ABC + DE + FG + HK + LMN with five-input function blocks.

SOLUTION: As it stands, the function requires five AND gates and one OR gate, as shown in Figure 9.32(a). The function can be factored as follows:

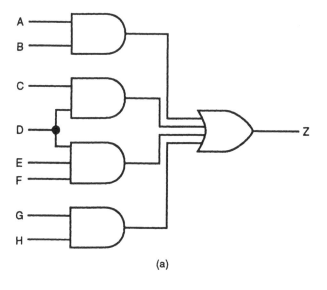

(a)

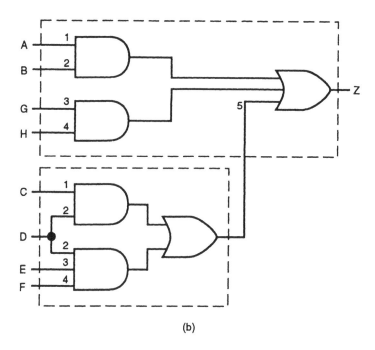

(b)

FIGURE 9.31. Realization of the function Z = AB + CD + DEF + GH (a) in AND-OR logic and (b) using a five- and a four-input function block.

Let $X = ABC + DE$ and X requires one five-input function block.
Let $Y = X + FG + HK$ and Y requires one five-input function block.
Then $Z = Y + LMN$ and Z requires one four-input function block.
The function has been factored to give the solution $Z = [(ABC + DE) + FG + HK] + LMN$ and it can be realized with two five-input and one four-input function blocks, as shown in Figure 9.32(b).

EXAMPLE 9.12: Realize the function $Z = ABC(F + G + H)IJK + DE(F + G + H)IJK + ABC(F + G + H)LM + DE(F + G + H)LM$ with the least number of five-input function blocks.

SOLUTION: The function can be factored into $Z = (ABC + DE) \cdot (F + G + H) \cdot (IJK + LM)$, as shown in Figure 9.33(a). It can be realized in a POS form with three five-input function blocks, as shown in Figure 9.33(b).

9.9.2   Multiplexer Architecture

In MUX-based architectures, the basic building block is a configuration of MUXs. The Actel architecture[12] has a basic building block composed of three two-input MUXs and one OR gate, as shown in Figure 9.34. This macrocell can realize functions with up to eight inputs. The rows of logic blocks are separated by routing channels consisting of routing tracks and a clock distribution network.

Given a MUX-based architecture, one must choose a set of 2-to-1-line MUXs as the basic function, and then program them appropriately. The same function can be realized in different forms, depending upon the choice of select control inputs and data inputs.

EXAMPLE 9.13: Realize the function $Y = A\overline{C}\,\overline{D}\,\overline{E} + BCE + BDE$ in MUX-based architecture.

SOLUTION: The function Y can be realized with inputs $S_{0A} = S_{0B} = E$, $S_1 = C + D$, and data inputs A, 0, 0, and B. This gives Y in the form $Y = A\overline{E}\left(\overline{C + D}\right) + BE(C + D)$. The function is shown in Figure 9.35.

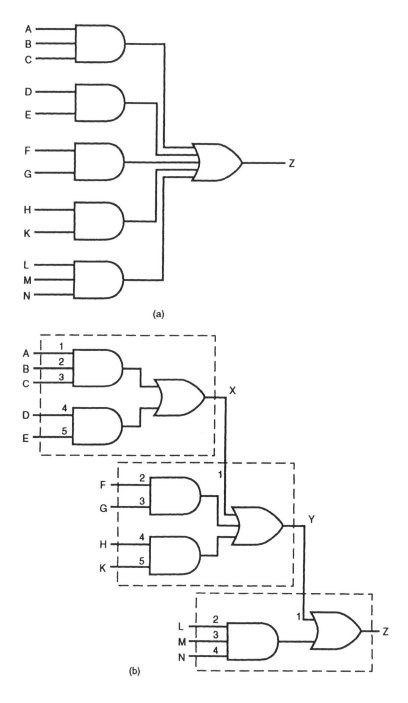

(a)

(b)

FIGURE 9.32. Realization of the function $Z = ABC + DE + FG + HK + LMN$ in (a) AND-OR logic and (b) using three logic function blocks.

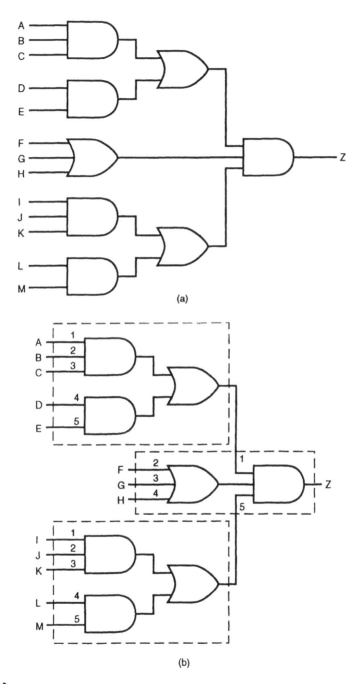

FIGURE 9.33.   Realization of the function $Z = (ABC + DE) \cdot (F + G + H) \cdot (IJK + LM)$ in (a) AND-OR-AND logic and (b) with three five-input logic blocks.

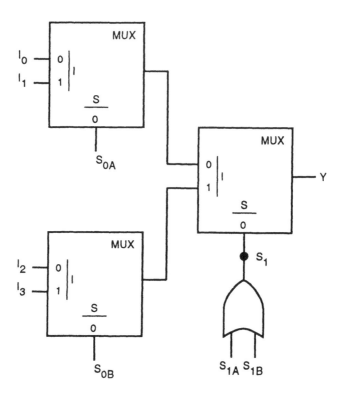

FIGURE 9.34.   One of the Actel multiplexer-based logic blocks.

Another architecture capable of realizing up to an eight-input function is achieved by adding an AND gate to the control input of the first two MUXs, as shown in Figure 9.36(a).[12] The function block performs a 4-to-1-line MUX operation, as shown in Figure 9.36(b), with control inputs $S_0 = I_4 \cdot I_5$ and $S_1 = I_6 + I_7$.

If the respective input assignments I_0 through I_7 are A, B, C, D, G, H, E, and F, the function realized by the circuit of either Figure 9.36(a) or (b) is

$$Y = A(\overline{E + F})(\overline{GH}) + B(\overline{E + F})(GH) + C(E + F)(\overline{GH}) + D(E + F)(GH)$$

$$= A(\overline{E}\,\overline{F})(\overline{G} + \overline{H}) + B(\overline{E}\,\overline{F})(GH) + C(E + F)(\overline{G} + \overline{H}) + D(E + F)(GH)$$

$$= A\overline{E}\,\overline{F}\overline{G} + A\overline{E}\,\overline{F}\overline{H} + B\overline{E}\,\overline{F}GH + CE\overline{G} + CE\overline{H} + CF\overline{G} + CF\overline{H}$$
$$+ DEGH + DFGH$$

9.9.3   The Multilevel NAND Structure

A third logic block structure, due to Altera,[13] is based on an AND-OR circuit, which can be implemented in AND-OR logic, feeding an EOR gate, as shown in

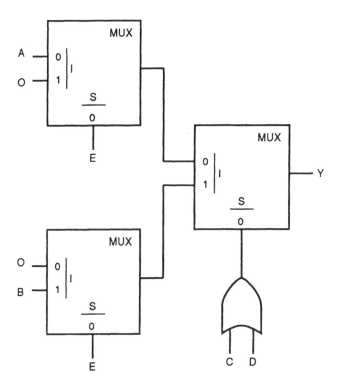

FIGURE 9.35. Realization of the function $Y = A\overline{E}(\overline{C+D}) + BE(C + D)$ in the MUX structure.

Figure 9.37(a). This basic circuit can be augmented by a flip-flop and a MUX, as shown in Figure 9.37(b). The MUX selects either the latched or the latch-bypassed output signal.

The EOR gate can be used to obtain programmable inversion. The function X = ABC + $\overline{A}$ BD and its complement can both be realized, as shown in Figure 9.38. X is the output of the OR gate. E is used as an input to the EOR gate. If E = 0, the output of the EOR gate is Y = X, and if E = 1, the output of the EOR gate is Y = $\overline{X}$.

If the inputs to an EOR gate are disjoint, it acts the same as an inclusive-OR gate. This allows the OR gate and the EOR gate to form a larger OR function, as seen in Example 9.14.

EXAMPLE 9.14: Realize a 4-to-1 MUX with inputs I_0, I_1, I_2, and I_3. The control inputs are A and B.

SOLUTION: The OR and EOR gates can be used to realize a 4-to-1 MUX, as shown in Figure 9.39. The output of the OR gate is $X = \overline{A}\,\overline{B}I_0 + \overline{A}\,BI_1 + A\overline{B}I_2$, and the output of the EOR gate is $Y = X \oplus ABI_3 = \overline{A}\,\overline{B}I_0 + \overline{A}\,BI_1 + A\overline{B}I_2 + ABI_3$.

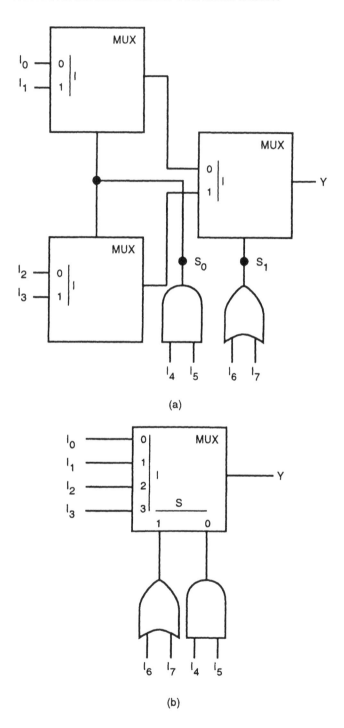

FIGURE 9.36.   A second Actel multiplexer-based logic block: (a) the basic architecture and (b) the equivalent 4-to-1-line MUX.

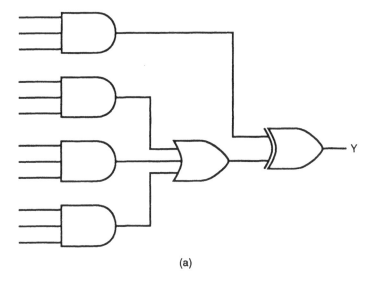

(a)

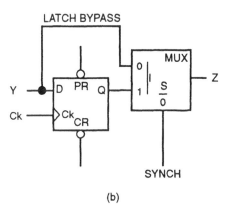

(b)

FIGURE 9.37. (a) An AND-OR-EOR logic block and (b) a synchronous/asynchronous output circuit.

   In Figure 9.40, three expander NAND gates have been added to realize functions in factored form. The D flip-flop and 2-to-1-line MUX have been included to give the complete macrocell. If a product of terms is fed into the NAND gate, its output will be a sum of the complemented terms. The expander performs the first-level logic, and the macrocell provides two more levels of logic before feeding the output of the third level to the EOR gate. One should avoid cascading expanders that feed expanders whenever possible or the multilevel logic becomes too slow.

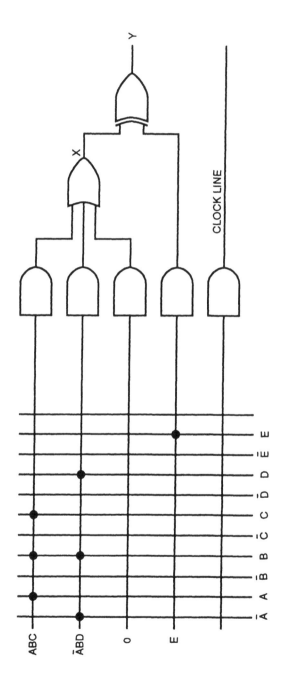

FIGURE 9.38. Realization of the function $X = ABC + \bar{A}BD$ or its complement, using the EOR gate as an inverter.

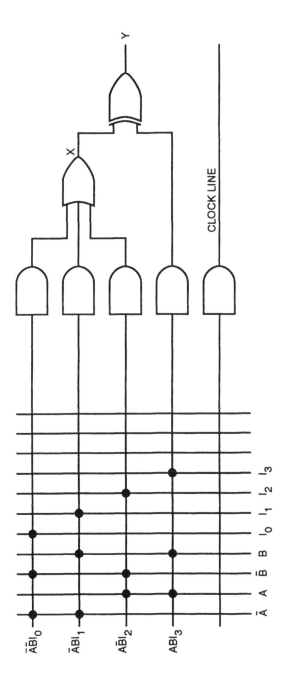

FIGURE 9.39.   Using the EOR gate as an inclusive-OR gate to implement a 4-to-1-line MUX.

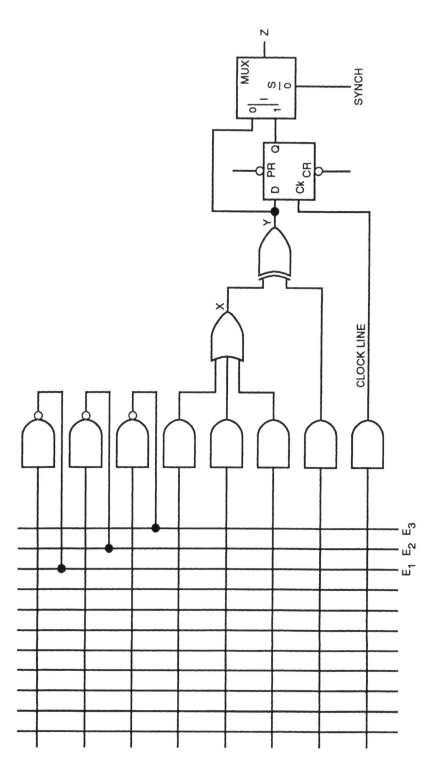

FIGURE 9.40. The Altera logic block augmented by expander NAND gates employing feedback.

EXAMPLE 9.15: Realize the function $Y = \overline{X}$, where

$$X = AD + \overline{B}D + CD + A\overline{E} + B\overline{E} + C\overline{E} + \overline{A}\,\overline{C}\,\overline{D}$$
$$+ \,\overline{B}\,\overline{C}\,\overline{D}$$

SOLUTION: The function X can be factored to give

$$X = (A + B + C)(D + \overline{E}) + (\overline{A} + \overline{B})\overline{C}\,\overline{D}$$

This function can be realized with three expander gates as follows. Feed $\overline{A}\,\overline{B}\,\overline{C}$ into the first expander gate and it outputs $E_1 = A + B + C$. Feed $\overline{D}E$ into the second expander gate and its output is $E_2 = D + \overline{E}$. Feed AB into the third expander gate and its output is $E_3 = \overline{A} + \overline{B}$. If expander outputs E_1 and E_2 are fed into an AND gate, its output is $(A + B + C)(D + \overline{E})$. If $\overline{C}\,\overline{D}$ and E_3 are fed into a second AND gate, its output is $(\overline{A} + \overline{B})\overline{C}\,\overline{D}$. If these two products are ORed, X is realized at the output of the OR gate. Inputting a 1 to the EOR gate causes it to output $Y = \overline{X}$, which is the desired function. The circuit is shown in Figure 9.41.

9.10 SUMMARY

Application-specific integrated circuits are tailored to meet a specific requirement, and they often require less space and less power than an equivalent fixed function integrated circuit. Most PLDs consist of a PLA AND plane-OR plane structure with various input buffers and output circuitry, including D and J-K flip-flops used as memory elements.

Four catagories of two-level PLDs were investigated: the FPLA, the PROM, the PAL, and the DLA.

Area, speed, and power are critical parameters for large PLA designs. The PLA should be reduced as much as possible, after which the physical layout can be done by PLA generation tools, which take input in boolean algebra and create PLA layouts. This enables design engineers to implement changes quickly and easily, with little impact upon the layout of the rest of the logic circuitry.

State-of-the-art PLDs have typical delays, including external buffers, of 2 to 3 ns per gate, and power dissipations of the order of 0.5 to 1.5 mW per equivalent gate. Speed-power products of this magnitude are competitive in today's market.

The OR plane outputs can be fed back as AND plane inputs to create a finite-state machine. Clocking Moore machines with two nonoverlapping clock phases prevents races. When flip-flops and/or shift registers are included in the feedback path, the FSM becomes a very effective sequential-machine design tool.

The importance of the PLD or FSM is due to its regularity, which makes it easy

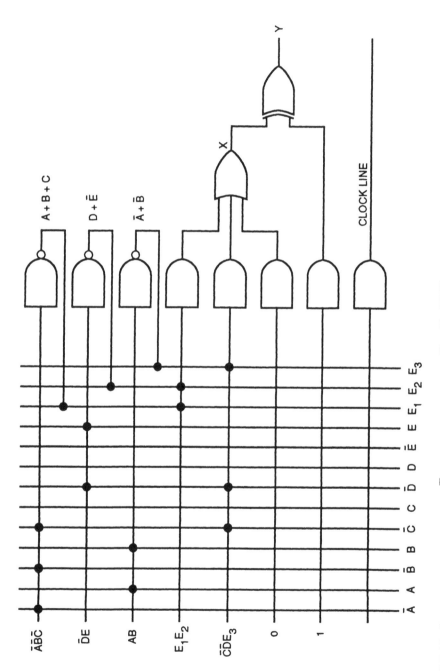

FIGURE 9.41. Realization of the function $Y = \bar{X}$ with $X = (A + B + C)(D + \bar{E}) + (\bar{A} + \bar{B})C\bar{D}$.

to design and easy to expand. Because of its modularity, it is possible to design hierarchical PLDs and FSMs into large sequential systems. It is reasonably compact, it is efficient for small circuits, and it is amenable to being computer generated.

A measure of the effective use of chip area by a PLA is the ratio of crosspoints containing switching devices to the total number of crosspoints required. When this ratio is low, the PLA is said to be sparse, and it does not utilize its chip area efficiently.

Direct implementation of an SOP realization of a function often leads to a very large PLA with few interconnecting transistors. This is very wasteful of chip area. Partitioning inputs into two-variable minterms or maxterms can minimize the AND plane.

PLA folding and sharing techniques tend to minimize the number of unused crosspoints in both the AND and the OR planes. This reduces the size of the PLA, which in turn reduces propagation delays. The row breaks and column breaks of a folded PLA allow more implicants on a single physical row and more outputs on a single physical column than are obtained with the original PLA structure. This results in a more compact external routing than in the original sparse PLA.

PLDs can be reprogrammed very quickly as compared to custom designed prototypes, which have rework cycles of 8 to 10 weeks. EPLDs can be erased and reprogrammed many times. They can be erased either electrically or using ultra-violet light, and they are very useful for prototyping.

The PLA-based architecture generates a two-level device which can realize any SOP or POS solution, but modern design often requires multilevel logic. The folded-NAND and folded-NOR programmable array is often superior when multiple logic levels are required. Each NAND of the folded-NAND structure and each NOR of the folded-NOR structure generates one level of logic. A single programmable array can generate N levels of logic with N passes through the array.

The latest addition to the arsenel of field programmable devices is the FPGA. An FPGA contains nonvolatile memory cells that allow the interconnection pattern to be loaded and changed after the device is manufactured. FPGAs are generally more complex and offer more logic functionality than PLDs. FPGAs come in three basic flavors, the Actel multiplexer, the Altera logic block, and the Zilinx look-up table architectures.

PROBLEMS

9.1. Four input signals are decoded to give the minterms of AB and of CD. Input F is not decoded with another literal.

   (a) Realize the implicant $A\overline{B}D\overline{F}$ with and without input partitioning.

   (b) Realize the implicant $\overline{A}C\overline{D}F$ with and without input partitioning.

(c) Realize the function $A\overline{B} + \overline{C}D + CD\overline{F}$ with and without input partitioning.

(d) Realize the function $A\overline{B} + CDF + B\overline{C}$ with and without input partitioning.

9.2. Repeat Problem 9.1 for the following:

(a) Implicant $AB\overline{C}\overline{D}$

(b) Implicant $A\overline{B}\overline{D}F$

(c) Function $AB + \overline{A}C + C\overline{D}F$

(d) Function $\overline{A}\,\overline{B}\,\overline{F} + AB + C\overline{D}$

9.3. Choose suitable decoding pairs to decode the following:

(a) Realize the implicant $C\overline{F}G\overline{H}$ with and without input partitioning.

(b) Realize the implicant $\overline{C}DF\overline{G}$ with and without input partitioning.

(c) Realize the function $G\overline{H} + C\overline{D} + CD\overline{F}$ with and without input partitioning.

(d) Realize the function $G\overline{H} + CGH + \overline{D}GH$ with and without input partitioning.

9.4. Repeat Problem 9.3 for the following:

(a) Implicant $BC\overline{E}\,\overline{F}$

(b) Implicant $B\overline{C}\,\overline{E}F$

(c) Function $BC + \overline{B}F + B\overline{E}F$

(d) Function $\overline{B}\,\overline{C}\,\overline{F} + BC + \overline{E}F$

In Problems 9.5 through 9.12 inclusive, design a PLA with and without a 3-to-8-line input decoder to realize the given functions. Compare the rows, columns, and crosspoint sites required for the two realizations.

9.5. $F(A,B,C) = \sum m\,(0,2,4,5,7)$ and $G(A,B,C) = \sum m\,(0,1,2,5,7)$.

9.6. $H(A,B,C) = \sum m\,(1,2,3,5,6)$ and $J(A,B,C) = \sum m\,(1,3,4,6,7)$.

9.7. $K(A,B,C) = \sum m\,(0,3,5,6,7)$ and $L(A,B,C) = \sum m\,(0,1,3,5,7)$.

9.8. $M(A,B,C) = \sum m\,(0,1,2,5,7)$ and $N(A,B,C) = \sum m\,(2,3,5,6,7)$.

9.9. $P(A,B,C) = \prod M\,(0,2,4,5,7)$ and $Q(A,B,C) = \prod M\,(0,1,2,5,7)$

9.10. $R(A,B,C) = \prod M\,(1,2,3,5,6)$ and $S(A,B,C) = \prod M\,(1,3,4,6,7)$.

9.11. $T(A,B,(C) = \prod M\,(0,3,5,6,7)$ and $U(A,B,(C) = \prod M\,(0,1,3,5,7)$.

9.12. $V(A,B,C) = \prod M\,(0,1,2,5,7)$ and $W(A,B,C) = \prod M\,(2,3,5,6,7)$.

9.13. Choose suitable input decoding to design a minimum size PLA for the functions $Y_1 = CD + \overline{C}\,\overline{D} + C\overline{D} + \overline{B}$, $Y_2 = A + C \oplus D$, and $Y_3 = AD(B \oplus C)$.

9.14. Choose suitable input decoding to design a minimum size PLA for the functions $Y_1 = AB + \overline{C}\,\overline{D} + B\overline{C} + \overline{D}$, $Y_2 = B + A \oplus D$, and $Y_3 = AB(C \oplus D)$.

In Problems 9.15 through 9.22 inclusive, do not use input partitioning.

9.15. Design a minimum size PLA to realize the counter of Example 8.1, Chapter 8.

9.16. Design a minimum size PLA to realize the counter of Example 8.2, Chapter 8.

9.17. Design a minimum size PLA to realize the counter of Example 8.4, Chapter 8.

9.18. Design a minimum size PLA to realize the counter of Example 8.5, Chapter 8.

9.19 (a) Design a minimum-size PLA to convert Gray code to weighted BCD. The truth table is given in Table 9.3.

   (b) How much larger would the PLA be if both true and complemented output lines are available?

   (c) If only one output per variable is provided, does it matter whether it is the true or complemented output?

9.20. Repeat Problem 9.19 for a weighted BCD to Gray code converter.

9.21. Repeat Problem 9.19 for a PLA that outputs a weighted BCD and a Johnson BCD code.

TABLE 9.3   Truth Tables

		Gray			8-4-2-1 BCD				Johnson BCD				
Decimal	A	B	C	D	W	X	Y	Z	R	S	T	U	V
0	1	1	1	1	0	0	0	0	0	0	0	0	0
1	0	1	1	1	0	0	0	1	0	0	0	0	1
2	0	0	1	1	0	0	1	0	0	0	0	1	1
3	1	0	1	1	0	0	1	1	0	0	1	1	1
4	1	0	0	1	0	1	0	0	0	1	1	1	1
5	0	0	0	1	0	1	0	1	1	1	1	1	1
6	0	1	0	1	0	1	1	0	1	1	1	1	0
7	1	1	0	1	0	1	1	1	1	1	1	0	0
8	1	1	0	0	1	0	0	0	1	1	0	0	0
9	0	1	0	0	1	0	0	1	1	0	0	0	0

9.22. Repeat Problem 9.19 for a PLA that outputs a Gray code and a Johnson BCD code.

9.23. Give the row-disjoint table for Figure 9.14.

9.24. Give the row-disjoint table for Figure 9.15(a).

9.25. Column fold the PLA whose personality matrix, Q, is

$$Q = \begin{bmatrix} 11000010 & 110 \\ 10001100 & 101 \\ 11011000 & 100 \\ 01100110 & 010 \\ 00100101 & 011 \\ 01000010 & 011 \end{bmatrix}$$

9.26. Row fold the PLA of Problem 9.25.

9.27. Column fold the PLA whose personality matrix, Q, is

$$Q = \begin{bmatrix} 11000010 & 110 \\ 00011101 & 101 \\ 11000000 & 100 \\ 01000111 & 010 \\ 00110100 & 011 \\ 01000011 & 011 \end{bmatrix}$$

9.28. Row fold the PLA of Problem 9.27.

9.29. Design a folded-NOR array to realize the function $Y = (A + B)(\overline{C} + \overline{D})(\overline{E} + F)$.

9.30. Design a folded-NOR array to realize the function $Z = \overline{A}[\overline{B} + \overline{C}(D + \overline{E})]$.

9.31. Design a folded-NAND array to realize the function $Y = \overline{A}\,\overline{B} + \overline{C}\,\overline{D} + EF$.

9.32. Design a folded-NAND array to realize the function $Z = \overline{A} + \overline{B}\,\overline{C} + D\overline{E}$.

Realize the functions in Problems 9.33 through 9.38 inclusive with the least number of function blocks. Four-input and five-input function blocks are available. Use the smallest combination that will do.

9.33. $U = ABC + DE + FGH$.

9.34. $V = AB\overline{C}D + \overline{B}CDE$.

9.35. $W = AB\overline{C}D + AB\overline{E}\,\overline{F} + C\overline{D}EF$.

9.36. $X = \overline{A}\,B\overline{C}D + CD\overline{E}\,\overline{F} + A\overline{C}DF$.

9.37. $Y = \overline{A}\,BC\overline{D}E + E\overline{F}GH$.

9.38. $Z = A\overline{B}E + CD\overline{E} + FG + HJ$.

Realize the functions in Problems 9.39 and 9.40 with the MUX macrocells of Figure 9.34.

9.39. $U = A\overline{B}(\overline{E} + \overline{F}) + CDEF$. Use B and D as the lower-level and $\overline{E} + \overline{F}$ as the higher-level select input.

9.40. $V = (A\overline{C} + \overline{B}C)(F + \overline{G}) + \overline{D}E\overline{F}G$. Use C and E as lower-level and $F + \overline{G}$ as the higher-level select inputs.

Realize the functions in Problems 9.41 through 9.44 inclusive with the MUX macrocells of Figure 9.36.

9.41. $W = (\overline{A} + \overline{B})(\overline{C}D + CF) + AB(\overline{C}\,\overline{E} + CG)$ and $S_0 = AB$, $S_1 = C$.

9.42. $X = (\overline{A}D + AE)(\overline{B} + \overline{C}) + (\overline{A}\,\overline{F} + AG)BC$ and $S_0 = A$, $S_1 = \overline{B} + \overline{C}$.

9.43. $Y = (\overline{A} + \overline{B})(\overline{C} + \overline{D})E + ABCDF$ and $S_0 = AB$, $S_1 = \overline{C} + \overline{D}$.

9.44. $Z = \overline{A}(C + \overline{D})E\overline{F} + (\overline{E} + F)H$ and $S_0 = \overline{C}D$, $S_1 = \overline{E} + F$.

Realize the functions in Problems 9.45 through 9.50 inclusive with the macrocells shown in Figure 9.40. Use the least number of expander gates needed. X is the output of the OR gate and Y is the output of the EOR gate.

9.45. $Y = X = \overline{A} + B + CD + C\overline{E} + \overline{F}H + GH$.

9.46. $Y = \overline{X}$ and $X = (\overline{A} + \overline{B})G + (\overline{C} + D)\overline{H} + E + \overline{F}$.

9.47. $Y = \overline{X}$ and $X = \overline{A}B(\overline{C} + D) + (\overline{A} + \overline{C})D$.

9.48. $Y = X \oplus (GH)$ and $X = \overline{A}\,\overline{B} + CD + EF$.

9.49. $Y = \overline{X}$ and $X = ABC + DG + GH[(\overline{A} + \overline{B})\overline{C}D + \overline{E} + F]$.

9.50. $X = (AB + CD)(\overline{E} + \overline{F})(\overline{G} + \overline{H})$ and $Y = \overline{AB + CD + EF + GH}$.

SPECIAL PROBLEMS

9.51. Design an ASIC to implement a modulo-16 synchronous binary counter that counts either up or down. Use parallel carry, as discussed in Chapter 8, Section 8.6, and an $\overline{UP}/DOWN$ control input as shown in Figure 8.14, Chapter 8.
   (a) Do the design with a PROM and a PLA.
   (b) Do the design in each of the three basic FPGA architectures.

9.52. Design an ASIC to implement a modulo-10 binary up counter using S-R, J-K, D, and T flip-flops. Do the design as a regular and as a folded PLA and compare their sizes.

REFERENCES

1. **Pellerin, D. and Holley, M.,** *Practical Design Using Programmable Logic,* Prentice-Hall, Englewood Cliffs, NJ, 1991, 65–74.
2. **Ligthart, M. M., Aarts, E. H. L., and Beenker, F. P. M.,** Design-for-Testability of PLAs Using Statistical Cooling, Proc. 23rd Design Automation Conf., June 29 to July 2, 1986, 339–345.
3. **Carr, W. N. and Mize, J. P.,** *MOS/LSI Design and Application,* Texas Instruments Electronics Series, McGraw-Hill, New York, 1972, 232–258.
4. **Hachtel, G. D., Newton, A. R., and Sangiovanni-Vincentelli, A. L.,** An Algorithm for Optimal PLA Folding, IEEE Trans. on Computer-Aided Design of Integrated Circuits and Systems, Vol. CAD-1, 1982, 63–76.
5. **Lewandowski, J. L. and Liu, C. L.,** A Branch and Bound Algorithm for Optimal PLA Folding, Proc. IEEE 21st Design Automation Conf., Albuquerque, NM, June 1984, 426–433.
6. **Hwang, A. S. Y., Dutton, R. W., and Blank, T.,** A Best-First Search Algorithm for Optimum PLA Folding, IEEE Trans. on Computer-Aided Design, Vol. CAD-5, 1986, 433–442.
7. **Wakerly, J. F.,** *Digital Design Principles and Practices,* Prentice-Hall, Englewood Cliffs, NJ, 1990, chap. 7.
8. **Pellerin, D. and Holley, M.,** *Practical Design Using Programmable Logic,* Prentice-Hall, Englewood Cliffs, NJ, 1991.
9. **Murgai, R., Nishizaki, Y., Shenoy, N., Brayton, R. K., and Sangiovanni-Vincentelli, A.,** Logic Synthesis for Programmable Gate Arrays, Proc. 27th ACM/IEEE Design Automation Conf., June 24 to 28, 1990, 620–625.
10. *Xilinx Programmable Gate Array User's Guide,* Xilinx, Inc., 1988.
11. **Francis, R., Rose, J., and Vranesic, Z.,** Chortle-crf: Fast Technology Mapping for Lookup Table-Based FPGAs, Proc. 28th ACM/IEEE Design Automation Conf., San Francisco, June 17 to 21, 1991, 227–233.
12. **Gamal, A., Greene, J., Reyneri, J., Rogoyski, E., El-Ayat, K. A., and Mohsen, A.,** An architecture for electrically configurable gate arrays, *IEEE J. Solid State Circuits,* 24(2), 394–398, 1989.
13. *Altera Corporation Maximalist Handbook,* Altera Corp., 1990.

CHAPTER 10

Multilevel Minimization

10.1 INTRODUCTION

A variety of methods have been developed for representing and manipulating boolean functions. Truth tables and Karnaugh maps are impractical for functions of many variables since an n-variable function has 2^n minterms or maxterms, and these representations grow exponentially with n. Even if a function has a simple representation, its complement may be very large. Also, reduced SOP and POS solutions are not in canonical form and can have more than one representation. This makes testing for equivalence or satisfiability difficult, and, in some cases, SOP and POS solutions can come up with prohibitively large fan-ins.

Logic synthesis is usually divided into two-level and multilevel synthesis. Two-level logic minimization is used to synthesize PLA-based PLDs for control logic and FSM design, and the optimization methods generally focus on minimizing the number of product terms. This minimizes the number of product lines in the PLD, which generally gives the minimum area. Two-level logic representations are special cases of multilevel representations, and a logic design approach should include multilevel synthesis in order to obtain better area and/or speed reduction. Multilevel logic is useful for control and data-flow logic. Its goal is to minimize overall area and critical path delay time while maximizing testability. Ideally, a complete set of test vectors would be generated as a byproduct of the optimization.

Many boolean networks are much smaller when represented in multilevel format. A common example is a parity checker, designed with EOR gates. The multilevel representation of the parity checker can be in the recursive form $A \oplus [B \oplus (C \oplus D)]$, or in the tree form $(A \oplus B) \oplus (C \oplus D)$. Both representations require n–1 EOR gates for n inputs. As a canonical SOP expansion the function requires many more gates. The EOR function is true for half the total number of minterms, and the two-level realization of a parity checker requires 2^{n-1} first-level gates.

When the number of variables in a function is very large it is necessary to have a more suitable representation of the function. There are two basic approaches to multilevel design: the boolean/algebraic approach and the graph-based approach. In the boolean/algebraic method, optimization is done by factoring and two-level minimization on each node of the factored form. In the graph-based approach, a function is represented as an acyclic graph whose nodes represent simple functions. Optimization is done using graph manipulation and data flow algorithms.

A factored form is isomorphic to a binary-decision-tree structure where each internal node is an AND or an OR operator. The factored form implicitly represents the complement of a function also, since DeMorgan's law is applied by simply interchanging the AND and OR operators and complementing all the literals.

A multilevel combinational circuit can be specified by a boolean network which is defined as a binary decision diagram (BDD) or an if-then-else (ITE) structure. Both the BDD and the ITE are directed acyclic graphs (DAGs). In both approaches, each node has two paths leading from the node, corresponding to a true and a false value of the node variable.

BDDs are ideally suited to realizing these functions with reduced DAGs, which, when ordered, give canonical representations of functions of many variables in a factored or multilevel form. The loss in speed due to many logic levels is partially offset by the greatly reduced fan-ins encountered in this approach. BDDs also can be quickly and easily converted to MUX realizations of the functions.

In the BDD approach, a boolean network is specified by a DAG with each node associated with a defining SOP expression and a unique reference variable. The edges of the graph are derived from the SOP expressions and reference variables of the nodes in the network. Nodes of a boolean network representing primary inputs to the network have no SOP expression. Each node of the network can be viewed as an incompletely specified, single-output boolean function.

This chapter offers an introduction to the techniques being investigated for implementation with computer-aided design of VLSI.

10.2 FACTORING BOOLEAN EXPRESSIONS

The factored form of an expression is the most useful for multilevel logic, and replaces the SOP and POS forms, which give minimum two-level logic solutions. The *factored form* is defined recursively. A literal is a factored form, a sum of factored forms is a factored form, and a product of factored forms is a factored form. A factored form is an SOP of an SOP of an SOP, etc.

Factored forms are not always unique. $ABC + ABD + CD$ can be factored into $AB(C + D) + CD$ or $ABC + (AB + C)D$. The minimum factored form contains the least number of literals. In the above example, $ABC + (AB + C)D$ has seven literals, while $AB(C + D) + CD$ has six literals and is the minimum factored form.

Factoring is the technique of converting a two-level SOP or POS form of a function to a multilevel function and *multilevel minimization* consists of obtaining a factored form which contains the least number of literals. The goal of factoring is the identification of common subexpressions. The primary objective in two-level SOP optimization is to minimize the number of cubes in the function. This is only of interest in multilevel optimization in so far as it correlates somewhat with the total number of literals. In general, optimization consists of finding a set of functions such that the resulting network optimizes a particular cost function, which normally would minimize layout area, routing, power dissipation, or propagation delays.

Multilevel logic can always reduce the number of literals, the number of gate inputs, and the maximum fan-in, MFI. It can often reduce the number of gates, but it will increase some path lengths and add signal delay to some inputs. Unless otherwise specified, the main purpose of optimizing functions in this chapter will be to minimize the number of literals.

Literal factoring is perhaps the easiest to do. An SOP realization of the expression $Y = AC + AD + AE + AG + BC + BD + BE + BF + CE + CF + DF + DG$ takes 12 2-input AND gates and 1 12-input OR gate. The function contains 24 literals, and factoring literals A, B, C, and D yields a function with 16 literals, viz: $Y_1 = A(C + D + E + G) + B(C + D + E + F) + C(E + F) + D(F + G)$. This form of the function can be realized in three-level logic and requires an MFI of four. The expression $C + D + E$ is a factor of the first 2 terms of the above function, and by factoring out this term, Y_1 can be reduced to a function having only 15 literals, $Y_2 = (A + B)(C + D + E) + AG + BF + C(E + F) + D(F + G)$.

Next, factoring literals F and G out of Y_2, one obtains a function with 14 literals, $Y_3 = (A + B)(C + D + E) + (B + C + D)F + (A + D)G + CE$. This is probably the solution with the least number of literals.

As was the case with SOP and POS solutions, techniques are required that guarantee an optimum solution to multilevel minimization. Unfortunately, there may not be unique solutions for complex functions or the solutions may require prohibitive computer time. For this reason many programs work toward a goal of "reasonably good" or "acceptably good" solutions.

Literal factoring is the same thing as factoring 0-cubes of the function. This can be generalized to cube factoring by counting the number of times a cube appears in an expression and how many literals can be saved by factoring the cube out. In the function $X = ABC + ABD + ABE + FGH + FGJ$ cube AB occurs three times and factoring it out can save four literals, while cube FG occurs twice and factoring it out can save two literals. When both cubes are extracted the function becomes $X = AB(C + D + E) + FG(H + J)$. It is seen that the literal count did indeed drop by 6, from 15 literals in the SOP form to 9 literals in the factored form.

This approach requires two steps to factor an expression as simple as $Y = AC + AD + BC + BD$, where each literal occurs twice. Factoring in alphabetical order $Y = A(C + D) + B(C + D)$ saves two literals. Factoring $C + D$, which now occurs twice, gives $Y = (A + B)(C + D)$, a savings of two more literals.

The process could have been more easily accomplished by examining the common factors of the function that are not cubes. A *kernel* of a boolean function (from cyrnel, the Anglo-Saxon word for little corn and meaning the essence or central part of anything) is a prime multicube divisor of the function. The determination of the factors common to two or more boolean cubes is referred to as kernel extraction and will be studied in depth in Section 10.7.

The above methods of factoring are algebraic and do not work for functions which have no algebraic factors. The function $Z = A + BC$ has no algebraic factors, but it has two boolean factors, $A + B$ and $A + C$. Factoring can thus be divided into algebraic factoring and boolean factoring. In *algebraic factoring,* the function is considered to be an algebraic polynomial with the true and complemented variables treated as different. In *boolean factoring,* the theorems and techniques of boolean algebra, including the use of don't care terms, are applied to find redundancies or simplify the function. Boolean factoring techniques are slower than algebraic methods, but generally give much better results.

Define the *support* of an expression as the set of variables that are elements of the cubes of the expression. The support of expression Y is written sup(Y), where $sup(Y) = \{X | X$ is a variable and either $X \in Y$ or $\overline{X} \in Y\}$. Sup($AB + \overline{A}C$) = {A, B, C} since the function $AB + \overline{A}C$ is composed of these three variables. The support of a function is proportional to the number of primary inputs, which is the number of wires that must be routed to the function, and is a measure of both the area needed and the routing difficulty.

If two expressions Y and Z have disjoint support, there is no overlap of variables in the two support sets and $sup(Y) \cap sup(Z) = \varnothing$, the null or empty set. When this is true, the product YZ is an *algebraic product* or *algebraic expression*. No boolean operations are required to obtain an algebraic product. If YZ is not an algebraic product, it is a *boolean product* or *boolean expression*. For example, if $Y = AB + AC$ and $Z = EF + H$, then YZ is an algebraic product because $sup(Y) \cap sup(Z) = \varnothing$, and Y and Z have disjoint support. If $Y = AB + AC$ and $X = DE + A$, then XY is not an algebraic product because $sup(X) \cap sup(Y) = A$. $(A + B)(C + D) = AC + AD + BC + BD$ is an algebraic product, while $(A + B)(A + C) = A + BC$ and $(A + B)(\overline{B} + C) = A\overline{B} + AC + BC$ are boolean products.

10.3 DECOMPOSITION AND EXTRACTION TECHNIQUES

Decomposition, restructuring, substitution and *resubstitution* are all terms for the process of re-expressing a single function as a collection of new functions. Decomposition is similar to factoring except that each divisor is converted into a new node variable which is substituted into the function being decomposed. Decomposition creates a directed acyclic graph, whereas factoring creates a tree. Decomposition can be algebraic or boolean, depending upon which type of factoring is used to achieve the answer. Decomposition requires the identification and resubstitution of common subexpressions of a function into the original function in order to reduce the number of literals.

A simple check will reveal whether or not literals are saved. For example, if $Y = A + B$ and $Z = A + BC = (A + B)(A + C)$, then $Z \rightarrow (A + C)Y$, and the literal count has risen from three to four, while if $Z = AC + AD + B\overline{C}\,\overline{D}$, and $Y = C+D$ then $Z \rightarrow AY + B\overline{Y}$, and the literal count has dropped from seven to four.

Extraction is the process of factoring multiple functions by identifying intermediate functions and variables, and re-expressing the original functions in terms of the original and the intermediate variables. Decomposition applies to single expressions, whereas extraction applies to multiple functions. The goal of extraction is to identify common subexpressions among the different logic functions forming a network and simultaneously to simplify the logic functions of the original network.

Decomposition of the function $W = \overline{A}C + BC + \overline{A}D + BD + \overline{A}E + BE = (\overline{A} + B)(C + D + E)$ results in $W = UV$, where $U = \overline{A} + B$ and $V = (C + D + E)$. Extraction applied to the three functions $X = (\overline{A} + B)CD + \overline{E}$, $Y = (\overline{A} + B)E$, and $Z = CD\overline{E}$ yields $X = UV + \overline{E}$, $Y = UE$, and $Z = V\overline{E}$, where $U = \overline{A} + B$ and $V = CD$.

EXAMPLE 10.1: Factor the functions X, Y, and Z and show the networks that realize the factored forms. $X = A\overline{B}C + A\overline{B}D + A\overline{B}E + \overline{F}GH + \overline{F}GJ$, $Y = AC + AD + \overline{B}C + \overline{B}D + H + J$, and $Z = A + \overline{B}C$.

SOLUTION: Algebraic factoring of X and Y yields $X = A\overline{B}(C + D + E) + \overline{F}G(H + J)$ and $Y = A(C + D) + \overline{B}(C + D) + H + J = (A + \overline{B})(C + D) + H + J$. Boolean factoring of Z yields $Z = (A + \overline{B})(A + C)$. The networks are shown in Figure 10.1.

EXAMPLE 10.2: Decompose the functions of Example 10.1 and show the circuits to realize them.

SOLUTION: Define the new functions: $S = C + D + E$, $T = H + J$, $U = A + \overline{B}$, $V = C + D$, and $W = A + C$. Then $X = A\overline{B}S + \overline{F}GT$, $Y = UV + T$, and $Z = UW$. The three functions are realized in Figure 10.2, with a total of 11 literals and 6 gates. The functions have been reduced to two-level SOP form, Z being a degenerate SOP form.

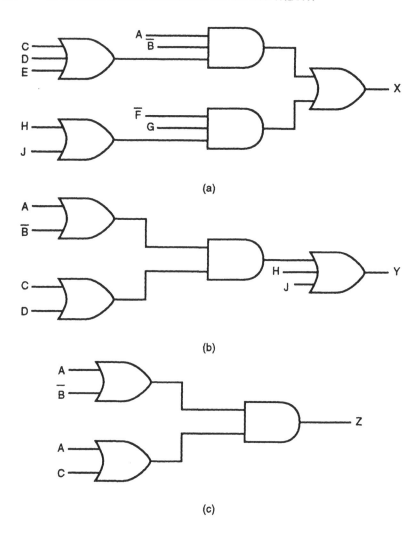

FIGURE 10.1. Circuit realizations of (a) $X = A\overline{B}(C + D + E) + \overline{F}G(H + J)$, (b) $Y = (A + \overline{B})(C + D) + H + J$, and (c) $Z = (A + \overline{B})(A + C)$.

EXAMPLE 10.3: Use extraction to realize the simultaneous solution to the three functions of Example 10.2.

SOLUTION: The same subfunctions are created, and Figure 10.3 shows the three functions realized simultaneously. The circuit requires a total of nine literals and six gates and is again reduced to SOP form.

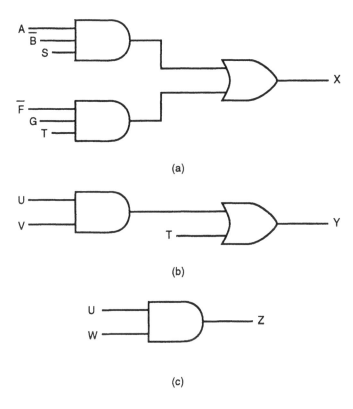

(a)

(b)

(c)

FIGURE 10.2. The functions of Figure 10.1 decomposed using subfunctions $S = C + D + E$, $T = H + J$, $U = A + \overline{B}$, $V = C + D$, and $W = A + C$, to give $X = A\overline{B}S + \overline{F}GT$, $Y = UV + T$, and $Z = UW$.

Collapsing, also referred to as *elimination* or *flattening,* is the converse of factoring or decomposition and reduces the number of levels of logic, thus flattening the graph. If Y is a subfunction of Z, then collapsing Y into Z re-expresses Z without Y. This undoes the operation of substituting Y into Z. If node Y is not an output, it may be eliminated, resulting in a boolean network with one less node. If $Z = (A + C)Y$ and $Y = A + B$, then Z has three nodes and requires one AND gate and two OR gates, whereas flattening gives $Z = A + BC$, which has two nodes and requires only one AND gate and one OR gate. Flattening is used to eliminate nodes that do not reduce the literal count of the function. A network that is completely flattened is reduced to an SOP or POS form.

EXAMPLE 10.4: Use node elimination to flatten the functions, $Y_1 = CD\overline{Y}_3$, $Y_2 = DE\overline{Y}_3$, $Y_3 = CDE$.

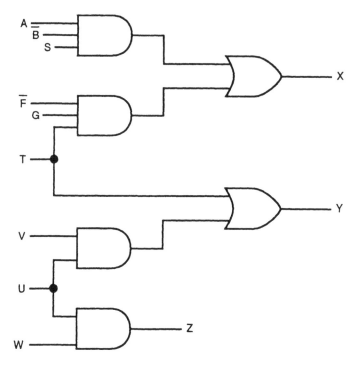

FIGURE 10.3.   The three functions of Figure 10.1 simultaneously reduced.

SOLUTION: $\overline{Y}_3 = \overline{C} + \overline{D} + \overline{E}$. Therefore, $Y_1 = CD(\overline{C} + \overline{D} + \overline{E}) = CD\overline{E}$ and $Y_2 = DE(\overline{C} + \overline{D} + \overline{E}) = \overline{C}DE$. The literal count has decreased from seven to six, and flattening was beneficial. The circuit is shown before and after flattening in Figure 10.4. It was flattened from a two-level circuit to a one-level circuit.

EXAMPLE 10.5:   Flatten the function $Y = (A + B)W$, where $W = (C + DE)$.

SOLUTION: Multiplying the function out, one obtains $Y = AC + ADE + BC + BDE$. The function has been flattened into an SOP form with 10 literals, twice the original count, as seen in Figure 10.5. The node count has increased from four to five also.

A literal is usually defined as each occurrence of a variable in either the true or complemented form. A literal can also be considered to be a variable combined with a *phase*. If the phase of the literal is positive, the literal is true, and if the phase is negative, the literal is complemented. A is a literal with positive phase, while $\overline{B}$ is a literal with negative phase. Phase assignment is a global procedure that helps simplify multilevel boolean networks by minimizing the total number of

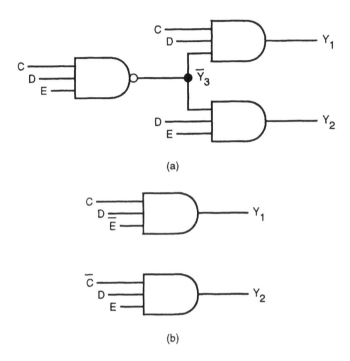

(a)

(b)

FIGURE 10.4. (a) Functions $Y_1 = CD\overline{Y}_3$, $Y_2 = DE\overline{Y}_3$, and $Y_3 = CDE$ flattened into (b) $Y_1 = CD(\overline{C} + \overline{D} + \overline{E}) = CD\overline{E}$ and $Y_2 = DE(\overline{C} + \overline{D} + \overline{E}) = \overline{C}DE$.

inverters required and defining the types of gates needed to implement the function.

EXAMPLE 10.6: Group the products of the function $\overline{Z} = ABC + DE + \overline{F}G + \overline{H}\,\overline{J}\,\overline{K}$ according to phase, and implement $\overline{Z}$ with two- and three-input NAND and NOR gates. All primary inputs are to have positive phase.

SOLUTION: Let $Z = \overline{Y_1 + Y_2 + Y_3}$. Y_1, with no complemented variables, consists of products ABC and DE and can be realized with three NANDgates. Y_2, with both true and complemented variables, consists of $\overline{F}G$ and can be realized with two NOR gates, one gate being used to form an inverter. Y_3, with complemented variables only, consists of product $\overline{H}\,\overline{J}\,\overline{K}$ and can be realized with one NOR gate. An additional NOR gate is needed to form the function Z. The circuit is shown in Figure 10.6.

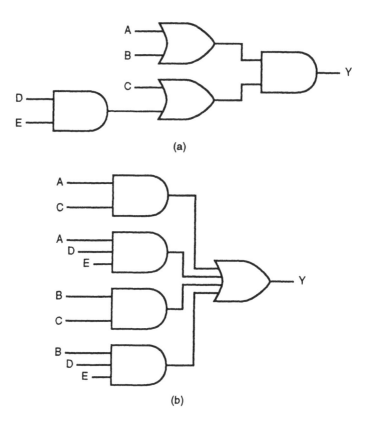

FIGURE 10.5.   The function $Y = (A + B)(C + DE)$ (a) in factored form and (b) flattened into $Y =$ $AC + ADE + BC + BDE$.

10.4   DIVISION OF SWITCHING FUNCTIONS

Factoring of a boolean expression is analogous to the operation of polynomial division, and collapsing is analogous to the operation of polynomial multiplication. In optimizing logic functions it is necessary to define operations which, when given Y and P, will find the functions Q and R such that $Y = PQ + R$. Because the operations required are similar to the division operation in other algebras, it is reasonable to refer to the function P as a *boolean divisor* of Y, if $R \neq 0$, and a *boolean factor* if $R = 0$. Q is called the quotient of Y divided by P, and R is called the remainder. Boolean division is not unique, since the values of Q and R may be dependent upon the particular representation of Y and P. The number of boolean factors and divisors of a given function can be very large.

If P and Q have disjoint support, then PQ is an algebraic product, P and Q are

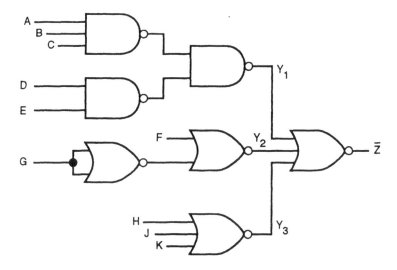

FIGURE 10.6. Implementation of the function $\overline{Z} = ABC + DE + \overline{F}G + \overline{H}\,\overline{J}\,\overline{K}$ factored by phase.

algebraic divisors of Y, and Y/P is a unique algebraic quotient. Otherwise P and Q are *boolean divisors* of Y, PQ is a boolean product, P and Q are boolean divisors of Y, and Y/P is a nonunique boolean quotient.

The candidate list of potential divisors can be reduced by observing that a function Y is not an algebraic divisor of another function Z if it contains any literals not in Z, if it contains more cubes than Z, or if it has any literal occurring more often than in Z.

EXAMPLE 10.7: Find the quotient and the remainder of X divided by both Y and Z, where X = AD + BCD + E, Y = A + B, and Z = A + BC. Are the divisions algebraic or boolean?

SOLUTION: The quotient of X/Y = [(A + B)(A + C)D + E]/(A + B) is (A + C)D, and the remainder is E. This is boolean division. The quotient of X/Z = [(A + BC)D + E]/(A + BC) is D, and the remainder is E. This is algebraic division.

10.5 ALGEBRAIC DIVISION

A function Y is an *algebraic expression* if Y is a set of cubes such that no one cube contains another. Cube A is said to contain cube B, A⊃B, if the set of minterms in cube A contains the set of minterms in cube B. Cube AB can be factored into minterms AB$\overline{C}$ and ABC in the boolean space of variables A, B, and C, and cube AB contains cubes ABC and AB$\overline{C}$.

The *product of two expressions* Y and Z is a set defined by $YZ = \{C_1 C_2 | C_1 \in Y,$ $C_2 \in Z$ and $C_1 C_2 \neq \emptyset\}$. YZ is an *algebraic product* if Y and Z are algebraic expressions and have disjoint support, that is, they have no input variables in common. If Y and Z have common support, then YZ is a *boolean product*. (A + B)(C + D) = AC + AD + BC + BD is an algebraic product, whereas (A + B)(A + D) = A + BD and $(A + B)(\overline{A} + D) = AD + \overline{A}B$ are boolean products.

An operation *DIV* is called *division* if, given two functions Y and P, it generates Q and R, such that Y = PQ + R, i.e., DIV(Y,P) = (Q,R). If PQ is an algebraic product, DIV is called an algebraic division, otherwise PQ is a boolean product and DIV is called a boolean division.

10.5.1   Weak Division

In working with a function of many variables, it is desirable to find good divisors first since they will be used many times in inner loops of any multilevel factoring program. One subset of algebraic division that accomplishes this is referred to as *weak division*.[1] Given two algebraic expressions Y and P, a division is said to be weak if it generates Q and R such that PQ is an algebraic product, R has a minimum number of cubes, and both PQ + R and Y have the same set of cubes and are therefore the same expression. Weak division is so named because it finds only the level-0 kernels of a function. Functions are factored in weak division by successively dividing them by subexpressions that appear more than once.

The first step is to identify the candidate subexpressions and a cost function to be optimized. The best candidate factor is chosen and divided out of each function. Substituting for the subexpression may allow new candidates to be generated, so the process is repeated until no more desirable candidates are found. Weak division of Y by P does give the largest set of cubes common to the result of dividing the numerator Y by each cube of the denominator P.

EXAMPLE 10.8:   Find the factors of Z_1 and Z_2 that minimize the total number of literals in the simultaneous solution. Z_1 = AEF + BEF + CEG and $Z_2 = \overline{A}$ EF + BEG + DEF.

SOLUTION: The candidate subexpressions (and the number of literals saved) are EF (6), and EG (2). EF is chosen first and gives Z_1 = (A + B)X + CEG and $Z_2 = (\overline{A} + D)X + BEG$, where X = EF. A second iteration using EG gives Z_1 = (A + B)X + CY and $Z_2 = (\overline{A} + D)X + BY$, where Y = EG. The factored solution is shown in Figure 10.7. The literal count has been reduced from 18 to 10.

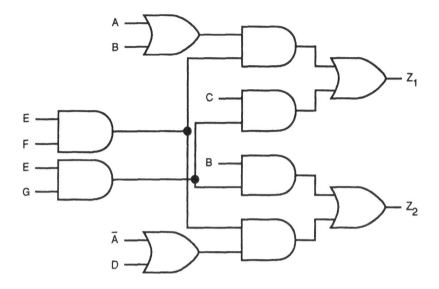

FIGURE 10.7. The functions $Z_1 = AEF + BEF + CEG$ and $Z_2 = \overline{A}\,EF + BEG + DEF$ factored as $Z_1 = (A + B)EF + CEG$ and $Z_2 = (\overline{A} + D)EF + BEG$.

Weak division incorporates a method of determining the candidate factors of an expression and systematically obtaining the largest factors of the function. Once a candidate factor P of a function Y is chosen, weak division of Y by P is accomplished by finding the set of cubes U in Y with the literals not in P deleted, and the set of cubes V in Y with the literals in P deleted. One at a time, the elements of V that multiply each variable in U are next obtained, and the intersection of all these terms yields the largest factor in the function Y.[1,2] The order in which the cubes are listed is immaterial. This can be stated symbolically as follows.

Given a function Y and a product expression P, where $P = \{p_i\}$,

$U = \{u_j | u_j$ are the cubes in Y with literals not in P deleted$\}$;

$V = \{v_j | v_j$ are the cubes in Y with literals in P deleted$\}$;

$V^i = \{v_j \in V | u_j = p_i\}$ $V^1 = \{v_j$ that are cofactors of p_1, etc.$\}$; and

$Q = \cap V_i$ and $R = Y - PQ$.

EXAMPLE 10.9: Use weak division to divide the function $Y = AC + AD + BC + BD + E$ by $P = A + B$.

SOLUTION: $U = \{A, A, B, B, 1\}$ and $V = \{C, D, C, D, E\}$. $V^A = \{C, D\}$ is the set of elements in V that are multiplied by A in the original function Y, and $V^B = \{C, D\}$ is the set of elements in V that are multiplied by B in the original function Y. Hence, $V^A \cap V^B$

$= \{C, D\}$, $Q = C + D$, $R = E$, and the function Y factors into
$(A + B)(C + D) + E$.

EXAMPLE 10.10:   Use weak division to divide the function $Y = AC + AD + AE + BC + BD + \overline{A} B$ by $P = A + B$.

SOLUTION: $U = \{A, A, A, B, B, B\}$ and $V = \{C, D, E, C, D, \overline{A}\}$. $V^A = \{C, D, E\}$ and $V^B = \{C, \underline{D}, \overline{A}\}$. Hence, $V^A \cap V^B = \{C, D\}$, $Q = \underline{C} + D$, and $R = AE + \overline{A} B$. Thus, $Y = (A + B)(C + D) + AE + \overline{A} B$.

10.5.2   Iterative Weak Division

Weak division can be used iteratively to produce better results. After weak division is applied to factor the level-0 kernel, the quotient Q is made cube free by dividing out any cube divisors of Q. If this yields a higher level kernel factor of the original expression, weak division is applied again using this kernel as a new factor. If there is an exact factorization, this method will find it.[2] The iteration successively factors higher level kernels of the function.

EXAMPLE 10.11:   Use the WEAK_DIV algorithm iteratively to factor $Y = AE + AFG + AFH + BCE + BCFG + BCFH + BDE + BDFG + BDFH$, given that $P_1 = C + D$ is a factor.

SOLUTION: WEAK_DIV (Y, P_1).

$U = \{1, 1, 1, C, C, C, D, D, D\}$
$V = \{AE, AFG, AFH, BE, BFG, BFH, BE, BFG, BFH\}$
$V^C = \{BE, BFG, BFH\}$
$V^D = \{BE, BFG, BFH\}$
$V^C \cap V^D = \{BE, BFG, BFH\}$
and
$Q_1 = BE + BFG + BFH = B(E + F(G + H))$
$R_1 = Y - P_1 Q_1 = AE + AFG + AFH$
and
$Y = (C + D)(BE + BFG + BFH) + AE + AFG + AFH$
Make Q_1 cube free by dividing out B, to obtain $E + FG + FH$, and let $P_2 = E + FG + FH$.

WEAK_DIV(Y, P_2)

$U = \{E, FG, FH, E, FG, FH, E, FG, FH\}$
$V = \{A, A, A, BC, BC, BC, BD, BD, BD\}$

$V^E = \{A, BC, BD\}$
$V^{FG} = \{A, BC, BD)$
$V^{FH} = \{A, BC, BD\}$
$V^E \cap V^{FG} \cap V^{FH} = \{A, BC, BD\}$
and
$Q_2 = A + BC + BD = A + B(C + D)$
$R_2 = Y - P_2Q_2 = 0$
and the final answer is
$Y = [A + B(C + D)][E + F(G + H)]$
The function has been factored into two level-1 kernels.

10.6   BOOLEAN DIVISION

In boolean division, expressions are treated as true logic functions, and both boolean identities and don't care terms are used to simplify the result. It is slower than algebraic division, but it can theoretically achieve the best possible results, whereas algebraic division can often give results that are not optimal but are acceptable. The trade-off is between computer run time and the quality of the result, and the challenge of multilevel logic optimization is the development of algorithms that can find "good" divisors. Frequently, a reasonably fast solution can be obtained by using algebraic division to give a "local algebraic minimum", after which boolean division is used to improve the answer. This approach can be iterated if necessary.

For any given function or functions there is a larger set of boolean divisors than algebraic divisors. Any function with at least one minterm in common with Y is a boolean divisor of Y, and the problem of choosing a "best" factor of Y is difficult.

All algebraic factors of a function are boolean factors of the function, but not all boolean factors of a function are algebraic factors of the function. Thus, algebraic division is a subset of boolean division. This is the reason algorithms using algebraic division run faster than algorithms using boolean division. Given the function $Y = AD + BCDE + F$, algebraic division yields one divisor $A + BCE$, giving the solution $Y = (A + BCE)D + F$, whereas boolean division yields the factors, $A + B$, $A + C$, and $A + E$, giving $Y = (A + B)(A + C)(A + E)D + F$. Boolean division gives more literals of this function, but it offers more possibilities of simplifying this and other functions simultaneously. A serendipitous combination of both boolean and algebraic operations can give relatively good results with little penalty in run time.[1,3,4]

Given the function $Y = \overline{A}(B + C) + \overline{B}(A + C) + \overline{C}(A + B)$ and potential divisor $Z = A + B + C$, it is seen that Z is not an algebraic divisor of Y. Boolean division gives $Y = (A + B + C)(\overline{A} + \overline{B} + \overline{C}) = Z(\overline{A} + \overline{B} + \overline{C})$, and Z is a boolean factor of Y. The original pair Y and Z had 12 literals; after boolean resubstitution, Y and Z have 6 literals, a saving of 50%.

A logic function Z is a boolean divisor of Y if $YZ \neq 0$, and Z is a boolean factor of a logic function Y if $Y\overline{Z} = 0$. In the above example, $YZ = A + B + C = Z \neq 0$, $\overline{Z} = \overline{A}\,\overline{B}\,\overline{C}$, and $Y\overline{Z} = 0$; hence, Z is a boolean factor of Y. Given $Y = AC + BC + AD + BD$ and $Z = A + B$, then $Y\overline{Z} = (A + B)(C + D)\overline{A}\,\overline{B} = 0$, but $YZ = Y$, and Z is a boolean factor of Y, whereas $Z = \overline{A}\,\overline{B}$ is not even a boolean divisor of Y since $YZ = 0$.

EXAMPLE 10.12:   Show whether or not $Z = AC$ is a boolean factor or divisor of $Y = ABC + ACD + E$.

SOLUTION: $Y\overline{Z} = (ABC + ACD + E)(\overline{A} + \overline{C}) = E(\overline{A} + \overline{C}) \neq 0$ and AC is not a boolean factor of Y. $YZ = (ABC + ACD + E)AC = ABC + ACD + ACE \neq 0$ and Z is a boolean divisor of Y. Thus $Y = AC(B + D) + E = Z(B + D)$ with a remainder E.

Boolean resubstitution uses boolean division to substitute one function into another function and reduce the number of literals in the function. Boolean resubstitution is generally better than algebraic resubstitution but, as with other boolean operations, it usually takes more machine time.[1] To decide whether it is advisable to use resubstitution, divide one function, X, by another function, Y, or its complement, $\overline{Y}$. Express the function in terms of Y or $\overline{Y}$ and see if literals are saved.

10.7   KERNELS AND CO-KERNELS OF AN EXPRESSION

The kernels of a function were defined in Section 10.2 as the cube-free primary divisors of the function. Kernels are used to find subexpressions common to two or more expressions. Conversely, a cube-free expression is a kernel. Kernel extraction is not canonical. A proper choice of kernels extracted will give a good network. In order to minimize multi-output boolean expressions once the set of kernels is obtained, one must determine the intersection of kernels of one function with the kernels of a second function.

The level of a kernel is defined recursively. A kernel that contains no kernels other than itself is said to be of level 0, and it cannot be factored. A kernel is of level n if the highest kernel contained in it is of level n–1. The function $Y = AE + AF + BCE + BCF + BDE + BDF + G$ can be factored to obtain $Y = [A + B(C + D)](E + F) + G$. $K^0 = C + D$ is a kernel of level 0, $K^1 = A + B(C + D)$ is a level-1 kernel, and Y itself is a level-2 kernel since it is cube-free. The function $Y = [A + B(C + D)](E + F)$ consists of three kernels. $C + D$ and $E + F$ cannot be factored, and are cube free; hence, they are kernels of level 0, and $A + B(C + D)$ is a kernel of level 1 since it contains a kernel of level 0. A given kernel can have more than one co-kernel.

Two functions Y and Z have a multiple-cube common divisor if, and only if, the intersection of a kernel from Y and a kernel from Z has more than one cube.[1,2] This provides a method of detecting whether two or more expressions have any common algebraic divisors other than single cubes.

Let C_i and C_j be cubes and let the cube C_{ij} denote the conjunction of the maximal set of literals present in both C_i and C_j. C_{ij} is defined as the *literal intersection* of cubes C_i and C_j. If $C_i = ABC$ and $C_j = ACD$, then $C_{ij} = AC$.

Let K_i and K_j be kernels of a function and K_{ij} denote the conjunction of the maximal set of cubes present in both K_i and K_j. K_{ij} is defined as the *cube intersection* of kernels K_i and K_j. The kernels of an algebraic expression are used to provide a means of finding subexpressions common to two or more expressions.[2] All operations used to find kernels are algebraic, and the division is weak. The cube C used to obtain the kernel Y/C is called the co-kernel of K.

EXAMPLE 10.13: Find all the kernels and co-kernels of the function Y = ADF + AEF + BDF + BEF + CDF + CEF + G.

SOLUTION: Y can be factored to obtain Y = (A + B + C)(D + E)F + G, from which the kernels and co-kernels can be obtained by inspection. They are

Kernels	Co-kernels
A + B + C	DF, EF
D + E	AF, BF, CF
(A + B + C)(D + E)	F
(A + B + C)(D + E)F + G	1

To locate common multiple-cube divisors, compute the set of kernels for each logic expression and form nontrivial (more than one term) intersections among kernels from different functions. It is not necessary to compute the set of all algebraic divisors for each expression since the set of kernels is smaller than the set of all algebraic divisors. If the intersection set of all kernels consists of single cubes or is empty, then one need only look for common divisors consisting of single cubes.[1]

EXAMPLE 10.14: Use kernel extraction to simplify the functions $Y_1 = ABCD + ABCE + ABF + ABG + H$ and $Y_2 = AJCD + AJCE + AJF + K$.

SOLUTION: $Y_1 = AB[C(D + E) + F + G] + H$, and $Y_2 = AJ[C(D + E) + F] + K$. The kernels of Y_1 are $K^0 = D + E$, $K^1 = C(D + E) + F + G$, and $K^2 = Y_1$. The kernels of Y_2 are $K^0 = D + E$, $K^1 = C(D + E) + F$, and $K^2 = Y_2$. The intersection of the two level-1 kernels is $A[C(D + E) + F] = X$. Thus, $Y_1 = BX$, and $Y_2 = JX$.

10.7.1   Kernel Determination by Tabular Means

A tabular method of computing the kernels and co-kernels of an expression Y is shown below by means of an example.[5] The function to be factored is a three-out-of-four majority function, $Y = ABC + ABD + ACD + BCD$. The procedure consists of first listing all the cubes of the function in both rows and columns, and then comparing each cube to each of the remaining cubes to find the largest common cube.

	ABC	ABD	ACD	BCD
ABC	*			
ABD	AB	*		
ACD	AC	AD	*	
BCD	BC	BD	CD	*

The table has generated smaller cubes common to two or more of the original cubes of the function. Form a new table for these cubes.

	AB	AC	AD	BC	BD	CD
AB	*					
AC	A	*				
AD	A	A	*			
BC	B	C	0	*		
BD	B	0	D	B	*	
CD	0	C	D	C	D	*

The co-kernels of Y are the set of cubes C = {AB, AC, AD, BC, BD, CD, A, B, C, D}. Next, the function Y is divided by each co-kernel and the reminders are suppressed to obtain the kernels of the function. One set of kernels is:

$$C_1 = AB, Y/AB = C + D = K_1$$
$$C_2 = AC, Y/AC = B + D = K_2$$
$$C_3 = AD, Y/AD = B + C = K_3$$
$$C_4 = BC, Y/BC = A + D = K_4$$
$$C_5 = BD, Y/BD = A + C = K_5$$
$$C_6 = CD, Y/CD = A + B = K_6$$
$$C_7 = A, Y/A = BC + BD + CD = B(C + D) + CD = K_7$$
$$C_8 = B, Y/B = AC + AD + CD = A(C + D) + CD = K_8$$
$$C_9 = C, Y/C = AB + BD + AD = D(A + B) + AB = K_9$$
$$C_{10} = D, Y/D = AB + BC + AC = C(A + B) + AB = K_{10}$$

The level-0 kernels of the function are $C + D$, $B + D$, $B + C$, $A + D$, $A + C$, and $A + B$; the level-1 kernels are $B(C + D) + CD$, $A(C + D) + CD$, $D(A + B) + AB$, and $C(A + B) + AB$. This is a highly symmetrical function and all four variables are interchangeable. Four isomorphic solutions are

$$Y_1 = A[B(C + D) + CD] + BCD \text{ and } Y_2 = B[A(C + D) + CD] + ACD$$
$$Y_3 = C[D(A + B) + AB] + ABD \text{ and } Y_4 = D[C(A + B) + AB] + ABC$$

Solution Y_1 is shown in Figure 10.8.

10.7.2 Kernel Determination by Rectangle Covering

Rectangles in a matrix provide an alternate method of interpreting the kernels and co-kernels of a boolean function, and many algorithms such as kernel intersections, cube extraction, and factoring can be expressed as rectangle covering problems. One such approach is the cube-literal matrix.[2]

The *cube-literal matrix* is formed by assigning one row for each cube in the SOP expression of the function and one column for each unique literal present in the function. The elements of the matrix are 1 and 0. For each cube of the function, a 1 is placed in the column of each literal which appears in that cube, and a 0 is placed in the column of each literal which does not appear in the cube.

Any rectangular array of elements in a matrix which are all 1s identifies a cube of the function represented by the matrix. The rectangle (R,C) of a matrix is a subset of rows R and a subset of columns C of the matrix. A *prime rectangle* (R, C) is a rectangle which is not completely contained in any other rectangle of the matrix. Rows and columns of a rectangle do not have to be adjacent, since a rearrangement of the rows or columns of the matrix can make them adjacent.

The *co-rectangle* of rectangle (R, C) is the pair (R, C′), where C′ is the set of columns that are in R but not in C and that have a 1 in them. That is, the remaining columns consisting of the same rows as in the cube form the kernel or co-rectangle associated with the cube. If one chooses the largest possible rectangle, it corresponds to a prime cube. The co-rectangles will all be level-0 kernels.

The rules for rectangle covering are equivalent to the covering rules for Karnaugh maps. Each 1 in matrix M must be covered by at least one rectangle from the cover, don't cares are optional, and a covering need not be disjoint since a 1 in M may be covered by more than one rectangle.

The function $Y = A\overline{B} + AC + \overline{B}DE + \overline{B}D\overline{F}$ has four cubes and six unique literals in it. It can be represented by a matrix of four rows and six columns, viz

	A	$\overline{B}$	C	D	E	$\overline{F}$
$A\overline{B}$	1	1	0	0	0	0
AC	1	0	1	0	0	0
$\overline{B}DE$	0	1	0	1	1	0
$\overline{B}D\overline{F}$	0	1	0	1	0	1

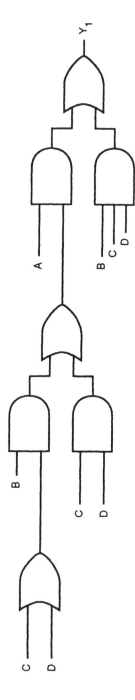

FIGURE 10.8. The 3-out-of-4 majority function factored as $Y^1 = A[B(C + D) + CD] + BCD$.

The matrix is redrawn below, with asterisks replacing the elements which are of no concern. This is not necessary, but it helps focus attention on the rectangles and corectangles of interest.

	A	$\overline{B}$	C	D	E	$\overline{F}$
$A\overline{B}$	1	1	0	*	*	*
AC	1	0	1	*	*	*
$\overline{B}DE$	*	1	*	1	1	0
$\overline{B}D\overline{F}$	*	1	*	1	0	1

Variable A forms a 2×1 rectangle and is a prime cube with corectangle or kernel $\overline{B}$ + C. Variables $\overline{B}$ and D form a 2×2 rectangle which is also a prime cube, with corectangle or kernel E + $\overline{F}$. The function has two level-0 kernels and can be factored as $Y = A(\overline{B} + C) + \overline{B}D(E + \overline{F})$. This can be stated formally by labeling the rows from top to bottom and labeling the columns from left to right, identifying the rectangles and corectangles as follows:

$$\{R_1, C_1\} = \{(1, 2), (1)\} \text{ and } \{R_1, C'_1\} = \{(1, 2), (2, 3)\}$$
$$\{R_2, C_2\} = \{(3, 4), (2, 4)\} \text{ and } \{R_2, C'_2\} = \{(3, 4), (5, 6)\}$$

EXAMPLE 10.15: Obtain the cube-literal matrix for the function Z and find its kernels. $Z = \overline{A}BC + \overline{A}BD + EFG\overline{H} + EFGJ$.

SOLUTION: The function consists of four cubes and nine unique literals and has the following 4×9 cube-literal matrix:

	$\overline{A}$	B	C	D	E	F	G	$\overline{H}$	J
$\overline{A}BC$	1	1	1	0	0	0	0	0	0
$\overline{A}BD$	1	1	0	1	0	0	0	0	0
$EFG\overline{H}$	0	0	0	0	1	1	1	1	0
EFGJ	0	0	0	0	1	1	1	0	1

Z has two prime cubes $\overline{A}B$ and EFG corresponding to rectangles $\{R_1, C_1\} = \{(1, 2), (1, 2)\}$ and $\{R_2, C_2\} = \{(3, 4), (5, 6, 7)\}$, respectively, and two kernels C + D and $\overline{H}$ + J, corresponding to corectangles $\{R_1, C'_1\} = \{(1, 2), (3, 4)\}$ and $\{R_2, C'_2\} = \{(3, 4), (8, 9)\}$, respectively. The function factors as $Z = \overline{A}B(C + D) + EFG(\overline{H} + J)$.

EXAMPLE 10.16: Use rectangle theory to find the cover of the function represented by the cube-literal matrix shown.

	$\overline{A}$	B	C	D	E	F	G	$\overline{H}$	J
$\overline{A}BC$	1	1	1	0	0	0	0	0	0
$\overline{A}BDEF$	1	1	0	1	1	1	0	0	0
$\overline{A}BDEG\overline{H}$	1	1	0	1	1	0	1	1	0
$\overline{A}BDEGJ$	1	1	0	1	1	0	1	0	1

SOLUTION: There are three cubes, $\overline{A}$ B, DE, and G corresponding to rectangles $\{R_1, C_1\} = \{(1, 2, 3, 4), (1, 2)\}$, $\{R_2, C_2\} = \{(2, 3, 4), (4, 5)\}$, and $\{R_3, C_3\} = \{(3, 4), (7)\}$. The cubes overlap, which indicates kernels higher than level-0. Remove cube G and level-0 kernel $K^0 = \overline{H} + J$ and replace them by literal X. Then $X = G(\overline{H} + J)$, $Z = \overline{A}BC + \overline{A}BDEF + \overline{A}BDEG(\overline{H} + J)$ $= \overline{A}BC + \overline{A}BDEF + \overline{A}BDEX$, and the matrix reduces to

	$\overline{A}$	B	C	D	E	F	X
$\overline{A}BC$	1	1	1	0	0	0	0
$\overline{A}BDEF$	1	1	0	1	1	1	0
$\overline{A}BDEX$	1	1	0	1	1	0	1

There are still two overlapping rectangles. Let $Y = DE(F + X)$, where $K^1 = F + X = F + G(\overline{H} + J)$ is a level-1 kernel, and the matrix further reduces to

	$\overline{A}$	B	C	Y
$\overline{A}BC$	1	1	1	0
$\overline{A}BY$	1	1	0	1

The function realized by the matrix is $Z = \overline{A}B(C + Y) =$ $\overline{A}B[C + DE(F + X)] = \overline{A}B[C + DE\{F + G(\overline{H} + J)\}]$, and Z has a level-2 kernel $C + Y = C + DE\{F + G(\overline{H} + J)\}$.

Rectangle theory can be used to determine the kernel intersections of two or more functions also. The best kernel intersection will result in the least number of literals in the network. The cokernel-cube matrix is defined as the matrix formed by using the cokernel cubes as rows and the cubes of the kernels as columns.[2] Element $B_{ij} = 1$ if the kernel associated with row i contains the cube associated with column j, and 0 otherwise.

EXAMPLE 10.17: Find the best kernel intersection of functions X, Y, and Z. $X = ABGH + CDGH + EFGH + K$, $Y = ABLM + CDLM + AN + BN$, and $Z = ABPQ + CDPQ + AR + BR$.

SOLUTION: The kernel of X is AB + CD + EF, the kernels of Y are AB + CD and A + B, and the kernels of Z are AB + CD and A + B. The co-kernel of X is GH, the co-kernels of Y are LM and N, and the co-kernels of Z are PQ and R. The nonredundant cubes of the three kernels are AB, CD, EF, A, and B, and the matrix is

Function	Co-kernel	AB	CD	EF	A	B
X	GH	1	1	1	0	0
Y	LM	1	1	0	0	0
	N	0	0	0	1	1
Z	PQ	1	1	0	0	0
	R	0	0	0	1	1

Rectangle $\{(1,2,4),(1,2)\}$ = AB + CD is common to all three functions, and rectangle $\{(3,5), (4,5)\}$ = A + B is common to Y and Z. The best solution to all three functions simultaneously is X = (AB + CD)GH + EFGH + K, Y = (AB + CD)LM + (A + B)N, and Z = (AB + CD)PQ + (A + B)R.

10.8 BOOLEAN TREES

Multilevel logic can be synthesized by mapping the function or functions into one or more trees. This can be done by extracting the kernels of the functions or by the use of Shannon's expansion theorem. Both methods will be considered.

Tree mapping techniques transform an optimized boolean network into a forest of trees with each tree rooted either in a primary output or in a multi fan-out gate. A multilevel function can be represented by a tree as follows. Each node of the tree is either an OR-gate, represented by an OR plus sign at the node, or an AND-gate, represented by an AND dot at the node. Tree mappings of the functions Y_1 = AB + CD and $Y_2 = (A + B)(C + D)$ are given in Figure 10.9. In multilevel logic, each common kernel selected becomes the root of a new tree.

In the tree graph, a branch connects node i to node j if the output of node i is part of the support of node j (if Y_i sup(Y_j). Primary inputs to a tree are the leaves, and primary outputs are the roots. Each node in a boolean network is a completely specified boolean function represented by an SOP form and a factored form. Multilevel synthesis of boolean functions gives a covering of the function tree, subdividing it into two-level gates.

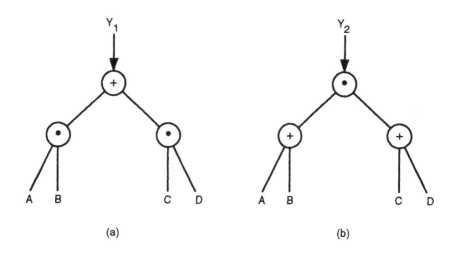

FIGURE 10.9. Tree mappings of the functions (a) $Y_1 = AB + CD$ and (b) $Y_2 = (A + B)(C + D)$.

10.8.1   Tree Mapping by Kernel Extraction

Once the kernels of an expression are obtained by whatever means, the function can be factored to yield a tree structure. The process is demonstrated in Examples 10.18 and 10.19.

EXAMPLE 10.18: The SOP representation of a function, Y_1, is $Y_1 = AB + A\overline{C} + AD + \overline{E}FG + \overline{E}FH + \overline{E}FJ$. Factor Y_1 and give a tree mapping of the function.

SOLUTION: The cokernels of Y_1 are found to be $C_1 = A$ and $C_2 = \overline{E}F$. The kernel of A is $K_1 = Y_1/A = B + \overline{C} + D$, and the kernel of $\overline{E}F$ is $K_2 = Y_1/\overline{E}F = G + H + J$. The factored form of Y_1 is thus $Y_1 = A(B + \overline{C} + D) + \overline{E}F(G + H + J)$. This maps into the tree shown in Figure 10.10.

EXAMPLE 10.19: Give a tree mapping of $\overline{Y}_1$, the complement of the function Y_1 in Example 10.18.

SOLUTION: $\overline{Y}_1 = (\overline{A} + \overline{B}C\overline{D})(E + \overline{F} + \overline{G}\overline{H}\overline{J})$. This maps into the tree shown in Figure 10.11. The kernels of $\overline{Y}_1$ are $K_3 = \overline{A} + \overline{B}C\overline{D}$ and $K_4 = E + \overline{F} + \overline{G}\overline{H}\overline{J}$, and algebraic resubstitution gives $\overline{Y}_1 = K_3K_4$. Notice that this is the same structure as the tree for Y_1 in Figure 10.10. The tree mapping maintains the symmetry of DeMorgan's law which is not obvious in the SOP and POS realizations.

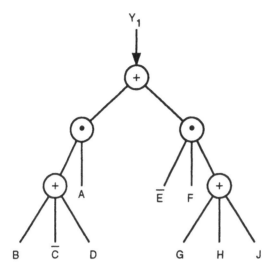

FIGURE 10.10. The tree mapping of the function $Y_1 = A(B + \overline{C} + D) + \overline{E}F(G + H + J)$.

Note in Figures 10.9, 10.10, and 10.11 that the kernels are visible by inspection of the tree. The function represented by each node with an OR sign is a kernel of the root function represented by the tree.

10.8.2   Tree Mapping by Shannon's Theorem

A function is said to be *positive unate (negative unate)* in X_i if $Y(X_1, X_2, ..., X_n)$ is a logic expression with X_i only occurring in the uncomplemented (complemented) form. The function, $Y = AB + B\overline{C}\overline{D} + \overline{A}BC + AC\overline{D}$ is positive unate in B, negative unate in D, and binate (nonunate) in A and C. A positive (negative) unate function is a function which is positive (negative) unate in all its variables. A *unate function* is one which is either positive unate or negative unate in each of its variables. $Y = ABC + AB\overline{D} + AC\overline{D} + E\overline{F} + \overline{F}G + \overline{F}H$ is positive unate in variables A, B, C, E, G, and H, and negative unate in variables D and F. The function Y is a unate function and has a unique minimum SOP and minimum POS realization.

A function Y that is binate in a variable X_i of its cover can be expanded about that variable by Shannon's theorem to give $Y = X_iY(X_i) + \overline{X}_iY(\overline{X}_i)$. Element X_i is referred to as the splitting variable, $Y(X_i)$ is the positive-phase cofactor of Y with respect to X_i, and $Y(\overline{X}_i)$ is the negative-phase cofactor of Y with respect to X_i. The function can be implemented in NAND-NAND logic as shown in Figure 10.12(a).

The complement of the function can be written in the form $\overline{Y} = [\overline{X}_i +$

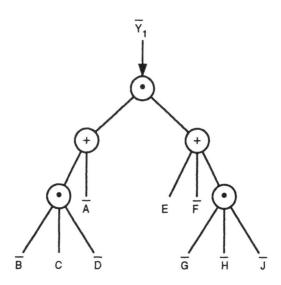

FIGURE 10.11. Tree mapping of the function $\overline{Y}_1 = (\overline{A} + \overline{B}\,C\overline{D})(E + \overline{F} + \overline{G}\,\overline{H}\,\overline{J})$.

$\overline{Y}(X_i)][X_i + \overline{Y}(\overline{X}_i)]$, and it can be implemented in NOR-NOR logic as shown in Figure 10.12(b).

The tree mapping of a function is obtained by recursively expanding the function by Shannon's theorem until a full binary tree is generated. This amounts to decomposing the function into unate cofactors. The cofactor of Y with respect to X is $Y_x = Y(X=1)$, obtained by replacing X by 1 in the function Y. The cofactor of Y with respect to $\overline{X}$ is $Y_{\overline{x}} = Y(\overline{X}=1) = Y(X = 0)$, obtained by replacing X by 0 in the function Y. The cofactors of a function are independent of the splitting variable.

EXAMPLE 10.20: Obtain the binary tree representation of the function $Y = ABD + A\overline{B}\,\overline{C} + \overline{A}\,B\overline{C} + \overline{A}\,\overline{B}CD$.

SOLUTION: Expanding in alphabetical order, one obtains $Y_A = BD + \overline{B}\,\overline{C}$ and $Y_{\overline{A}} = B\overline{C} + \overline{B}CD$. Expanding Y_A, $Y_{AB} = D$ and $Y_{A\overline{B}} = \overline{C}$. Expanding $Y_{\overline{A}}$, $Y_{\overline{A}B} = \overline{C}$ and $Y_{\overline{A}\,\overline{B}} = CD$. The expanded function can be written in factored form as $Y = A(BD + \overline{B}\,\overline{C}) + \overline{A}(B\overline{C} + \overline{B}CD)$. The tree is shown in Figure 10.13.

10.9 THE BERKELEY ESPRESSO PROGRAM

ESPRESSO is a program originally developed for two-level PLA minimization by the University of California at Berkeley.[6,7] It can be applied as part of a routine for simplifying multilevel logic. The basic idea is that minimizing the total

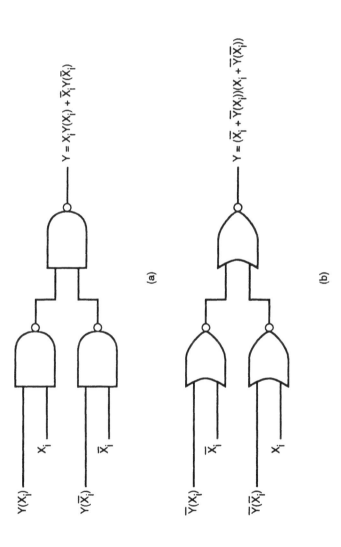

(a)

(b)

FIGURE 10.12   Shannon's theorem represented as (a) a NAND tree and (b) a NOR tree.

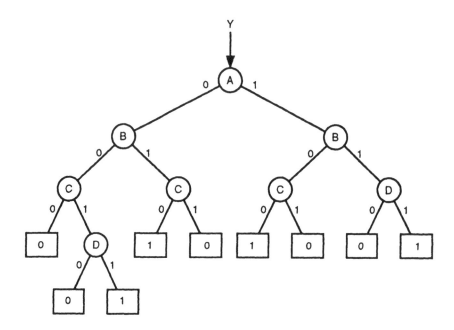

FIGURE 10.13.   The boolean tree realization of $Y = A(BD + \overline{B}\,\overline{C}) + \overline{A}(B\overline{C} + \overline{B}CD)$.

number of literals in an SOP form is a first approximation to minimizing the number of literals in the factored form. ESPRESSO proceeds in three steps to simplify an SOP expression. The steps are labeled *Reduce, Expand,* and *Irredundant Cover*. The algorithm can be stated as follows (DC is the don't care set):

```
ESPRESSO (Y, DC){
    For (F∈ Y) {
        do {
                G ← REDUCE (F)
                G ← EXPAND (G)
                G ← IRR_COV (G)
                if (NUM_LITERALS(G) < NUM_LITERALS(F) continue
                F ← G
        }
        return Y
    }
```

The input to Reduce is an irredundant SOP expression. Reduce then selects a set of candidate reduction variables and an ordering for processing the product terms. It then takes the product terms one at a time and adds as many literals as possible to each product term. By shrinking the prime implicants to their smallest size that still covers the logic function, Reduce tries to find a better cover. The order in which the variables are processed is important.

EXAMPLE 10.21: Reduce $Y = AB + AC$.

SOLUTION: Y is first converted to $Y = (AB\overline{C} + ABC) + (A\overline{B}C + ABC)$. If AB is processed first, ABC is deleted from AB, since $ABC \subset AC$, giving $Y = AB\overline{C} + AC$. If AC is processed first, ABC is deleted from AC since $ABC \subset AB$, giving $Y = AB + A\overline{B}C$.

Reduce gives a cover that is generally not prime. Expand is next used to expand implicants to their maximum size and remove implicants which are not prime. Expand selects a variable processing order and tries to delete that variable from the product term. The result is an SOP expression from which no literals can be deleted and still maintain equivalence with the original expression.

EXAMPLE 10.22: Expand $Y = AB\overline{C}D$ with don't care terms $AB\overline{D}$, $\overline{B}\,\overline{C}D$ and ABC, by processing literals first in the order B, C, D, A, then in the order C, D, A, B.

SOLUTION: Taking the order B, C, D, and A, delete B by adding $A\overline{B}\,\overline{C}D$ to Y, since $A\overline{B}\,\overline{C}D \subset \overline{B}\,\overline{C}D$. Then $Y = AB\overline{C}D + A\overline{B}\,\overline{C}D = A\overline{C}D$. No more literals can be deleted. Delete C by adding $ABCD$ to Y, since $ABCD \subset ABC$. Then $Y = AB\overline{C}D + ABCD = ABD$. Next, add $AB\overline{D}$ to Y, since $AB\overline{D}$ is a don't care. Then $Y = ABD + AB\overline{D} = AB$. No more literals can be deleted. One now has an SOP expression consisting of one prime cube. Taking the order C, D, A, and B, delete C by adding $ABCD \subset ABC$, and $Y = ABD$. Next, delete D by adding $A\overline{B}D$, and $Y = AB$. No more literals can be deleted.

Irredundant Cover is next used to reduce the expression by deleting as many cubes as possible, until an irredundant SOP solution is obtained. The order in which the cubes are taken determines the output. Irredundant Cover does not reverse the work of Expand; it always finds a new reduction to compare to the previous reduced expression.

EXAMPLE 10.23: Reduce $Y = ABD + A\overline{C}D + BCD + A\overline{B}\,\overline{C} + \overline{A}BC$ by processing the cubes in the order given, then in the order $A\overline{C}D$, BCD, ABD, $A\overline{B}\,\overline{C}$, $\overline{A}BC$.

SOLUTION: Delete ABD since $ABD \subset (A\overline{C}D + BCD)$, and $Y = A\overline{C}D + BCD + A\overline{B}\,\overline{C} + \overline{A}BC$. No more cubes can be deleted. Taking the cubes in the second order, delete $A\overline{C}D$ since $A\overline{C}D \subset (ABD + A\overline{B}\,\overline{C})$, and $Y = ABD + BCD + A\overline{B}\,\overline{C} + \overline{A}BC$. Next delete BCD since $BCD \subset (ABD + \overline{A}BC)$, and $Y = ABD + A\overline{B}\,\overline{C} + \overline{A}BC$. No more cubes can be deleted.

10.10 BINARY DECISION DIAGRAMS AND DIRECTED ACYCLIC GRAPHS

The second major approach to multilevel logic is the graphic approach as exemplified by *binary decision diagrams* (*BDDs*) and *if-then-else* (*ITE*) diagrams. A *boolean network* can be treated as a directed acyclic graph (DAG), each node of which is associated with a variable X_i and a logic function Y_i. A directed arc from node i to node j is in the graph if the variable X_i is an element in the support of function Y_j, node j being closer to the root than node i.

Binary decision diagrams were proposed by Akers[8] in 1978, and directed acyclic graphs were proposed by Bryant[9] in 1986, both as alternative approaches to truth table solutions. A BDD is a DAG which starts at a *root node* that represents the function and terminates on one of two nodes called leaves and labeled 0 and 1. The leaves of the graph are labeled 0 if the function is not asserted, or 1 if the function is asserted, and they are represented by a square containing the appropriate value. Each non-leaf node has a variable associated with it represented by a circle containing the label of the variable at that node, called its index. There are two paths leading from it to its two successors, one for the node variable asserted true and pointing to a successor node representing the function cofactored with respect to the positive phase of the node variable, and one for the node variable asserted false and pointing to a successor node representing the function cofactored with respect to the negative phase of the node variable.

Figure 10.14 shows the BDD for the trivial functions $Y = A$ and $Y = \overline{A}$. The only difference between A and its complement is in the ordering of the leaves. This is true in general, and complementation can be accomplished at any node by simply interchanging the leaves. Figure 10.15 shows the BDDs for the elementary functions $Y = AB$, $Y = A + B$, and $Y = A \oplus B$.

10.10.1 The Ordered Binary Decision Diagram

An *ordered binary decision diagram (OBDD)* is a BDD with the constraint that the input variables are ordered, and every path from the root to a leaf in the OBDD visits the input variables in the same order. Unless otherwise specified, the order will be alphabetical starting from the root. A *reduced ordered binary decision diagram (ROBDD)* is an OBDD for which each node represents a distinct logic function. It is a canonical representation of a completely specified logic function and is suitable as a replacement for a truth table representation of the function. Changing the ordering of the node variables produces a new canonical function, but there is only one ROBDD for a given variable ordering. It is easy to identify reducable nodes in OBDDs, and the reduced OBDD can always be made to terminate in one of two leaves, 0 or 1.

Any path from the root to either leaf of an ROBDD traverses the nodes whose variables are in the proper order, although a variable may be skipped if it is missing. In the ROBDD for $Y = AB + C$, shown in Figure 10.16, the two paths

Y = A

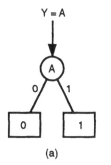

(a)

Y = $\overline{A}$

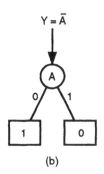

(b)

FIGURE 10.14. The BDDs for functions (a) Y = A and (b) Y = $\overline{A}$.

from the root Y to the leaf 0 are A true-B false-C false, and A false-C false. If A is false, variable B is a don't care and it is missing in the ROBDD path, but the variables are still traversed in alphabetical order.

The BDD is reduced by observing that if any two internal nodes represent the same function, they can be merged into one node. Finding an optimum ordering of the node variables, though, is a major problem in multilevel synthesis, and it will be addressed in Section 10.10.6.

Since an ordered BDD is a canonical representation of a function, one application of BDDs is in verifying whether or not two multilevel networks are equivalent. Each output of a network is converted to an ROBDD over the input variables, and the output functions are equivalent if, and only if, the two OBDDs are *isomorphic*. Checking isomorphism of ROBDDs with a computer is very fast.

To transform any circuit to its equivalent ROBDD, start by creating the BDD of the root, then create BDDs for each node of the circuit working down to the leaves. Merge and reduce each node to obtain the ROBDD.

Consider the function Y = $\overline{A}\overline{B}C + \overline{D} + E$ and take the variables in alphabetical order. If A and B are false and C is true, Y = 1; hence, there must be a path from A false to B false to C true to leaf 1. If this cube is false, control transfers to variable D. If D is false, Y = 1. This path will be from either A true, B true, or C false to D false to leaf 1. If D is true, control passes to variable E, and the function is true if E is false, and false if E is true. To summarize, Y = 1 if ABC = 001 or AD = 10, ABD = 010, ABCD = 0000, or ADE = 110, ABDE = 0110, ABCDE = 00010. The circuit is shown in Figure 10.17(a), and the ROBDD is shown in Figure 10.17(b).

10.10.2 Inversion and Reduced Binary Decision Diagrams

Inversion can be introduced into the diagram by using an invert "bubble" at either a fan-in to a node or at one or both fan-outs from a node.[8] The addition of this *negate pointer* to a BDD allows the combining of a function and its comple-

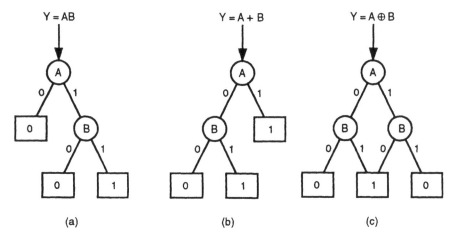

FIGURE 10.15.   The BDDs for the functions (a) Y = AB, (b) Y = A + B, and (c) Y = A ⊕ B.

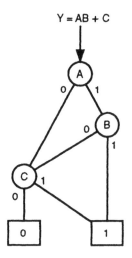

FIGURE 10.16.   The BDD for the function Y = AB + C.

ment into the same DAG. If one node has a successor Y, and another node has a successor $\overline{Y}$, then only one node is necessary if one of the predecessors has a negate pointer associated with it.

The invert bubble tells the user to complement the binary value that is ultimately obtained. This is done by simply interchanging the 1 and 0 path labels, as shown in Figure 10.14. Invert bubbles cancel in pairs; an even number of invert bubbles along a path will not cause an inversion of the root function, and an odd number of invert bubbles along a path will invert the root function. The use of inverters further reduces the number of nodes required in a BDD, as will be seen in the next section.

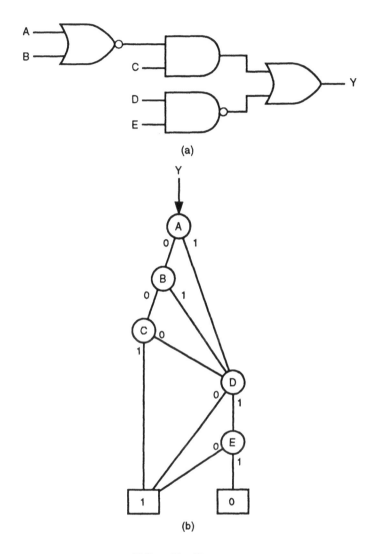

(a)

(b)

FIGURE 10.17. The function $Y = \overline{A}\,\overline{B}C + \overline{D} + \overline{E}$ (a) realized in factored form and (b) the BDD for Y.

10.10.3 MUX Realizations of Binary Decision Diagrams

Given the diagram for a boolean function, one must construct a circuit that behaves as indicated in the diagram. Since each node starting from the root of the tree has two possible paths leaving it, the node can be represented by a 2-to-1 multiplexer with the node variable as the MUX select variable. Any node C can be represented by a function of the form $Y = A\overline{C} + BC$, which can be realized with a 2-to-1 line MUX having C as the control or select variable, A and B as the two inputs, and Y as the output. Thus, there is a one-to-one correspondence between any BDD and a MUX tree. In the MUX implementation, however, the signals flow from the leaves of the tree to the root. A MUX with inverted and noninverted outputs can realize a network having inverters in it.

Since BDDs and ROBDDs are representations of boolean functions which can be implemented directly in MUXs they can be used to realize MUX-based architectures. This makes BDDs very convenient to the implementation of MUX-based FPGAs.

Since the BDD can be implemented with a multiplexer, any valid MUX operation is a valid BDD operation. Note that these are not the same rules as for gates, unless AND and OR operations are interchanged. In particular, any rules for inverting bubbles that apply to multiplexers also apply to BDDs. Bubbles can be applied at the root of a node (output of a MUX) or at either or both of the successors of the node (MUX inputs). An inverter at the input to one node is the same as an inverter at the output of that fan-in node, a bubble at a node input and a bubble at a fan-in node gives double complementation and can be cancelled, etc.

An output inverter indicates complementation of the subfunction realized at that particular node. The use of multiplexers with output inverters can reduce the size of a BDD by up to half, and execute a NOT operation without generating another graph. A MUX input inverter performs the logical complementation of an input variable, which is equivalent to the operation of swapping a 0-edge with a 1-edge at the next lower node of the BDD. The use of input inverters can also reduce the size of a BDD by up to half. Examples of some of the rules for BDDs with inverters are shown in Figures 10.18 and 10.19.

EXAMPLE 10.24: Draw a merged BDD for the functions $Y_1 = A + B$, and $Y_2 = \overline{A}\,\overline{B}$. Use inverters to simplify the result. Show a single 2-to-1 MUX implementation.

SOLUTION: The function Y_2 is the complement of Y_1. Figure 10.18(a), (b), (c), (d) and (e) show the BDD for Y_1, the reduced BDD for Y_1 obtained by including an inverter, the merged realization of both functions, and the MUX realization of Y_1 and of both functions, respectively.

EXAMPLE 10.25: Find a reduced BDD for $Y_1 = A \oplus B$, and $Y_2 = \overline{A \oplus B}$. Use inverters as needed. Give the MUX implementation.

SOLUTION: The function Y_2 is again the complement of Y_1. The BDD, the merged realization of both Y_1 and Y_2, the MUX implementation of Y_1, and of both functions are shown in Figure 10.19.

The reduced BDD for the EOR function in Figure 10.19(b) is read as follows. There are four paths, viz

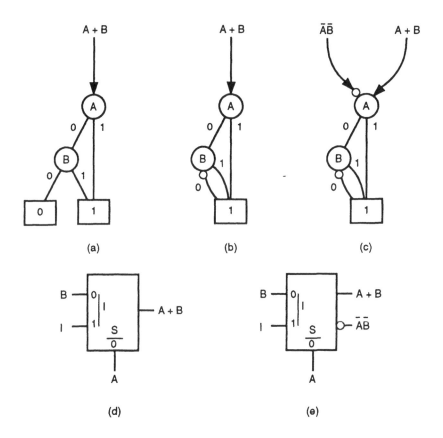

FIGURE 10.18. (a) The BDD for $Y_1 = A + B$, (b) the ROBDD for Y_1, (c) the merged realization of Y_1 and $\overline{Y}_1$, and the MUX implementations of (d) Y_1, and (e) Y_1 and $\overline{Y}_1$.

Path 1: $A \rightarrow B \rightarrow \overline{1} = 0$, or $AB = 11$, $Y_1 = 0$

Path 2: $A \rightarrow \overline{B} \rightarrow 1$, or $AB = 10$, $Y_1 = 1$

Path 3: $\overline{A} \rightarrow B \rightarrow 1$, or $AB = 01$, $Y_1 = 1$

Path 4: $\overline{A} \rightarrow \overline{B} \rightarrow \overline{1} = 0$, or $AB = 00$, $Y_1 = 0$

The correct answer can be obtained directly from the BDD by counting the number of inversions in any path. Paths 1 and 4 have one inversion, and the answer is the complement of the 1 leaf, while Paths 2 and 3 have 2 and 0 inversions respectively, and the answer is uncomplemented, or 1.

The proof of equivalence consists of showing that both output functions are the same for the DAG or for the MUX which implements the function. In Figures 10.20 and 10.21, the MUX implementations of the given pair of graphs is shown. Several more equivalence examples are left as an exercise.

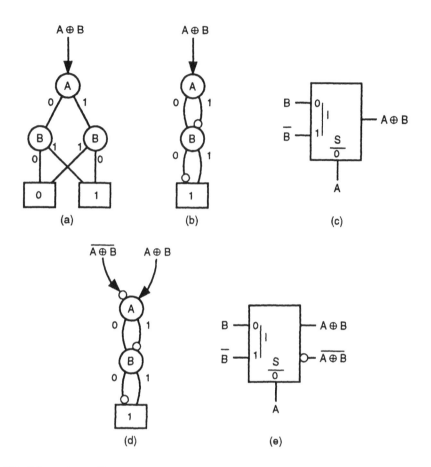

FIGURE 10.19.   (a) The BDD for $Y_1 = A \oplus B$, (b) the ROBDD for Y_1, (c) the MUX implementation of Y_1, (d) the merged BDD for Y_1 and $\overline{Y}_1$, and (e) the MUX for Y_1 and $\overline{Y}_1$.

EXAMPLE 10.26: Show that Figures 10.20(a) and (b) are equivalent by showing that their two MUX implementations have the same outputs.

SOLUTION: For Figures 10.20(a) and 10.20(c), $Y = \overline{A} B + AC$, while for Figures 10.20(b) and (d), $Y = \overline{\overline{A} \overline{B}} + AC = (A + B)(\overline{A} + C) = AC + \overline{A} B + BC = AC + \overline{A} B$, since BC is the consensus of the other two terms and is redundant.

EXAMPLE 10.27: Show that Figures 10.21a and b are equivalent by showing that their two MUX implementations have the same outputs.

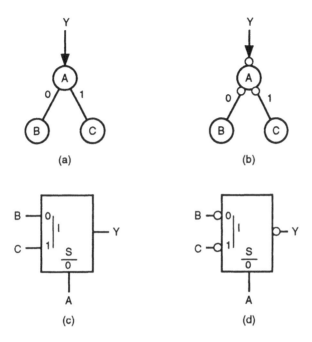

FIGURE 10.20. (a) and (b) Two BDD representations of $Y = \overline{A}\,B + AC$ and (c) and (d) the equivalent MUX realizations of Y.

SOLUTION: For Figure 10.21 a and c, $Y = \overline{A}\,\overline{B} + A\overline{C}$, while for b and d,
$$Y = \overline{\overline{A}\,B + AC} = (A + \overline{B})(\overline{A} + \overline{C}) = A\overline{C} + \overline{A}\,\overline{B} + \overline{B}\,\overline{C} =$$
$$A\overline{C} + \overline{A}\,\overline{B}, \text{ since } \overline{B}\,\overline{C} \text{ is the consensus of the other two terms.}$$

In all the diagrams discussed until now, it has been assumed that the node variables were primary input variables of the function to be realized. This can be generalized to allow the node variable to represent a subfunction of the primary or root function, which has its own BDD. When a node containing a subfunction is encountered, one must go to the BDD for that subfunction, determine its value, and then take the appropriate path on the original BDD. The primary advantage of using subfunctions as node variables is the ease with which diagrams can then be connected when merging two or more BDDs.

EXAMPLE 10.28: Construct the BDD for $Y = A + B + C$, using subfunction $Z = A + B$. Find the value of Y when $A = B = 0$ and $C = 1$. Show a MUX realization.

SOLUTION: The BDDs for $Y = Z + C$ and $Z = A + B$ are shown in Figure 10.22. In reading the diagram for Y, one first comes to node Z. The value of Z can be obtained from its BDD. If either A or B are asserted, then Z and Y are asserted. If $Z = 0$, then $Y = C$.

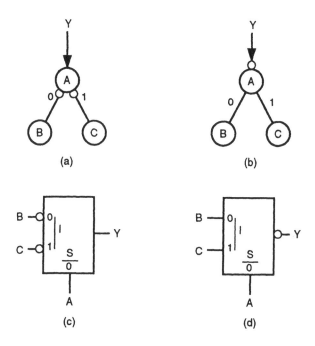

FIGURE 10.21.   (a) and (b) The BDDs for $Y = \overline{A}\,\overline{B} + A\overline{C}$ and $Y = \overline{\overline{AB} + AC}$, and (c) and (d) the MUX realizations of each form of Y.

10.10.4 Temporal Binary Decision Diagrams

Latches and flip-flops can be modeled as BDDs also. Representations of sequential positive-edge-triggered delay, toggle, and J-K flip-flops are shown in Figure 10.23. These BDDs are read in the same manner as the combinational network diagrams already discussed. The characteristic equation of each flip-flop is given as the function to be realized. In these diagrams, Q^+ is the state of the flip-flop after the next clock pulse or positive-edge transition P, D is the input data to the delay flip-flop, T is the input to the toggle flip-flop, and J and K are inputs to the J-K flip-flop. The same BDDs represent equivalent latches if P represents a positive clock level, and if P is replaced by $\overline{P}$, the same BDDs represent either negative-edge-triggered flip-flops or negative-level-triggered latches.

10.10.5 Binary Decision Diagrams From Binary Decision Trees

Any switching function can be realized in the form of a binary decision tree by applying Shannon's expansion theorem to the binate variables of the function. A binary decision tree can also be constructed from a truth table description of the function or equivalently from a minterm expansion, as shown in Figure 10.24. In this approach the minterms are the leaves of the tree. BDDs share many properties

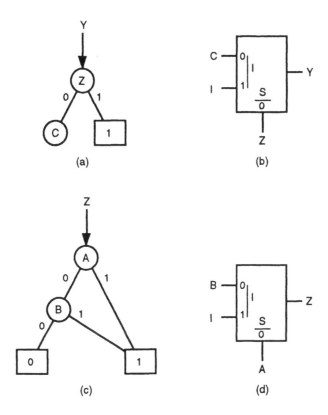

FIGURE 10.22. The BDD and MUX for (a) and (b) $Y = Z + C$, and (c) and (d) $Z = A + B$.

with binary decision trees, but they differ from the binary decision tree in that a node of a BDD can have more than one branch directed into it. This occurs when the tree is reduced to a BDD.[8]

Consider the EOR of three inputs, A, B, and C, which consists of minterms 1, 2, 4, and 7. The binary tree representation of the function is shown in Figure 10.25(a), and the nodes are labeled 1 through 7. Nodes 5 and 6 of the tree have the same output values and can be combined into one node, labeled 5,6 in Figure 10.25(b). Nodes 4 and 7 also have the same output values for the function and can be combined into one node, labeled 4,7 in Figure 10.25(b). The tree has been pruned to give a ROBDD. Each node has two output branches as in the tree, but nodes 4,7 and 5,6 have two input branches, as do the leaves 0 and 1.

From every node of a BDD there is only one active path from the root function to an output leaf corresponding to any given input vector. For the EOR function $Y = A \oplus B \oplus C$ shown in Figure 10.25(b), and an input vector ABC = 111, the path from Y to output 1 is shown in dark lines in Figure 10.26(a).

There are a total of four input vectors that result in $Y = 0$, and four that result in $Y = 1$. The number of paths from the function to an output can be obtained by assigning a number to each output branch which is equal to the sum of the

$$Q^+ = PD + \bar{P}Q$$

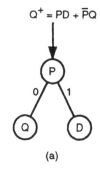

(a)

$$Q^+ = P(T\bar{Q} + \bar{T}Q) + \bar{P}Q$$

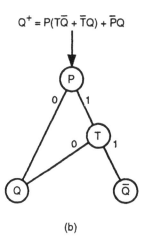

(b)

$$Q^+ = P(Q\bar{K} + \bar{Q}J) + \bar{P}Q$$

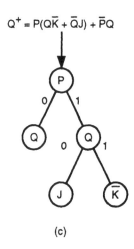

(c)

FIGURE 10.23. The BDDs for (a) delay, (b) toggle, and (c) J-K flip-flops.

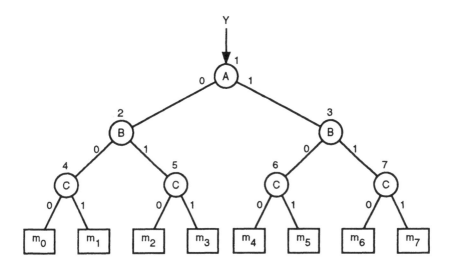

FIGURE 10.24. The binary tree diagram of a three-variable function has eight minterms for leaves.

numbers on all the input branches.[8] For the EOR function of Figure 10.25(b), there is only one input to node A; hence, the outputs of node A are assigned the number 1. Since both nodes B have one input their output lines are labeled 1 also. There are two inputs of weight 1 entering nodes C, however, and each node-C output is assigned the number 2. This is shown in Figure 10.26(b). Both the true and the false output leaves have two inputs of weight 2 each, and there are four paths from Y to either a true or a false output. The true output paths yield the SOP solution consisting of minterms 1, 2, 4, and 7.

$$Y = \overline{A}\,\overline{B}C + \overline{A}B\overline{C} + A\overline{B}\,\overline{C} + ABC = \Sigma m(1,2,4,7)$$

and the four paths to a false output give the POS solution, consisting of maxterms 0, 3, 5, and 6.

$$Y = (A + B + C)(A + \overline{B} + \overline{C})(\overline{A} + B + \overline{C})(\overline{A} + \overline{B} + C) = \Pi M(0,3,5,6)$$

By counting the weights at the true and the false output one can determine whether an SOP or a POS solution has fewer terms in it. In the above example there were four paths to each output, giving four terms in the SOP and the POS solution.

Consider next the function whose minterm expansion is 0, 2, 6, and 7. The binary tree representation of this function is shown in Figure 10.27(a). Examination of the tree shows that nodes 4 and 5 have the same outputs and can be merged. When this is done node B is seen to be redundant, and it can be removed, as shown in Figure 10.27(b). Also, the output of gate 6 is 0 and the output of gate 7 is 1

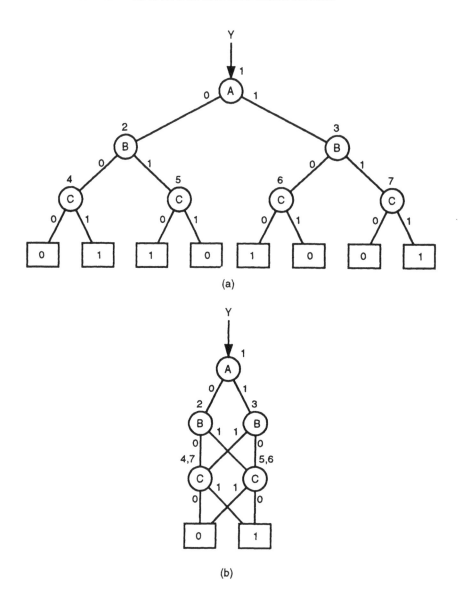

FIGURE 10.25. The function $A \oplus B \oplus C = \Sigma m(1,2,4,7)$. (a) The binary tree and (b) the ROBDD.

regardless of the value of C. In both cases, C is superfluous and can be deleted from the diagram. When these three reductions are applied, one obtains the BDD shown in Figure 10.27(c). The function realized in the BDD is $Y = \overline{A}\,\overline{C} + AB$, which is the result obtained by reducing the function to a minimum SOP form.

The above function was reduced to two prime implicants. The cubes obtained

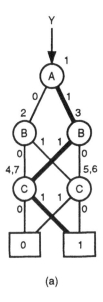
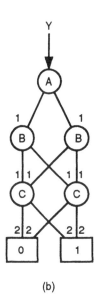

(a)   (b)

FIGURE 10.26. The BDD for $A \oplus B \oplus C$ (a) with the path of input vector $ABC = 111$ represented by dark lines and (b) the tabulation of paths to each leaf.

by the BDD method are not necessarily prime cubes, but each different path must have at least one branch that is different from the branches used in any alternative solution. This guarantees that the implicants obtained are disjoint and they are essential in the sense that each minterm is covered by one and only one implicant.

Consider the function $Y(A,B,D) = AB + D = \Sigma m(1,3,5,6,7)$. The binary tree is shown in Figure 10.28(a). The BDD is simplified by noting that nodes 4 and 5 have the same outputs and can be merged, making node 2 redundant. Node 7 is redundant also, since $Y = 1$ regardless of the value of D, and the binary tree is reduced as shown in Figure 10.28(b). Finally, since the outputs of node 4,5 and node 6 are the same, they can be merged, giving the BDD shown in Figure 10.28(c). Figure 10.28(d) reveals that there are three paths to a true output and two paths to a complemented output. The true output is $Y = AB + A\overline{B}D + \overline{A}D$ and the complemented output is $\overline{Y} = A\overline{B}\,\overline{D} + \overline{A}\,\overline{D}$.

The true and complemented outputs can be simplified to give $Y = A(B + \overline{B}D)$ $+ \overline{A}D = A(B + D) + \overline{A}D = AB + D$, which was the original function, and $\overline{Y} =$ $(\overline{A} + A\overline{B})\overline{D} = (\overline{A} + \overline{B})\overline{D} = \overline{A}\,\overline{D} + \overline{B}\,\overline{D}$, which is the complement of $AB + D$. Neither the true nor the complemented output is given in terms of prime cubes, although the cubes obtained from the graph are disjoint since $AB \cap A\overline{B}D =$ $AB \cap \overline{A}D = A\overline{B}D \cap \overline{A}D = \emptyset$, and $A\overline{B}\,\overline{D} \cap \overline{A}D = \emptyset$. While the SOP solution has three terms in it and the POS solution only has two, the SOP realization can be simplified to two terms also, as shown above.

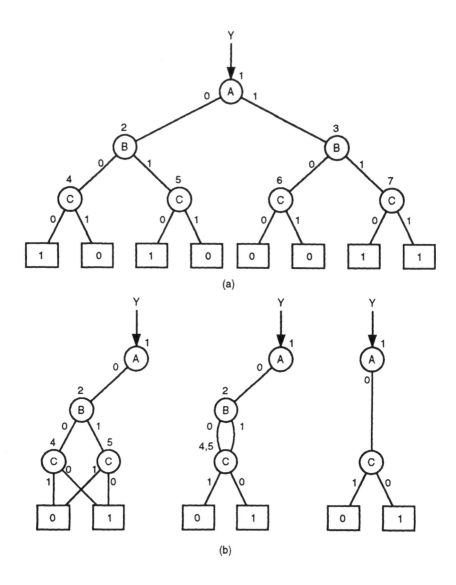

FIGURE 10.27. The function $Y = \sum m\,(0,2,6,7)$. (a) The binary tree, (b) the elimination of node 2, and (c) the ROBDD.

10.10.6 Ordering of Binary Decision Diagrams

The ordering of the input variables can make a great difference in the size of a BDD.[8] If there are binate variables in the function, a simpler graph is achieved by taking the binate variable(s) first. This follows from a consideration of the binary decision tree or the Shannon expansion theorem. A simple function such as $Y = ABC + \overline{B}\,\overline{D} + \overline{C}\,\overline{D}$ can be realized in alphabetical order with a BDD having

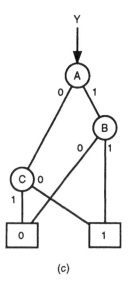

(c)

FIGURE 10.27. (continued).

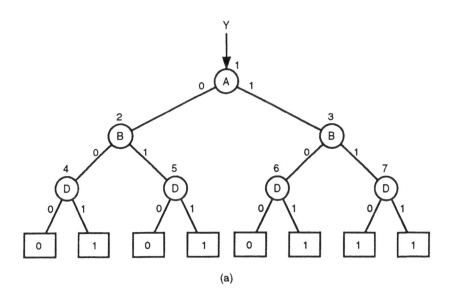

(a)

FIGURE 10.28. The function $Y = \Sigma \, m \, (1,3,5,6,7)$. (a) The binary tree, (b) the merging of nodes 4 and 5 and elimination of node 7, (c) the ROBDD, and (d) the number of paths to either output.

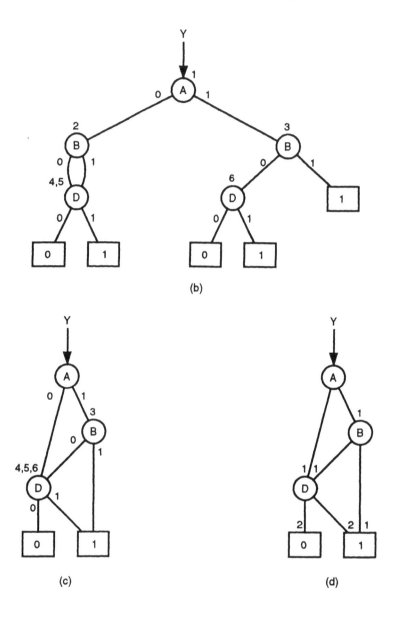

(b)

(c)   (d)

FIGURE 10.28. (continued).

six nodes as shown in Figure 10.29(a). Expanding first about the binate variable B in the order B, C, A, D gives a BDD with only four nodes, as shown in Figure 10.29(b).

ROBDDs have fewer nodes if control inputs are ordered prior to data inputs.[10] This can be seen very easily by considering the truth table of a 4-to-1-line MUX with data inputs A, B, C, and D, select inputs S_1 and S_0, and an enable input EN

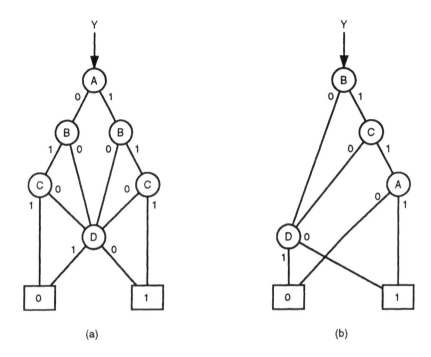

(a)   (b)

FIGURE 10.29. The function $Y = ABC + \overline{B}\,\overline{D} + \overline{C}\,\overline{D}$, (a) realized in order A, B, C, D and (b) expanded in order B, C, A, D.

as shown in Figure 10.30(a). The truth table can be reduced from $2^7 = 128$ rows to the 5 rows shown in Table 10.1, and the BDD can be reduced to four nodes as shown in Figure 10.30(b). A, B, C, and D are treated as leaves since they are data inputs that take on the value 0 or 1. The BDD can be reduced to the normal two leaves once values of the data inputs are given.

Obviously, the truth table and the BDD would be much more complicated if any other ordering were chosen. Even a poor choice of control input sequence will give a more complicated truth table and BDD. The minimum-size truth table for the same MUX with select variables chosen prior to the enable variable is expanded from five rows to eight rows as shown in Table 10.2, and the BDD is increased from four nodes to seven nodes, as shown in Figure 10.30(c).

TABLE 10.1

EN	S_1	S_0	Y
1	x	x	0
0	0	0	A
0	0	1	B
0	1	0	C
0	1	1	D

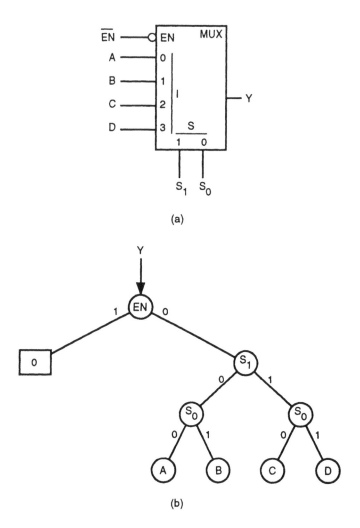

FIGURE 10.30. A 4-to-1-line MUX with enable. (a) The circuit, (b) the minimum BDD, and (c) a poorly ordered BDD.

TABLE 10.2

S_1	S_0	EN	Y
0	0	0	A
0	0	1	0
0	1	0	B
0	1	1	0
1	0	0	C
1	0	1	0
1	1	0	D
1	1	1	0

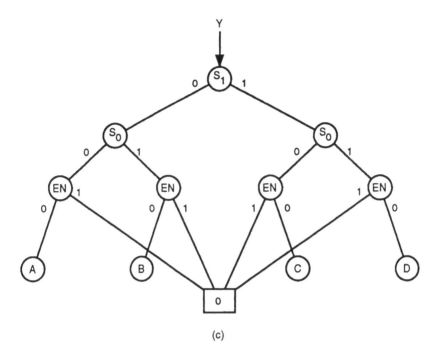

(c)

FIGURE 10.30.   (continued).

10.10.7 Shared Binary Decision Diagrams

Multiple BDDs that possess shared subgraphs can be merged into a single graph, called a *shared binary decision diagram (SBDD)*.[11,12] SBDDs do not contain two or more isomorphic subgraphs. Humans can create SBDDs by inspection. To accomplish this with a program, a *node table* can be formed, and every time a new node is to be created it can be checked against the node table. When another node with the same two edges and index is found in the node table, an edge of the graph is directed to the existing node. This guarantees that no redundant nodes are created, and the graph need not be reduced later.

Shared BDDs are discussed in more detail in Minato, et al.[11] and Sato, et al.[12] Both groups also include don't cares in their BDD representations. See also Ochi et al.[13]

EXAMPLE 10.29: Combine the following three functions into an SBDD.

$$Z_1 = AC + AD + BC + BD = (A + B)Y$$
$$Z_2 = BC + BD + \overline{G} = BY + \overline{G}$$
$$Z_3 = CE\overline{F} + DE\overline{F} = YE\overline{F}$$
where $Y = C + D$.

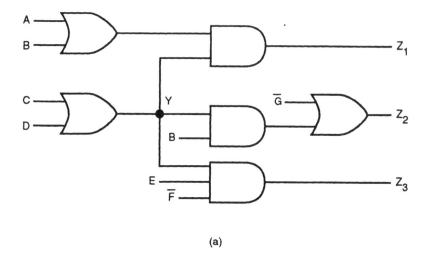

(a)

FIGURE 10.31. (a) Circuit realization of functions $Z_1 = AC + AD + BC + BD = (A + B)Y$, $Z_2 = BC + BD + \overline{G} = BY + \overline{G}$, $Z_3 = CE\overline{F} + DE\overline{F} = YE\overline{F}$, and $Y = C + D$. (b) Realization of BDDs for (b) Z_1, (c) Z_2, (d) Z_3, (e) subfunction Y.

SOLUTION: The network is shown in Figure 10.31(a). The BDDs for Z_1, Z_2, Z_3, and Y, are shown in Figures 10.31(b), (c), (d), and (e). These diagrams can be merged into one SBDD, as shown in Figure 10.32.

10.11 THE IF-THEN-ELSE DIRECTED ACYCLIC GRAPH

Another generalization of BDDs proposed by Karplus[14] is called the *if-then-else DAG (ITE DAG)*. The ITE operator is a universal operator which directly represents arbitrary networks of two-input gates. A representation of an ITE node is shown in Figure 10.33. The *if path* of the DAG is, in general, another ITE DAG representing an arbitrary boolean expression. The *then path* is the one taken whenever the if expression evaluates to true, and the *else path* is the one taken whenever the if expression evaluates to false.

The node of an ITE DAG can be a single variable or an arbitrary boolean expression. In general, ITE diagrams are not canonical. Karplus gives seven rules for constructing an ITE DAG which is canonical.[14] A simple example is the two statements "if A then B else C" and "if $\overline{A}$ then C else B." They both represent the algebraic statement $AB + \overline{A}C$, but they are not equivalent ITE statements. One of Karplus' rules is that the if statement and the then statement must never contain complemented variables, negation only being allowed for the else term or for the entire expression.

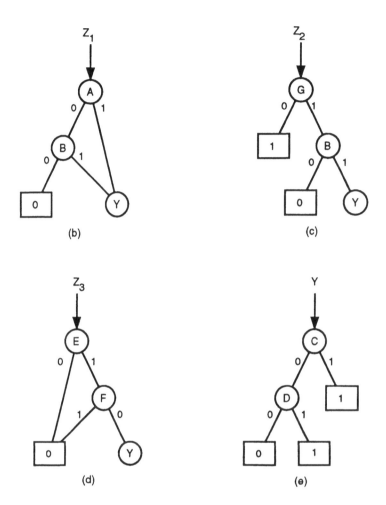

FIGURE 10.31. (continued).

An ITE is reduced when no two nodes represent the same function. ITE graphs reduce to ordinary BDDs when the conditional expression is a single variable. The ITE operator can replace the standard boolean operations of ANDing, ORing, and INVERTing. The ITE operators corresponding to several simple boolean functions are shown in Table 10.3.

ITE diagrams of AB, $A + B$, $A \oplus B$, and $AB + \overline{A}C$ are shown in Figure 10.34. Diagrams of $A + B + C$ and ABC each contain two nodes that represent two tree roots. The root function represented at each ITE node is labeled as shown in Figure 10.35.

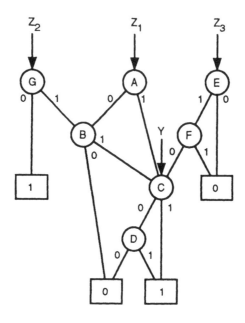

FIGURE 10.32.   The SBDD for Z_1, Z_2, and Z_3.

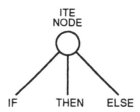

FIGURE 10.33.   The basic ITE node, with three successors, if, then, and else.

TABLE 10.3

A•B	If A then B else FALSE
A + B	If A then TRUE else B
A ⊕ B	If A then $\overline{B}$ else B
ABD	If (if A then B else FALSE) then D else FALSE
AB + $\overline{A}$C	If A then B else C
C(D + E)	If C then (if D then TRUE else E) else FALSE
AB(D + E)	If (if A then B else FALSE) then (if D then TRUE else E) else FALSE

To use ITEs with a computer program, the ITE operation can be defined recursively using Shannon's theorem. Let Y be a function of w, T, and E, where w is the node, T (then) is the path if the node variable is 1, and E (else) is the path if the node variable is 0. Let node v be no further from the root than node w; this

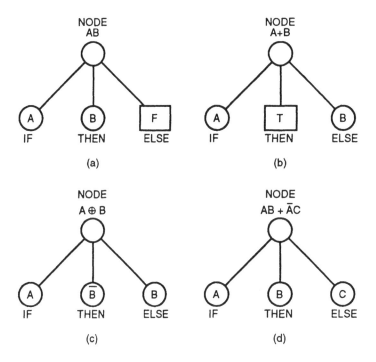

FIGURE 10.34. ITE graphs for the functions (a) AB, (b) A + B, (c) A ⊕ B, and (d) AB + $\overline{A}$C.

can be written as $v \le w$. Then $Y_v = Y$ if $v < w$ or T if $v = w$, and $Y_{\overline{v}} = Y$ if $v < w$ or E if $v = w$. The following recursive formulation will compute ITE(P, Q, R), for functions P, Q, and R represented in ROBDD form. Let $Y = $ ITE(P, Q, R) and let v be the top variable in P, Q, R. Then,

$$Y = vY_v + \overline{v}Y_{\overline{v}}$$
$$Y_v = (PQ + \overline{P}R)_v = P_vQ_v + \overline{P}_vR_v,$$
and
$$Y_{\overline{v}} = (PQ + \overline{P}R)_{\overline{v}} = P_{\overline{v}}Q_{\overline{v}} + \overline{P}_{\overline{v}}R_{\overline{v}}$$
Therefore,
$$Y = v(P_vQ_v + \overline{P}_vR_v) + \overline{v}(P_{\overline{v}}Q_{\overline{v}} + \overline{P}_{\overline{v}}R_{\overline{v}})$$
Now,
$$\text{ITE}(P, Q, R)_v = \overline{P}_vQ_v + \overline{P}_vR_v,$$
and
$$\text{ITE}(P, Q, R)_{\overline{v}} = P_{\overline{v}}Q_{\overline{v}} + \overline{P}_{\overline{v}}R_{\overline{v}}$$
Therefore,
$$Y = v \cdot \text{ITE}(P_v, Q_v, R_v) + \overline{v} \cdot \text{ITE}(P_{\overline{v}}, Q_{\overline{v}}, R_{\overline{v}})$$
$$= \text{ITE}(v, \text{ITE}(P_v, Q_v, R_v), \text{ITE}(P_{\overline{v}}, Q_{\overline{v}}, R_{\overline{v}}))$$

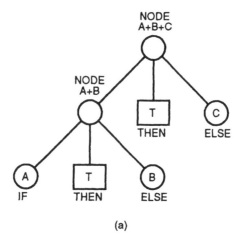

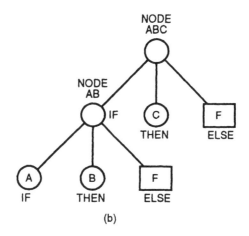

FIGURE 10.35. ITE graphs for the functions (a) A + B + C and (b) ABC.

The recursion terminates when the root is reached, and ITE(Y, 1, 0) = Y. The outline of a computer ITE algorithm is then[15]

```
ITE(P, Q, R){
        If (terminal case) {
                return result;
        } else if (computed-table has entry {P, Q, R}) {
                return result;
        } else {
                let v be the top variable of {P, Q, R};
                T = ITE(P_v, Q_v, R_v);
```

$E = ITE(P_{\overline{v}}, Q_{\overline{v}}, R_{\overline{v}})$;
if T = E, return T;
S = find_or_add_unique_table (v, T, E);
insert_computed_table({P, Q, R}, S);
return S;

EXAMPLE 10.30: Convert the ROBDD shown in Figure 10.16 to an iterative ITE statement.

SOLUTION: Let Y = ITE(P, Q, R). Then,
at node A, P = A, Q = B, R = C, and Y = ITE(A, B, C)
at node B, P = B, Q = TRUE, R = C,
and Y_B = ITE(B, TRUE, C)
at node C, P = C, Q = TRUE, R = FALSE, and
Y_C = ITE(C, TRUE, FALSE)
Y = if A then (if B then TRUE else C) else C
= if A then (if B then TRUE else (if C then TRUE else FALSE))
else (if C then TRUE else FALSE)
Y = ITE(A, ITE(B, TRUE, C),C)
= ITE[A, ITE(B, TRUE, ITE(C, TRUE, FALSE)), ITE(C, TRUE, FALSE)]

The situation can occur where ITE(A, B, C) = ITE(D, E, F), but A ≠ D, B ≠ E, and C ≠ F. In this event, a *standard triple* can be defined and stored in memory. For example, A + B and AB can be written as

$A + B \rightarrow ITE(A, 1, B) = ITE(A, A, B) = ITE(B, 1, A) = ITE(B, B, A)$
$AB \rightarrow ITE(A, B, 0) = ITE(A, B, A) = ITE(B, A, 0) = ITE(B, A, B)$

When complemented edges are used, the following equivalences arise:

$ITE(A, B, C) = ITE(\overline{A}, C, B) = \overline{ITE(A, \overline{B}, \overline{C})} = \overline{ITE(\overline{A}, \overline{C}, \overline{B})}$

The proof of the above equivalence is left as an assignment, as are the proofs of the following statements.

ITE(A, A, B) = ITE(A, 1, B)	ITE(A, 1, B) = ITE(B, 1, A)
ITE(A, B, $\underline{A}$) = ITE(A, B, 0)	ITE(A, B, 0) = ITE(B, $\underline{A}$, 0)
ITE(A, $\underline{B}$, $\overline{A}$) = ITE(A, B, 1)	ITE(A, B, 1) = ITE($\overline{B}$, $\overline{A}$, 1)
ITE(A, $\overline{A}$, B) = ITE(A, 0, B)	ITE(A, 0, B) = ITE($\overline{B}$, 0, $\overline{A}$)
	ITE(A, B, $\overline{B}$) = ITE(B, A, $\overline{A}$)

EXAMPLE 10.31: Show that DeMorgan's law can be written as
$$\overline{\text{ITE}(A, 1, B)} = \text{ITE}(\overline{A}, \overline{B}, 0).$$

SOLUTION: $\text{ITE}(A, 1, B) = A + B$, and $\text{ITE}(\overline{A}, \overline{B}, 0) = \overline{A}\,\overline{B} = \overline{A + B} = \overline{\text{ITE}(A, 1, B)}$.

EXAMPLE 10.32: Show that $A + \overline{A} = 1$ with ITE statements.

SOLUTION: $A + \overline{A} = \text{ITE}(A, 1, \overline{A}) = \text{ITE}(A, 1, 1) = 1$.

There are 16 different output combinations of 2 variables, as shown in Table 10.4. The first column of the table gives the vector representation of each particular output combination. This is a shorthand representation of the OR of the four minterms of the two variables A and B, namely, $\overline{A}\,\overline{B}$, $\overline{A}B$, $A\overline{B}$, and AB. Thus, the second row of the table is $AB = [0001]$, the third row is $A\overline{B} = [0010]$, and the fourth row is $A\overline{B} + AB = A = [0011]$, etc.

TABLE 10.4   The 16 ITE Functions of 2 Variables

Vector	Name	Expression	Equivalent form
0000	0	0	0
0001	AND(A,B)	$A{\cdot}B$	$\text{ITE}(A, B, 0)$
0010	$A > B$	$A{\cdot}\overline{B}$	$\text{ITE}(A, \overline{B}, 0)$
0011	A	A	A
0100	$A < B$	$\overline{A}{\cdot}B$	$\text{ITE}(A, 0, B)$
0101	B	B	B
0110	EOR(A, B)	$A \oplus B$	$\text{ITE}(A, \overline{B}, B)$
0111	OR(A, B)	$A + B$	$\text{ITE}(A, 1, B)$
1000	NOR(A, B)	$\overline{A + B}$	$\text{ITE}(A, 0, \overline{B})$
1001	ENOR(A, B)	$\overline{A \oplus B}$	$\text{ITE}(A, B, \overline{B})$
1010	NOT(B)	$\overline{B}$	$\text{ITE}(B, 0, 1)$
1011	$A \geq B$	$A + \overline{B}$	$\text{ITE}(A, 1, \overline{B})$
1100	NOT(A)	$\overline{A}$	$\text{ITE}(A, 0, 1)$
1101	$A \leq B$	$\overline{A} + B$	$\text{ITE}(A, B, 1)$
1110	NAND(A,B)	$\overline{A{\cdot}B}$	$\text{ITE}(A, \overline{B}, 1)$
1111	1	1	1

10.12 SUMMARY

Multilevel logic minimization transforms a boolean function into an equivalent representation that can be implemented as a circuit that is smaller, faster, or more testable than a circuit built in an SOP or POS form. The function may or may not have multiple outputs. A large effort is currently being expended in the construction of software algorithms which will synthesize multilevel circuits with one or more outputs, a special case of two-level multi-output circuits being the PLA.

Multilevel minimization requires boolean expressions to be in factored form, which generates a tree structure. A function is said to be in a *canonical form* when there is only one correct representation of the function. Truth tables, minterm expansions, and maxterm expansions are canonical as there is only one correct representation of the function. For a given ordering of the variables, the OBDD is cannonical. If certain simple restrictions are applied to ITE forms they are canonical also.

The SOP form is the natural representation for two-level implementations such as PLAs, and factored forms are the natural representation for multilevel networks. In two-level realizations the factored form is the same as the POS form, which may or may not be simpler than the SOP form. In multilevel realizations the factored form is isomorphic to a binary tree structure, where each internal node is an AND or an OR operator and each leaf is a literal.

The factored form implicitly represents both a function and its complement, since the complemented factored form can be obtained directly by applying DeMorgan's law to the factored form. This is accomplished by interchanging all AND and OR nodes and complementing all the literals. This produces a factored form of the complement of the function with the same literal count as that of the uncomplemented function. Also, the complemented form of a factored function can be obtained by complementing the output of the tree structure. Thus, a function and its complement are equally complex, except for, at most, one inverter. The factored form gives a tree structure, and is not in general a canonical representation of a function.

A major problem in algebraic operations is the identification of divisors. Kernels are the primary multicube divisors of a function and form a good set of divisors both for factoring (or decomposition) and extraction. The kernel approach favors the development of fast and effective algorithms. This makes kernel extraction, intersection, and phase assignment very important algebraic operations in multilevel synthesis and optimization of combinational networks.

The synthesis of a complex boolean expression into multilevel logic is done by obtaining and selecting those kernels that reduce the literal count of the network. These kernels become new subfunctions of the network, and the process is repeated until no more literal reduction is possible. Kernel intersections determine which factors are common to two or more output functions. A common factor of two or more functions can be implemented with one gate feeding two or more outputs.

Binary decision diagrams and if-then-else diagrams are directed acyclic graphs which offer an alternative approach to the synthesis of multilevel logic. BDDs define a digital function in terms of a diagram which represents the function and contains the information necessary to implement the function. Nodes of the BDD are either variables or subfunctions. Ultimately, nodes can be reduced to single variables. The root of the tree is the function to be implemented, and the leaves are either 0s or 1s. The diagram is entered at the arrow pointing to the function realized by using that node as a root. The value of the node variable is indicated on the two branches leaving the node, and the nodes are directed downward.

There is a one-to-one correspondence between a BDD and a MUX realization; hence, BDDs provide a multilevel decomposition of a network. Each node of a BDD maps into a MUX controlled by the node variable, with the other two inputs coming from the successor node. This makes the BDD a convenient tool for designing MUX-based FPGAs.

As pointed out by Bryant,[9] BDDs can be written in the form of ITE boolean operations, formed by using the Shannon expansion theorem and recognizing that this is a function of the form $AB + \overline{A}C$. In an ITE representation, the out-edge labeled 1 is relabeled *then*, and the out-edge labeled 0 is relabeled *else*. Thus, the ITE operator is a boolean function defined for three inputs: A, B, C, as ITE(A, B, C) is identical to if A then B else C, which is equivalent to the statement $AB + \overline{A}C$. The ITE operator can be used to implement all two-variable boolean opeations.

PROBLEMS

For Problems 10.1 through 10.12 determine which form gives a minimum cost. Let G be the number of gates, L the number of literals, MFI the maximum fan-in, and MPL the maximum path length. Use a cost factor CF of $2G + L + MFI + MPL$.

10.1. Factor $Y = ABC + AD + AE$ and compare the cost factors of the SOP and the factored forms.

10.2. Factor $Z = ABCD + ABEF$ (a) using three-input gates and (b) using two-input gates. Compare all three cost factors.

10.3. Factor $Y = AE + AF + AG + BCE + BCF + BCG + BDE + BDF + BDG$ and compare cost factors if the SOP form is realized with three-input gates.

10.4. Factor $Z = A\overline{C}DE + B\overline{C}DE + AC\overline{D}E + BC\overline{D}E + C\overline{D}EF$ into three-level logic and four-level logic and compare all three cost factors.

10.5. Factor the two functions Y and Z simultaneously. $Y = A\overline{B}CD + A\overline{B}CE + A\overline{B}F + A\overline{B}G + H$, and $Z = ACDJ + ACEJ + AFJ + AJK + L$.

10.6. Factor the two functions Y and Z simultaneously. $Y = AB\overline{C}D + AB\overline{C}E + AB\overline{F} + ABG + H$, and $Z = A\overline{C}DJ + A\overline{C}EJ + A\overline{F}J + AJK + L$.

10.7. Use algebraic resubstitution of the largest common kernel to reduce the two functions $Y_1 = A\overline{C} + AD + AE$, and $Y_2 = B\overline{C} + BD + BE$.

10.8. Use algebraic resubstitution of the largest common kernel to reduce the two functions $Z_1 = A\overline{B} + AC + AD$, and $Z_2 = \overline{B}E + CE + DE$.

10.9. Flatten the function $Y = \underline{A}(BC + D)\overline{E}F + (G + \overline{H})K$.

10.10. Flatten the function $Z = \overline{A}(BC + D)EF + (\underline{G} + H)J$.

10.11. Flatten the function $W = (AB + CD + EF)(\overline{A}\,\overline{B} + \underline{C}\overline{D} + \overline{E}\,\overline{F})$.

10.12. Flatten the function $X = (AC + BE + DF)(\overline{A}C + \overline{B}E + DF)$.

10.13. Group the products of $Y = ABC + ADE + \overline{A}\,\overline{B} + A\overline{B}E + \overline{A}D + CD + \overline{A}\,\overline{F}$ by positive, negative, and mixed phase.

10.14. Group the products of $Z = \overline{A}\,\overline{B}\,\overline{C} + ABC + A\overline{B}\,\overline{E} + A\overline{B}DF + AD + CF + \overline{C}\,\overline{G}$ by positive, negative, and mixed phase.

10.15. Find the support of the functions $X = ABCE + \overline{A}BD + \overline{B}D$, $Y = BC + AFG + HJ + K$, and $Z = EFG + HJ$.

10.16. Find the support of the functions XY, XZ, and YZ of Problem 10.15.

10.17. Find the support of the functions $X = \overline{A}CDE + \overline{A}DF + E\overline{G}$, $Y = AB + DEF + GH + J$, and $Z = AB + CDE$.

10.18. Find the support of the functions XY, XZ, and YZ of Problem 10.17.

10.19. $U = AB + CD$, $V = EF + GH$, and $W = EF + JK$. Examine UV, UW, and VW and determine which are algebraic expressions and which are not.

10.20. $U = AC + DE$, $V = AB + EF$, and $W = EF + JK$. Examine UV, UW, and VW and determine which are algebraic expressions and which are not.

10.21. Find the kernels and co-kernels of the function Y, where $Y = ADF + AEF + BDF + BEF + CDF + CEF + G$.

10.22. Find all the kernels of Z, where $Z = ACD + AEF + BCD + BEF + CDE + CEF + H$.

10.23. Realize as much of the function $Y = ABC + DEF + GHK + L\overline{M}N + PQR + \overline{S}\overline{T}\overline{U} + \overline{V}\overline{W}\overline{X}$ with as few AND, OR and NOR gates as possible, if the AND-gates and OR-gates are restricted to three inputs each, and all the input literals are to have positive phase. (Hint: Group cubes by unateness, and use DeMorgan's law to change literals of negative to positive phase.)

10.24. Repeat Problem 10.23 for the function $Y = A\overline{B}C + \overline{D}\overline{E}\overline{F} + GHK + LMN + \overline{P}\overline{Q}\overline{R} + STU + VWX$.

10.25. Use the method of Section 10.7 to find all the kernels and co-kernels of the functions Y_1 and Y_2, and show the multilevel circuit for the simultaneous realization of the two functions. Compare the number of gates and literals in the SOP form and the factored form of the solution. $Y_1 = ABC + ABD + ABE + A\overline{B}F + A\overline{B}G$, and $Y_2 = \overline{A}CF + \overline{A}DF + \overline{A}EF + C\overline{D}F + C\overline{D}G$.

10.26. Repeat Problem 10.25 for the two functions $Z_1 = \overline{A}BD + \overline{A}BE + \overline{A}BF + \overline{A}\,\overline{B}H + \overline{A}\,\overline{B}J$, and $Z_2 = ADH + AEH + AFH + D\overline{E}H + D\overline{E}J$.

10.27. For the binary tree in Figure 10.36:
  (a) give the Shannon cofactors represented by each node of the tree.
  (b) Obtain the expression for the root Y.
  (c) Find the number of paths to each 0 and 1 leaf.

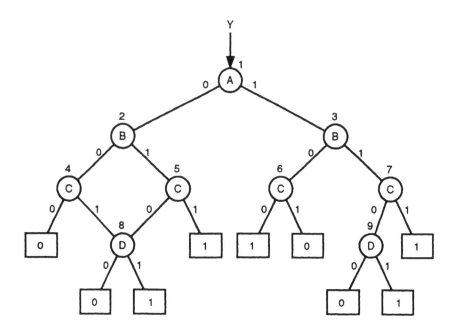

FIGURE 10.36.

10.28. Repeat Problem 10.27 for the tree in Figure 10.37.
10.29. Repeat Problem 10.27 for the tree in Figure 10.38.
10.30. Repeat Problem 10.27 for the tree in Figure 10.39.
10.31. For the tree in Figure 10.36 show the paths represented by the vectors ABCD = 0101 and 1100.
10.32. For the tree in Figure 10.37 show the paths represented by the vectors ABCD = 1101 and 0011.
10.33. For the tree in Figure 10.38 show the paths represented by the vectors ABCD = 01x1 and 110x.
10.34. For the tree in Figure 10.39 show the paths represented by the vectors ABCD = 101x and 01x1.
10.35. (a) Show the binary tree and the ROBDD for the function Y(A,B,C) = Σm(3,4,5,6,7).
   (b) Map the function and compare a minimum SOP to the BDD solution.
10.36. (a) Show the binary tree and the ROBDD for the function Y(A,B,C) = Σm(0,2,4,6,7).
   (b) Map the function and compare a minimum SOP to the BDD solution.
10.37. (a) Show the binary tree and the ROBDD for the function Y(A,B,C) = Σm(1,2,4,5,7).
   (b) Map the function and compare a minimum SOP to the BDD solution.

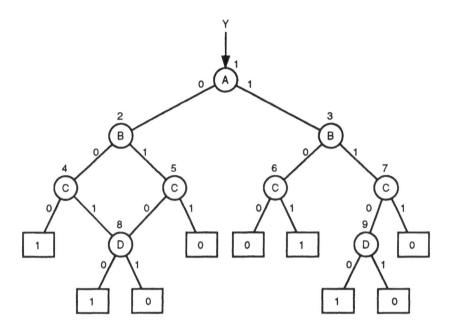

FIGURE 10.37.

10.38. (a) Show the binary tree and the ROBDD for the function Y(A,B,C) = Σm(1,2,3,5,7).

(b) Map the function and compare a minimum SOP to the BDD solution.

10.39. Obtain the ROBDD for the function Y = $\overline{A}$ BC$\overline{D}$E.

10.40. Obtain the ROBDD for the function Y = A$\overline{B}$CD$\overline{E}$.

10.41. Obtain the ROBDD for the function Y = A$\overline{B}$ $\overline{C}$ + DE.

10.42. Obtain the ROBDD for the function Y = A$\overline{B}$ + $\overline{C}$DE.

10.43. Obtain the ROBDD for the function Y = A$\overline{B}$C + $\overline{D}$EF.

10.44. Obtain the ROBDD for the function Y = A$\overline{B}$ $\overline{C}$ + DEF.

10.45. Obtain the two ROBDDs for a half adder with sum S = A $\oplus$ B and Carry-out C$_{out}$ = AB.

10.46. Obtain the two ROBDDs for a full adder with sum S = A $\oplus$ B $\oplus$ C and Carry-out C$_{out}$ = AB + C$_{in}$(A $\oplus$ B).

10.47. Obtain the ROBDDs for a full adder using A $\oplus$ B as a subfunction.

10.48. Obtain the ROBDD for a 3-out-of-5 majority function.

10.49. Obtain the ROBDD for the function A $\oplus$ B $\oplus$ C $\oplus$ D using invert bubbles.

10.50. Obtain the ROBDD for the function $\overline{A}$ $\oplus$ B $\oplus$ C $\oplus$ D using invert bubbles.

10.51. Give the MUX representations of the two BDDs in Figure 10.40 and find the output function Y in each case.

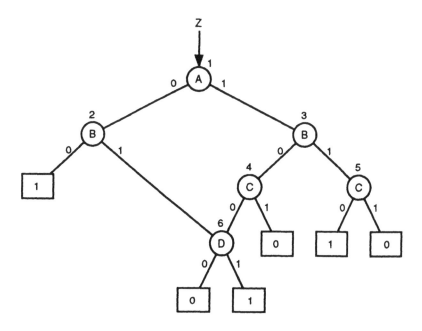

FIGURE 10.38.

10.52. Give the MUX representations of the two BDDs in Figure 10.41 and find the output function Y in each case.

10.53. If the carry-in of a full adder is treated as a control variable, the sum and carry-out are given by: $S = A \oplus B$ and $C_{out} = AB$ when $C_{in} = 0$, and $S = \overline{A} \oplus B$ and $C_{out} = A + B$ when $C_{in} = 1$. Obtain the ROBDD with C_{in} the most significant variable. Use invert bubbles to reduce the BDD.

10.54. Obtain the best ROBDD for the function given by the table below.

A	B	C	D	Y
0	0	x	x	CD
0	1	x	x	$C \oplus D$
1	x	x	x	C + D

10.55. Obtain the shared ROBDD for the functions $A \oplus B$, $A\overline{B}$, and $A + \overline{B}$.

10.56. Obtain the shared ROBDD for the functions $\overline{A} \oplus B$, AB, and $A + \overline{B}$.

10.57. Give the four possible ROBDD topologies for a tree of three two-input MUXs.

10.58. Give the ROBDD for the MUX tree shown in Figure 10.42.

10.59. Obtain the ROBDD for the function $Y = \Sigma m(2,3,5,7,9,11,12,13,14,15)$. Map the function and compare the SOP solution to the ROBDD solution.

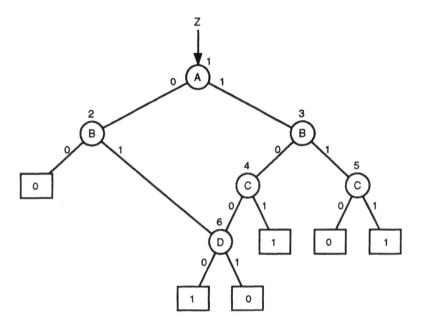

FIGURE 10.39.

10.60. Obtain the ROBDD for the function $Z = \Sigma m(0,1,4,6,8,10)$. Map the function and compare the SOP solution to the ROBDD solution.

10.61. For the function $Y = \Sigma m(0,2,3,6,7)$,

(a) Give the binary decision tree for Y, and implement it with an 8-to-1-line MUX.

(b) Give the ROBDD for Y and implement it with a 4-to-1 line MUX.

10.62. Repeat Problem 10.61 for the function $Z = \Sigma m(1,3,4,5,6)$.

10.63. Let $Y = AB\overline{D} + ACD + BCD$.

(a) Obtain the cube-literal matrix for Y.

(b) Give a minimum factored-form solution for Y.

10.64. Repeat Problem 10.63 for the function $Z = ABC + ABD + EFG + EFH + JK + JL$.

10.65. Repeat Problem 10.63 for the function $Y = ABC + ABD + ABE + ACE + ACF$.

10.66. Repeat Problem 10.63 for the function $Z = CDE + CDF + ADH + ADJ$.

10.67. Let $U = A\overline{B} + ACDE + ACDF + G$ and $V = A\overline{B} + \overline{B}CDE + \overline{B}CDF$.

(a) Obtain the cube-literal matrix for functions U and V.

(b) Give a minimum factored-form solution for U and V separately.

(c) Give a minimum factored-form solution for U and V simultaneously.

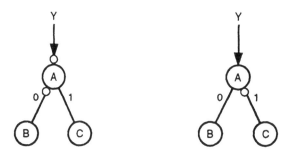

FIGURE 10.40.

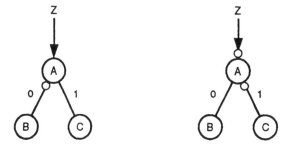

FIGURE 10.41.

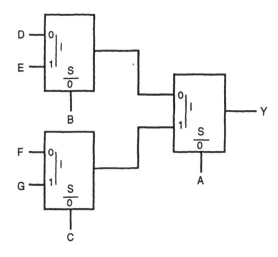

FIGURE 10.42.

10.68. Repeat Problem 10.67 for functions $W = AB + BC\overline{D}E + BC\overline{D}F + G$ and $X = AB + AC\overline{D}E + AC\overline{D}F$.

10.69. Repeat Problem 10.67 for functions $X = ABC + ABD + EFG$ and $Y = AC + AE + CEF + DEF$.

10.70. Repeat Problem 10.67 for functions $U = AB + ACDE + ACDF + G$ and $V = AB + BCDE + BCDF$.

10.71. Give the ITE diagram for $Y = ABC + D$.

10.72. Give the ITE diagram for $Z = AB + CD$.

10.73. Show that $A + B \rightarrow ITE(A, 1, B) = ITE(A, A, B) = ITE(B, 1, A) = ITE(B, B, A)$.

10.74. Show that $AB \rightarrow ITE(A, B, 0) = ITE(A, B, A) = ITE(B, A, 0) = ITE(B, A, B)$.

10.75. Show that $ITE(A, A, B) = ITE(A, 1, B)$ and $ITE(A, 1, B) = ITE(B, 1, A)$.

10.76. Show that $ITE(A, B, A) = ITE(A, B, 0)$ and $ITE(A, B, 0) = ITE(B, A, 0)$.

10.77. Show that $ITE(A, B, \overline{A}) = ITE(A, B, 1)$ and $ITE(A, B, 1) = ITE(\overline{B}, \overline{A}, 1)$.

10.78. Show that $ITE(A, \overline{A}, B) = ITE(A, 0, B)$ and $ITE(A, 0, B) = ITE(\overline{B}, 0, \overline{A})$.

SPECIAL PROBLEMS

10.79 (a) Do the three steps of ESPRESSO (Reduce, Expand, and Irredundant Cover) algebraically to simplify the functions $Y = \overline{A}\,\overline{C} + \overline{A}\,B + AC + A\overline{B}$ and $Z = \overline{B}C + \overline{A}C + A\overline{B}\,\overline{C}$.

(b) Show each step of part a on a Karnaugh map.

10.80. Use iterative weak division to simplify the functions

(a) $Y = AB + ACD + ACE + BF + CDF + CEF$ and $P_1 = D + E$.

(b) $Z = AC + AD + AE + AG + BC + BD + BE + BF + CE + CF + DF + DG$ and $P_1 = A + B$.

REFERENCES

1. **Brayton, R. K., Rudell, R., Sangiovanni-Vincentelli, A., and Wang, A. R.,** MIS: a multiple-level logic optimization system, IEEE Trans. on Computer-Aided Design, Vol. CAD-6, November 1987, 1062–1081.

2. **Brayton, R. K., Hachtel, G. D., and Sangiovanni-Vincentelli, A. L.,** Multilevel logic synthesis, Proc. *IEEE,* 78(2), 1990, 264–300.

3. **Bostick, D., Hachtel, G., Jacoby, R., Lightner, M., Moceyunas, P., Morrison, C., and Ravenscroft, D.,** The boulder optimal logic design system, Proc. Int. Conf. Computer Aided Design, ICCAD-87, 1987.

4. **Gregory, D., Bartlett, K., de Geus, A., and Hachtel, G.,** Socrates: A system for automatically synthesizing and optimating combinational logic, Proc. 23rd Design Automation Conf., June 1986, 79–85.

5. **Hachtel, G. D. and Lightner, M. R.**, A tutorial on logic synthesis: algebraic decomposition, paper presented at the IEEE Conf. on CAD, ICCAD-88, Santa Clara, CA, November 7, 1988.

6. **Katz, R. H.**, *Modern Logic Design for Rapid Hardware Prototyping*, Benjamin Cummings, in press.

7. **Brayton, R. K., Hachtel, G. D., McMullen, C. T., and Sangiovanni-Vincentelli, A. L.**, Logic Minimization Algorithms for VLSI Synthesis, Kluwer Academic Publishers, Boston, 1984.

8. **Akers, S. B.**, Binary decision diagrams, IEEE Trans. *Comput.*, C-27(6), 509–516, 1978.

9. **Bryant, R. E.**, Graph-based algorithms for boolean function manipulation, *IEEE Trans. Comput.*, C-35(8), 677–691, 1986.

10. **Ohmura, M., Yasuura, H., and Tamaru, K.**, Extraction of functional information from combinational circuits, Proc. IEEE Int. Conf. Computer-Aided Design, ICCAD-90, Santa Clara, CA, November 11 to 15, 1990, 176–179.

11. **Minato, S., Ishiura, N., and Yajima, S.**, Shared binary decision diagram with attributed edges for efficient boolean function manipulation, Proc. 27th ACM/IEEE Design Automation Conf., Orlando, FL, June 24 to 28, 1990, 52–57.

12. **Sato, H., Yasue, Y., Matsunaga, Y., and Fujita, M.**, Boolean resubstitution with permissible functions and binary decision diagrams, Proc. 27th ACM/IEEE Design Automation Conf., Orlando, FL, June 24 to 28, 1990, 284–289.

13. **Ochi, H., Ishiura, N., and Yajima, S.**, Breadth-first manipulation of SBDD of boolean functions for vector processing, Proc. 28th ACM/IEEE Design Automation Conf., San Francisco, June 17 to 21, 1991, 413–416.

14. **Karplus, K.**, Using if-then-else DAGs for multi-level logic minimization, Proc. 1989 Decennial Caltech Conf., March 20 to 22, 1989, 101–117.

15. **Brace, K. S., Rudell, R. L., and Bryant, R. E.**, Efficient implementation of a BDD package, Proc. 27th ACM/IEEE Design Automation Conf., Orlando, FL, June 24 to 28, 1990, 40–45.

Index

Milton Keynes UK
Ingram Content Group UK Ltd.
UKHW031124141024
449569UK00006B/447